The Property Masters

The Property Masters

A HISTORY OF THE BRITISH COMMERCIAL PROPERTY SECTOR

Peter Scott
Business History Unit
University of Portsmouth Business School
UK

LONDON AND NEW YORK

Published by Taylor & Francis

Published 2008 by Routledge
2 Park Square, Milton Park, Abingdon, Oxfordshire OX14 4RN
711 Third Avenue, New York, NY 10017

First edition 1996

First issued in paperback 2015

Routledge is an imprint of the Taylor and Francis Group, an informa business

© 1996 Peter Scott

Typeset in Great Britain by Saxon Graphics Ltd, Derby

ISBN 0 419 20950 6

Apart from any fair dealing for the purposes of research or private study, or criticism or review, as permitted under the UK Copyright Designs and Patents Act, 1988, this publication may not be reproduced, stored, or transmitted, in any form or by any means, without the prior permission in writing of the publishers, or in the case of reprographic reproduction only in accordance with the terms of the licences issued by the Copyright Licensing Agency in the UK, or inaccordance with the terms of licences issued by the appropriate Reproduction Rights Organization outside the UK. Enquiries concerning reproduction outside the terms stated here should be sent to the publishers at the London address printed on this page.
  The publisher makes no representation, express or implied, with regard to the accuracy of the information contained in this book and cannot accept any legal responsibility or liability for any errors or omissions that may be made.

A catalogue record for this book is available from the British Library

Publisher's Note
The publisher has gone to great lengths to ensure the quality of this reprint but points out that some imperfections in the original may be apparent.

ISBN 13: 978-1-138-98399-1 (pbk)
ISBN 13: 978-0-419-20950-8 (hbk)

To my parents

Contents

Acknowledgements	x
Glossary of abbreviations and technical terms	xiii
List of tables	xvii
List of figures	xix
Foreword *Jack Rose*	xxi

Chapter 1 Introduction — 1
    1.1 The aims of this study — 1
    1.2 Scope, sources and methodology — 4
    References and notes — 7

SECTION ONE THE EVOLUTION OF THE BRITISH COMMERCIAL PROPERTY MARKET SINCE 1800 — 9

Chapter 2 The emergence of a commercial property investment market 1800–1918 — 11
    2.1 Introduction — 11
    2.2 'Traditional' landowners and the nineteenth century urban property market — 11
    2.3 Insurance company investment in the property sector 1800–1870 — 14
    2.4 The changing nature of the property investment market 1850–1914 — 18
    2.5 City office development and the property market — 21
    2.6 Insurance companies and the property market 1870–1918 — 26
    References and notes — 34

Chapter 3 The growth of a national commercial property market, 1919–1939 — 38
    3.1 Introduction — 38
    3.2 The growth of market intermediaries — 39
    3.3 The 'multiple revolution' and the property investment market — 48

Appendix: The mathematical derivation of the steady state
of corporate growth equation 63
References and notes 64

Chapter 4 Property investment, development and the capital market between the wars 68
4.1 The pattern of institutional property investment 1919–1939 68
4.2 Property companies and the property development market 73
4.3 Property development finance 83
4.4 Entrepreneurship in the inter-war property investment market 90
References and notes 95

Chapter 5 War and recovery 1939–1954 99
5.1 The Second World War 99
5.2 Property investment and the capital market 1945–1954 104
5.3 The financial institutions and the property entrepreneurs: 1945–1954 117
References and notes 128

Chapter 6 The property boom 1955–1964 132
6.1 Introduction 132
6.2 Property investment and the capital market 1955–1964 133
6.3 Institutional investors and the property developers 1955–1964 141
References and notes 162

Chapter 7 From Brown Ban to Barber Boom 1965–1973 166
7.1 Introduction 166
7.2 The property investment boom 166
7.3 The property development market 1965–1973 174
References and notes 187

Chapter 8 The property crash, aftermath and recovery 1974–1980 193
8.1 Introduction 193
8.2 The property crash 194
8.3 Recovery and boom, 1975–1980 201
References and notes 209

Chapter 9 The best of times and the worst of times – the commercial property market since 1980 213
9.1 Introduction 213
9.2 The retreat of the institutional investors 1981–1984 213
9.3 The changing face of commercial property 215
9.4 The merchant developers 221
9.5 The Big Bang Boom 1985–1989 223

9.6	The second property crash	229
9.7	The green shoots of recovery	236
	References and notes	239

SECTION TWO A STATISTICAL ANALYSIS OF THE COMMERCIAL PROPERTY MARKET — 245

Chapter 10 A statistical overview of the property investment market — 247

10.1	Introduction	247
10.2	Net investment in property 1922–1993	247
10.3	The composition of institutional property portfolios 1870–1990	260
10.4	Initial yields on investment property 1920–1993	266
10.5	Market rents	271
10.6	The rate of return on investment property 1921–1993	274
	References and notes	284

Chapter 11 An econometric analysis of property returns and the volume of institutional property investment — 286

11.1	The long-term determinants of rates of return and net investment in UK property	286
11.2	The determinants of investment property capital values	291
11.3	The determinants of the volume of institutional property investment	294
	References and notes	296

CHAPTER 12 CONCLUSIONS — 297
References and notes — 306

Bibliography — 308

Index — 321

Acknowledgements

Conducting the research for this study would not have been possible without the help of a number of institutional investors and other organizations which allowed me access to their private business papers. I am particularly indebted to the archival and investment staff of the Church Commissioners, Clerical Medical, Legal & General, the National Provident Institution and Standard Life, whose generous and patient assistance enabled me to assemble comprehensive case studies of their property investment histories. A number of other organizations have also provided me with access to their archives; I would particularly like to thank Brixton Estate plc, Colliers Erdman Lewis, Eagle Star, the Estates Gazette, Healey & Baker, Hillier, Parker, Marks & Spencer plc, Pearl Assurance and the Prudential, plus several other institutions which let me check through their archival papers, but had no surviving records covering the information I was seeking.

My thanks are also due to a number of record offices which assisted me with this project, particularly the Church of England Record Centre, the Guildhall Library, the Public Record Office and the West Yorkshire Archive Service. Stanley H. Burton gave me access to sections of the Montague Burton papers which are not normally open to the public and provided me with a copy of an additional unpublished paper, for which I would like to thank him. Richard Barkham, John Dunbabin, John Lawrence, T.W. LaPier, Ralph Turvey, J. Whitley, J.H. Treble and John Butt generously provided me with access to work they had undertaken which was unpublished at the time of writing. I would also like to thank Alison Sharpe for her help with my research on Legal & General, and Stephanie Jones for help with my work on Hillier, Parker.

Archival evidence can only provide an incomplete picture of the past; my research was greatly assisted by being able to conduct a number of interviews with people who are, or have been, active in the property investment and/or development markets. I am grateful to John Crickmay, Arthur Green, Michael Hallett, Arthur Hemens, Bernard High, Paul Orchard-Lisle, David Ormerod and Tony Overall, who granted me interviews. Particular thanks are due to Edward Erdman, Jim Pegler and Jack Rose, who gave me access to their unpublished memoirs and other written work, in addition to the interviews I conducted with them, and to J. Max Keyworth for allowing me to quote from his privately published history of the Covent Garden property companies.

ACKNOWLEDGEMENTS

I would like to give special thanks to the Investment Property Databank for their help with some of the statistical aspects of my research, and to Allsop & Co., DTZ Debenham Thorpe Research and Jones Lang Wootton for allowing me to make use of their data. My thanks are also due to James Foreman-Peck, Ken Mayhew, Sean Pascoe and Andrew Ryan for their advice regarding my various attempts at econometric analysis of the property market, and to Paul Gandy for his help with the mathematical derivation of the steady state of corporate growth equation, outlined in Chapter 3.

The original idea of writing a history of the property investment market was conceived in 1989 at the London School of Economics, and stemmed from an MSc dissertation I undertook there in the Department of Economic History. My thanks are due to Leslie Hannah, Paul Johnson and the Prudential Property Research Team for their help with this earlier project. The first incarnation of this study was assembled as a DPhil thesis in Modern History at the University of Oxford, the research for which involved considerable travel to London and elsewhere. I am grateful to the Economic and Social Research Council, the trustees of the Michael Postan Award, and Merton and Pembroke Colleges, Oxford, for their financial support with this research. I would also like to thank Tom Elliott, Shan Mitra, Rupert Nabarro, Katie Peters, Andrew Ryan and Richard Truscott for helping me to arrange accommodation in London during my research, and my sister Ruth for helping me with accommodation in Edinburgh.

My work on the recent history of the property market was further developed after the completion of my thesis, while I was working as a part-time consultant with London Economics. Particular thanks are due to John Kay, Nick Morris, Bill Robinson, Ron Smith and Nick Stern for their help and advice during this period. The long process of transforming my thesis into a book has largely been undertaken during the last year; I wish to thank my colleagues in the Economics Department, University of Portsmouth, for all their encouragement and support.

A number of people provided me with useful advice after reading through sections of this study, the thesis from which it was developed and a number of papers I produced in connection with this project; thanks are due to Richard Barkham, Virginia Bainbridge, Sue Bowden, Francesca Carnevali, Christina Fowler, Andrew Godley, Cliff Gulvin, Mark Hampton, Leslie Hannah, Carol Heim, Katrina Honeyman, Lester Hunt, David Kells, Katie Peters, Chris Reid, Tim Rooth, Paul Walker and Oliver Westall. I also wish to thank Nigel Wratten for his help and advice with my research, and Kate Pattullo, whose help with the intricacies of WordPerfect 5.1 and Excel saved me a great deal of time and frustration.

Thanks are due to the following people and organizations for allowing me to reproduce the photographs in this book: Jeremy Burton (Montague Burton and Burton store); City of Coventry Central Library (Coventry precinct); Clerical Medical (Sir Andrew Rowell); Greater London Library (Elephant and Castle Centre); Guildhall Library (Fountain House); Healey & Baker (Aubrey

ACKNOWLEDGEMENTS

Orchard-Lisle and Douglas Tovey); Lloyds of London (Lloyds of London Building); Marks & Spencer (Pantheon, Oxford Street); Portsmouth City Council (Odeon Southsea); Jack Rose (personal photograph); Westminster City Archives (Monico proposal and Bush House); Edward Erdman (all other photographs). I would also like to thank Alan Bailey for permission to reproduce his 1985 'STACKUP' cartoon.

My DPhil thesis examiners, Avner Offer and Barry Supple, provided me with a number of very useful comments and criticisms, for which I would like to thank them. I am also grateful to my editor, Madeleine Metcalfe, for her help and advice. Finally, and most importantly, I wish to thank my DPhil supervisor, Charles Feinstein, whose help in guiding me through this project has been of immense value, both to this study and to my skills as a historical researcher. Any errors and omissions are, of course, my own.

Glossary of abbreviations and technical terms

ABBREVIATIONS

CEDIC	Church Estates Development and Improvement Company Ltd.
CIC	Capital Issues Committee.
City Offices	City Offices Co. Ltd.
CLRP	City of London Real Property Co. Ltd.
Clerical Medical (Clerical)[1]	Clerical, Medical and General Life Assurance Society.
Hillier, Parker	Hillier, Parker, May & Rowden.
IPD	Investment Property Databank.
LCC	London County Council.
LDDC	London Docklands Development Corporation.
Legal & General	Legal and General Life Assurance Society.
NCB	National Coal Board.
NPI	National Provident Institution.
ODP	Office Development Permit.
Pearl	Pearl Assurance plc.
PFPUT	Pension Fund Property Unit Trust.
PRO	Public Record Office.
Prudential	Prudential Assurance Company.
REA	Royal Exchange Assurance.
Standard Life (Standard)	The Standard Life Assurance Company.
UDC	Urban Development Corporation.

GLOSSARY OF ABBREVIATIONS AND TECHNICAL TERMS

TECHNICAL TERMS

Book value	The value of a capital asset as recorded in the accounts of the institution owning the asset.
Building lease	A long-term lease, imposing an obligation on the lessee to erect one or more buildings on the leased land, which will become the property of the landlord after the lease expires.[2]
Deep discount bond	A bond which provides at least part of its return as a capital gain upon redemption, in addition to interest payments.
Equivalent yield	A yield figure which represents the present income from a property, plus the discounted value of expected future increases in income.
Feu duty	A property security, existing only in Scotland, which takes the form of an annual payment, similar to a perpetual rent charge.
Full repairing and insuring lease	A lease which places the obligation for maintaining and insuring the property on the tenant.
Ground rent	Rent payable to the landowner, following the granting of a building lease, representing the value of the land only.
Initial yield	The net rental income from a property at the time of purchase, as a percentage of the property's cost.
Interest rate cap	An arrangement by which a property company fixes a ceiling on the rate of interest paid on a loan, via the payment of a capital premium.
Interest rate swap	An arrangement by which a borrower of floating-rate funds exchanges its interest rate obligations with a borrower of fixed-rate funds, in order to provide fixed interest finance at a rate below that the borrower could arrange on its own account.[3]
Leveraged buy-out	A company take-over financed mainly via borrowed funds.
Limited-recourse loans	A form of off-balance sheet finance in which the parent company does not guarantee the sum borrowed, but provides some guarantees regarding the completion of the project and/or interest payments.[4]
Market value	The estimated current value of a capital asset, if sold in the open market, with a willing buyer and seller.
Marriage value	The excess value of two associated interests in land when held by the same person, or institution, over their value when held separately. This can apply both to adjoining portions of land and to different interests in the same property, such as a ground rent and a leasehold interest.

GLOSSARY OF ABBREVIATIONS AND TECHNICAL TERMS

Non-recourse loans	A form of off-balance sheet finance in which the loan is secured on the development rather than on the entire capital assets of the developer.
Off-balance sheet finance	Financial techniques which allow companies to incur debt, usually via associated companies or joint ventures, without the debt appearing in the group's consolidated accounts.
Peppercorn rent	A token rent, set well below a property's true rental value.
Prime property	A term used to define property which is of particular interest to investors. Definitions vary, but a prime property can be broadly defined as a modern or recently refurbished building, finished to a high specification, well situated in a commercially strong geographical location and let to a good tenant.[5]
Rate of return	A measure of the performance of a capital asset over a given time period, representing capital appreciation (or depreciation) during that period, plus net income from the asset over the period as a proportion of its initial value.[6]
Rent charge	A payment which is not rent but is supported by a power of distress and charged upon land.[7]
Sale and leaseback	An arrangement whereby a property is sold, the vendor simultaneously being granted a lease on the property by the purchaser.
Secondary property	A term used for property which is defective in one, or possibly two, of the characteristics which define prime property.
Securitization	The conversion of assets into tradeable securities.
Sterling commercial paper	A form of short-term finance, involving companies issuing securities at a discount to their face value, which are redeemed at the end of their term at their full value (the excess of full value over discounted value acting as the interest rate).
Unitization	A form of securitization, which involves the division of property rights in commercial buildings into a number of tradeable securities.
Years purchase	The amount by which the net income from a capital asset is multiplied to arrive at the capital value; the inverse of the yield.
Zero-coupon bond	A bond which pays no interest, but provides a return to the investor via its redemption at the end of its term at a price higher than its issue value.

REFERENCES AND NOTES

1. Where two abbreviations have been used, the shorter of the two is given in brackets.
2. Jones Lang Wootton (1989) *The Glossary of Property Terms*, Estates Gazette, London, p. 24
3. See M. Brett (1990) *Property and Money*, Estates Gazette, London, pp. 169–72
4. M. Brett (1990) *Property and Money*, Estates Gazette, London, p. 94.
5. D.P. Hager and D.J. Lord (1985) The property market, property valuations and property performance measurement, *Journal of the Institute of Actuaries,* **112**, 21.
6. Both capital appreciation and income/initial value are corrected for capital expenditure on the asset during the analysis period.
7. Jones Lang Wootton (1989) *The Glossary of Property Terms,* Estates Gazette, London, p. 155.

List of tables

Table 2.1	New public property companies 1845–1913	20
Table 2.2	The percentage distribution of life assurance company assets 1870–1913	27
Table 3.1	Business conducted by Hillier, Parker, May & Rowden, 1922–1937	42
Table 3.2	Indices of sales and lettings by Hillier, Parker, May & Rowden, corrected for changes in retail prices: 1922–1937	43
Table 3.3	Average annual net increase in number of branches of multiple retailers with 25 or more branches: 1876–1950	49
Table 3.4	Corporate growth with mortgage finance	56
Table 3.5	Corporate growth with mortgage finance and 5% cash inflow	57
Table 3.6	An outline of Marks & Spencer's growth: 1926–1939	61
Table 4.1	Net (inflow–outflow) new insurance company funds and net investment in mortgages and property 1922–1937	69
Table 4.2	New public property companies 1870–1938	73
Table 4.3	The distribution of NPI's mortgages between property companies and other borrowers: 1920–1938	87
Table 5.1	Net yields on institutional property holdings: 1938–1945	100
Table 5.2	A profile of the property investment market in 1949–1950	107
Table 5.3	Original occupations of property millionaires, 1945–1965	118
Table 6.1	New investment by Legal & General during 1961	137
Table 6.2	The Church Commissioners' assets, 31 March 1959	139
Table 6.3	Major leaseback transactions and funding arrangements undertaken by Legal & General to November 1960	150
Table 6.4	Mortgages held by members of the British Insurance Association in 1957	155
Table 6.5	Equity partnership arrangements between insurance companies and property companies 1939–1963	156
Table 6.6	Equity partnerships between various organizations and property companies 1955–1963	157
Table 6.7	The Income Conversion Factor for properties valued at 10 and 15 years purchase	161

LIST OF TABLES

Table 7.1	The impact of different rent review periods on effective property yields	170
Table 10.1	Net direct property investment by UK financial institutions (£M), nominal values, 1923–1993	248
Table 10.2	Net direct property investment by UK financial institutions (£M), real (1990) values, 1923–1993	252
Table 10.3	Property investment as a proportion of total investment for insurance companies and pension funds, 1963–1993	256
Table 10.4	An estimate of overall property investment activity (£M, annual averages for each period), 1946–1966	257
Table 10.5	An estimate of the percentage distribution of property investment activity by category of investor, 1946–1966	257
Table 10.6	The distribution of insurance company direct property holdings between rack-rented property and other property interests, 1870–1937	260
Table 10.7	The percentage distribution of institutional property holdings by property type, 1945–1990	261
Table 10.8	The geographical distribution of investment property (%), 1945–1990	264
Table 10.9	Regional distribution of investment property as percentage of regional GDP distribution, 1965–1990	265
Table 10.10	Regional distribution of offices, shops and industrial property as percentage of regional GDP distribution, 1970–1990	265
Table 10.11	Initial yields on investment property (%), 1920–1993	267
Table 10.12	Market rental indices for investment property 1962–1993 (1975 = 100)	272
Table 10.13	Average rates of growth of property market rents (nominal values), 1962–1993	274
Table 10.14	The data used to derive the 1921–1948 property returns series	277
Table 10.15	Capital, income and total returns on investment property (%), 1921–1993	277
Table 10.16	Annual returns on investment property (%), by sector, 1959–1993	280
Table 10.17	Average return and standard deviation of return on investment property, 1921–1993.	282
Table 10.18	Average rate of return for property by sector (%), 1959–1993	282
Table 10.19	Average return and standard deviation of return for property, shares and gilts (%), 1921–1993	283
Table 11.1	Current and average times to rent review for investment property, 1955–1993	287

List of figures

Figure 3.1	Sales and lettings arranged by Hillier, Parker, May & Rowden, corrected for changes in retail prices: 1922–1937	43
Figure 3.2	Maddox Street in the 1930s	46
Figure 4.1	Annual percentage changes in the volume of retail sales and retail construction 1930–1938, in real (1930) prices	76
Figure 6.1	The administrative structure of Legal & General's Estates Dept. at the beginning of 1957	136
Figure 6.2	The distribution of post-war office space in England and Wales (over 200 000 sq. ft. built or under construction) 1945–1962	142
Figure 6.3	The market capitalization of quoted property companies and the average market capitalization per company (£M), 1958–1967	159
Figure 9.1	The *Estates Times*' 'STACKUP' cartoon of 24th May 1985, reflecting the new trend away from modernism in commercial architecture	216
Figure 9.2	Bank property company debt as a percentage of all commercial loans, 1970–1994	231
Figure 9.3	Property company gearing (debt as a percentage of net asset value), 1987–1991 (median gearing for each year)	233
Figure 9.4	Vacancy rates for City and West End offices, 1985–1994	238
Figure 10.1	Real net institutional property investment 1923–1937	254
Figure 10.2	Real net institutional property investment 1946–1993	254
Figure 10.3	Real net property investment by category of institution ~ 1954–1993	255
Figure 10.4	Net property investment by financial institutions, property companies and overseas investors, 1980–1993	259
Figure 10.5	Net yields on investment property 1920–1938	269
Figure 10.6	Net yields on investment property 1946–1993	269
Figure 10.7	Yield differentials for property and gilts 1920–1993	270
Figure 10.8	Yield differentials for property and equities 1920–1993	270

LIST OF FIGURES

Figure 10.9	Annual percentage changes in market rents for investment property, 1963–1993	272
Figure 10.10	Annual percentage change in market rents for investment property, in nominal and real terms, 1963–1993	273
Figure 10.11	The capital and income components of property investment returns, 1921–1993	279
Figure 11.1	Current and average times to rent review, 1955–1977	288
Figure 11.2	Returns on property, equities and gilts, 1955–1977, expressed in terms of nine-year moving averages	288
Figure 11.3	Five-year moving average of property investment returns 1963–1993	290
Figure 11.4	The actual and fitted values of LCAP, 1965–1993	293
Figure 11.5	The actual and fitted values of LINV, 1965–1993	295

Foreword

Sixty-three years spent as an active estate agent, dealer and developer is perhaps the reason why I am asked to introduce this book.

I predict *The Property Masters* will prevail as the leading definitive and comprehensive history of the commercial property market from 1800 to the present time. The book is divided into two sections – the first providing a masterful account of the development of the British commercial property sector, recounting the activities of successive generations of property developers and investors against a background of the economic and other conditions under which they operated. It presents an accessible and highly readable account of the property sector's history, together with an authoritative analysis of the factors that have shaped events in the property investment and development market over the last two centuries.

Peter Scott has gained unprecedented access to the private business papers of a number of leading players in the property market, and using evidence from these he charts the factors which led these organizations to collectively develop one of the most sophisticated property markets in the world. The early chapters of the book show how the foundations for this market were laid during the years before 1945 – techniques developed during the inter-war years provided the basis on which property developers such as Jack Cotton, Charles Clore and Harold Samuel made their post-war fortunes. This led, in part, to the choice of the title of the book, since leading property companies were merely the names under which their masters operated.

The post-war property boom is discussed at length, with chapters outlining the investment boom in 'second-hand' property during the austerity years of 1945–1954, and the following property development boom in which over 100 people became property millionaires.

The more recent history of the property sector – a roller coaster ride of alternating booms and slumps – is also dealt with admirably. The factors which precipitated the property crashes of 1974 and 1990 are entertainingly described, together with a convincing analysis of the wider economic impact of the two crashes, particularly the most recent, the consequences of which the property market, and the wider economy, are still experiencing.

Section Two provides a statistical picture of the property investment market, discussing long run trends in property values, rents, yields, and investment levels. These are then used to analyse the factors which have driven property values and investment over the last 30 years, confirming the arguments presented in the first section of the book. Technically-minded readers should find this analysis provides a valuable supplement to the earlier chapters, while discussion of the findings of this Section in the earlier chapters, and the Conclusion, enables other readers to skip this part of the book while still being fully informed of its findings.

I find Chapter 12, which deals with the conclusions Peter Scott has reached, of particular interest in that he recognizes that the property market relies almost entirely on financial support from the banks and institutions and through them the investing public. The history of post-war property development and investment has been that of boom and bust. It is suggested that property finance should by-pass the banks entirely, relying instead on the securitization of property assets and property company debt. Advocates of this method of funding property development argue that as the market is far more sensitive to changes and conditions influencing future profitability than the banks, any indication of the likely future downturn in conditions would lead to a fall in the price of property securities which would reduce the magnitude of any slump by providing an early warning of a change in conditions.

This statement alone presages a change in attitude that may well have significant effect on the future of property development and investment, and Peter Scott's book could well be studied by banks and other institutions as well as property investors and developers.

Jack Rose

Introduction   1

1.1 THE AIMS OF THIS STUDY

The commercial property sector is of vital importance to the British economy. While the property industry makes a considerable direct contribution to national output, representing about 6% of Gross Domestic Product (GDP),[1] the overall significance of the sector to the prosperity of British industry is far greater than this figure suggests. Property forms a substantial element of the cost-base of the service and manufacturing sectors, accounting for 44% of the non-financial assets of UK companies.[2] Even this figure understates the true importance of Britain's non-residential property stock to its occupants, since a substantial proportion of properties occupied by corporate tenants are rented rather than owner-occupied, thus not constituting tangible company assets.

Property assets play a vital role in corporate finance. The British banking system has always had a marked preference for loans backed by collateral security, property forming one of the most important classes of acceptable collateral held by corporate borrowers. Indeed, it has been estimated that three quarters of all UK bank lending is dependent on property.[3] Property also offers opportunities for gearing which have allowed companies in property-rich sectors, such as retailing, to expand more rapidly than would have been possible with other forms of finance.[4] The growth of an active commercial property market has had an important influence on property's acceptability as collateral for corporate borrowing, by facilitating the valuation of property assets and increasing their marketability and liquidity. This link between the commercial property market and corporate investment has a long lineage, property having formed an important source of collateral for industrial and commercial finance since at least the nineteenth century.

Furthermore, in addition to constituting a major corporate asset, the character of Britain's commercial property stock also has an important influence on the technological and organizational flexibility of the work environment, which in turn has a substantial impact on efficiency in many service-sector industries. A large number of technological innovations in the retailing and office-based sectors are most easily introduced by their embodiment in new buildings, the ease with which companies can move to new premises, or redevelop their existing buildings, being a major determinant of the pace of diffusion for such innovations.

Property also plays an important role as an investment medium for the financial institutions and other long-term investors; the total value of British commercial property in 1989 amounted to more than twice that of UK government stock and about half that of the British equity market.[5] Much of the 'prime' high-quality segment of this stock is held by insurance companies and pension funds, which rely on its rental income and long-term capital appreciation to pay a significant proportion of their financial obligations to policyholders. The commercial property sector is also of considerable importance in employment terms; 236 000 people were employed in the real estate service sector in 1990, employment having increased at an average rate of 4.2% during the previous 30 years.[6] This is in addition to the very substantial construction employment which the sector generates.

However, while the property sector has made an important contribution to the British economy, it has also proved particularly prone to the boom–bust cycle which has been a hallmark of Britain's post-war economic growth. Indeed the property development sector has played a leading role in that cycle, particularly during the last 25 years. The 1974 property crash led to the most severe financial crisis the British banking system has witnessed during the twentieth century, while the aftermath of the 1990 crash may have significantly obstructed Britain's subsequent economic recovery by reducing the value of corporate assets and restricting bank lending to the corporate sector.[7]

Despite its considerable economic importance, the commercial property industry has received much less attention from economists and economic historians than other sectors of similar magnitude. This study seeks to begin to redress this balance by providing the first comprehensive economic history of the British commercial property market, from the nineteenth century to the present. This will focus on the evolution of the commercial property investment market and its interlinkages with the property development industry, institutional investment, corporate finance, government policy, and the changing character of office, retail and industrial property in Great Britain.

The history of the British commercial property sector also casts light on a number of more general issues which are currently subjects of debate in economics and economic history. Questions such as the efficiency of capital markets, the nature and economic role of entrepreneurship, the adoption of inflationary expectations into investment decision-making, and the nature and causes of financial crises are, therefore, discussed in relation to the experience of the commercial property investment and development markets. The analysis of these more abstract issues in the light of the experience of the property market will, it is hoped, both cast light on the validity of different theoretical models which have been used to analyse them and facilitate a better

THE AIMS OF THIS STUDY

understanding of the economic forces which have shaped the evolution of Britain's commercial property stock.

Many aspects of the British property market – such as the development cycle, the framework of government regulation, town planning legislation, property taxation and the changing nature of user-demand – have been subject to either relatively long-term fluctuations or infrequent shocks. Furthermore, fundamental influences on the growth of the property investment market during this century, such as the expansion of the tertiary sector, the concentration of retailing activity in prime 'High Street' pitches, growing geographical concentration of other types of commercial property and the emergence of persistent inflation after 1945, have been even longer-term in nature. The study of the commercial property market therefore lends itself to long-term, historical analysis, since the impact of such influences on the property investment and development markets can only be fully appreciated when examined from a historical perspective, spanning at least several decades.

In addition, as Britain's current property stock has a genesis which, in many cases, originates before the Second World War (and in some cases, dates well before 1900) a study of the history of the property market over the last two centuries is necessary to explain the character of today's built environment. In this study, as in many other areas of academic investigation, the historical approach is justified as only by examining the past can we fully understand the present.

This study will outline the emergence of a commercial property market in Britain during the nineteenth century, and its slow displacement of the previous system of estate management by the landed gentry and 'tradition institutions', which had hitherto characterized urban property development. The rise of the multiple retailers during the present century provided a considerable boost to the growth of the property sector, by necessitating the emergence of a national market in commercial property, and providing a substantial flow of attractive, secure, investment properties, as is outlined in Chapters 3 and 4. The cheap-money period of the 1930s led to the first boom in institutional property investment, property providing a high-yielding asset, but one which offered little capital appreciation as rents were typically fixed for 99, or even 999, years. This period also saw the growth of substantial links between developers, long-term investors and occupiers of commercial property, which were to form the basis of the substantial expansion of the sector following the Second World War.

During the immediate post-war years institutional investors remained happy to buy property on long leases at fixed rents. Selling properties to institutions on this basis, and simultaneously taking out long leases on them, proved an ideal means of raising capital during the 1940s and 1950s when other forms of finance were being severely rationed by government. Money raised by such activity formed the basis of the initial fortunes of many leading property developers, and allowed Britain's first hostile take-over bidders, such as Charles Clore, Isaac Wolfson and Hugh Fraser, to finance their activities.

Eventually, by the mid to late 1950s, institutional investors woke up to the dangers of inflation, and began to press for the transformation of property into an 'equity' security. The introduction of the key post-war innovation in the property investment market, the rent review, allowed rents paid on property to be periodically adjusted to market levels at regular intervals specified in leases. Gradual reductions in the intervals between these adjustments during the following years led to a steady upward movement in returns to investment in commercial property, any oversupply of new developments being prevented by the imposition of the 'Brown Ban' on office development in and around London in 1964.

However, by the late 1960s the property market was becoming 'overheated' with institutional and other funds. The relaxation of controls on office development, and the deregulation of the banking sector, by the Heath government during the early 1970s led to a massive boom in new development. Eventually a reversal of government monetary policy precipitated Britain's first property crash in December 1973, marking the end of the long post-war property investment boom. It was replaced by a regular boom–bust property investment and development cycle, which (intensified during the mid to late 1980s by a variety of short-sighted government policies) produced a further catastrophic collapse in commercial property values in 1990, from which the property sector has yet to recover.

This book looks in detail at the factors contributing to the early growth of the property market, the operation of the property investment market in the non-inflationary world of the 1930s, the symbiotic relationship which emerged between the financial institutions and the property developers during the 1945–1965 property boom, and the causes of the instability which has characterized the recent history of the commercial property sector.

1.2 SCOPE, SOURCES AND METHODOLOGY

This study is confined to investment in, and development of, UK commercial property. It omits any detailed analysis of investment in residential property, overseas property and agricultural land. Residential, agricultural and overseas property are avoided since the economic, and other, factors determining conditions in these markets are very different from those influencing commercial property. In addition, the main investors in, and developers of, commercial property generally undertake little or no activity in these sectors (with the exception of some major property development companies which have expanded into overseas markets). These sectors are, therefore, only discussed where a clear interlinkage between them and the commercial property market (or major 'players' in that market) exists.

This book also omits any detailed discussion of the industrial property market, other than with regard to the evolution of institutional investment and

SCOPE, SOURCES AND METHODOLOGY

property development funding in this sector. While factories and warehouses have been developed on a speculative basis since at least the 1920s, to adequately discuss the evolution of the industrial property development market is beyond the scope of the present study. Such a discussion would require an examination of issues such as the changing framework of government regional, and location of industry, policy; the evolution of the industrial estate; the interaction between industrial location and transport infrastructure; and the histories of the specialist industrial property development companies which have dominated this sector, doing justice to which would be the topic of a book in itself.

There is very little published material covering the history of the commercial property market, the growth of institutional property investment, the property development process and the financial returns to investment in property, prior to the recent past. Secondary sources on the commercial property market fall into four broad categories. Firstly, there are a number of books and articles written by financial journalists, such as Oliver Marriott's classic *The Property Boom* (1967). These works contain some extremely useful information, but are generally based on interview evidence, the accuracy of which is necessarily questionable with regard to events which occurred many years prior to the time they were written.[8]

The second group of sources consist of books written by people who have worked in the property industry, such as Charles Gordon's *The Two Tycoons* (1984) and Jack Rose's *The Dynamics of Urban Property Development* (1985). These studies sometimes provide considerable insight into the inner workings of the property investment and development sectors, though once again in many cases evidence is largely based on personal recollection, or the recollections of others.

Thirdly, there are a number of academic books, written by urban economists and others, covering various aspects of the property sector. Very few of these discuss the historical development of the sector in any detail, a notable exception being Hedley Smyth's book *Property Companies and the Construction Industry in Britain* (1985). There are also a handful of published company histories of property companies, property investing institutions and commercial chartered surveyors; two of the best books in this category are R. Redden's *The Pension Fund Property Unit Trust: A History* (1984) and Michael Cassell's history of Slough Estates, *Long Lease* (1991). However, compared to other industries of similar economic importance the property sector boasts very few published corporate biographies.

Finally, there are a range of statistical sources published by chartered surveyors, property research companies and other organizations, which provide data on property investment returns, market rents, yields and other market indicators. Few of these date before the 1970s, the most notable exception being the Allsop series of initial yields on investment property, which provides annual yield data from 1932.

The absence of extensive secondary material made it necessary to undertake a substantial amount of primary research. Two major avenues of investigation have been adopted. The first involved assembling a number of detailed historical case studies of institutions which have played a leading role in the property investment market. Case studies were compiled for four insurance companies which have been active in the commercial property market since the nineteenth century: Clerical Medical, Legal & General, the National Provident Institution and Standard Life, and one other property investing institution, the Church Commissioners. Primary research of a less comprehensive nature has also been undertaken into a number of other organizations, including the National Coal Board Pension Fund, Eagle Star, the Prudential, City Offices Ltd, Brixton Estate Ltd, Montague Burton Ltd, Marks & Spencer, Edward Erdman, Healey & Baker and Hillier Parker. This information has been supplemented by a number of published histories of particular companies, together with several more general works on the property industry. A number of interviews have also been conducted with people who are working, or have worked, in the property sector, either as institutional investors, chartered surveyors or developers.

The case studies, and other information relating to particular companies, provided a wealth of information regarding contemporary perceptions of property market conditions and opportunities, the timing and causes of long-term changes in investment behaviour, and the reaction of players in the property market to legislative and other 'shocks' to the sector. In assembling and selecting the case study, and other primary, evidence relating to particular firms, attention has concentrated on those companies which were among the leading, most innovative, institutions. This was necessary in order to gain a detailed picture of the timing and determinants of innovation in the sector, though it incurs the danger of giving a distorted picture of the overall market. In order to counter this, and provide a further avenue of inquiry to back up qualitative evidence from the case studies, statistical data has been collected on the aggregate behaviour of the property investment market.

Some statistical evidence is presented in Chapters 2–9, which examine the history of the commercial property market from the nineteenth century to the present. However, several of the questions outlined in section 1.1 required a more thorough statistical analysis than could be conducted within the confines of Chapters 2–9, and this is provided in Section Two. As statistical evidence regarding most aspects of the property investment market is extremely poor prior to the recent past it was necessary to compile series covering the years prior to the commencement of reliable official or commercially-produced indices. Series covering returns on capital invested, the distribution of investment property by geographical region and property type, the volume of annual property investment by various classes of investor, and initial yields on investment property, were extended further back into the past, using evidence from the business records of institutional investors together with a variety of other sources. The compilation of these series is outlined in Chapter 10, together with an examination of the major trends they reveal. Chapter 11 provides a discussion of the main long-run factors influencing returns to investment in property, together with an econometric analysis of the principal long-term determinants of rates of return and levels of investment in the sector.

The statistical section aims to provide an aggregate picture of the long-term behaviour of the property investment market, testing hypotheses which are put forward in the chronological chapters. Finally, Chapter 12 draws upon the evidence presented in both the chronological and statistical sections of this book to derive conclusions regarding the economic impact of the property cycle, the causes of the property sector's recent instability, and the role of government, the banks, the financial institutions and the property developers in preventing a recurrence of the boom–bust cycle which has resulted in two catastrophic property market crashes during the last 25 years.

REFERENCES AND NOTES

1. Commercial property values: forecasts from the London Business School–Royal Institution of Chartered Surveyors model (1992) *Economic Outlook*, October, 28.
2. Douglas McWilliams (1992) *Commercial Property and Company Borrowing*, Royal Institution of Chartered Surveyors, Paper No. 22, November, 3.
3. V. Houlder (1992) Why bricks are no longer bankable. *Financial Times*, 10 April, 16.
4. See Chapter 3.
5. Commercial property values: forecasts from the London Business School–Royal Institution of Chartered Surveyors model (1992) *Economic Outlook*, October, 28. (Not all UK commercial property can be regarded as part of the property investment market.)
6. Information supplied by T.W. LaPier, based on official sources.
7. See Chapter 9.
8. Although in the case of *The Property Boom* [O. Marriott (1967), Hamish Hamilton, London] the archival and other evidence examined in the course of this study has very largely served to confirm, rather than challenge, Mariott's analysis.

SECTION ONE

The evolution of the British commercial property market since 1800

The emergence of a commercial property investment market 1800–1918 | 2

2.1 INTRODUCTION

The nineteenth century witnessed the emergence of a commercial property investment market in Great Britain, replacing the previous system of 'estate management' by aristocratic and other 'traditional' landlords, whose property acquisition and development strategies were strongly influenced by social and political, as well as monetary, considerations. This chapter examines the factors which contributed to this transition, the ways in which barriers to an active market in commercial property were, to a limited extent, overcome, and the reasons behind the lack of significant direct investment in commercial property by the financial institutions prior to the First World War.

2.2 'TRADITIONAL' LANDOWNERS AND THE NINETEENTH CENTURY URBAN PROPERTY MARKET

The 'traditional' landowners (principally the aristocracy, Crown and Church, plus educational, social and charitable institutions such as Oxford and Cambridge colleges, public schools, London livery companies and hospitals) which had dominated the urban property market during previous centuries remained central players during the Victorian period. Their policies towards urban property underwent only minor adaption from the pattern which had emerged by the end of the eighteenth century, involving the development of urban landholdings, when opportunity arose, by the granting of building leases.

From the 1840s the pace of estate development was increased, due to changes in the law relating to settled estates and to charity and ecclesiastical land, allowing its development via building leases.[1] The accelerating growth of a large number of British towns and cities also facilitated development by bringing more aristocratic estates within urban boundaries.

The great London estates to the west of the City, owned by the Russell, Grosvenor, Portman and Cavendish-Bentinck families, continued to be run using estate management practices established during the seventeenth and eighteenth centuries, albeit with some refinements. This involved letting properties on 99 year building leases and 21 year repairing leases, both including restrictive convenants designed to prevent alterations which would mar the original building plan. The same was true of smaller estates such as those of the Howard and Cecil families and the Foundling Hospital.[2] In additon to these estates, which had already been extensively developed by the nineteenth century, those 'traditional' estates on the outskirts of London, which became 'ripe' for development as its boundaries expanded during the Victorian period, adopted similar estate management policies to those of their more central neighbours.[3]

Aristocratic and other 'traditional' landlords would often invest substantial sums in the laying-out of roads and the provision of other amenities on estates before granting building leases, thus making them more attractive to speculative developers by reducing their costs. The extent of such expenditure tended to vary according to the distance from the nearest developed area; it was particularly high in the case of relatively isolated seaside resorts such as Eastbourne and Skegness. Other ways in which landlords might indirectly contribute to, or reduce, the capital cost of speculative development included granting peppercorn or reduced rents for the first several years of leases, and, as Olsen has noted, failing to impose or enforce stringent convenants regarding the quality of building.[4]

Successful estate development sometimes entailed more direct financial assistance to developers on the part of the ground-landlord. In the late 1850s the Duke of Devonshire advanced £37 000 to speculative builders on his Eastbourne estate, at a time of crisis in the building trade, so that they could finish houses where construction had already commenced. Other instances include loans provided by the Duke of Norfolk, at 4% interest, for the rebuilding of his estate and the financial assistance provided to Thomas Cubitt by the Duke of Bedford in the late 1820s. While Cannadine claimed that such instances were sufficiently unusual to be described as 'extraordinary and unique' occurrences,[5] Olsen argued that their use was 'a common, but far from universal device'.[4]

It was rare for 'traditional' investors to purchase urban land or rack-rented property, other than limited acquisitions to consolidate existing estates. Their capital was usually tied up in their current landedholdings, the sale of which was prohibited 'almost without exception' by elaborate statutes or settlements.[6] In a few cases, where there were no statutes preventing sales, the redeployment of capital did occasionally occur; for example, during the 1880s the Marquess of Salisbury's trustees invested heavily in London ground rents, in addition to railway stock and debentures, using the proceeds from sales of consols and land.[7]

The legal difficulties involved in the sale of land were also a factor behind the characteristic mode of aristocratic estate development – the 99 year building lease. The development of estates via this means, rather than outright sale, was attractive as it did not fall foul of settlements preventing sale, and allowed the owner to retain the prestige associated with landownership.

Prestige and political power are factors which are not possible to quantify when assessing the investment patterns of traditional landowners, but were clearly of considerable importance. In 1823 Lord Calthorpe had planned to sell the Edgebaston estate in Birmingham, and his family's other urban property, and purchase an 11 000 acre agricultural estate in Culford, Suffolk. This was expected to result in a fall in income, but would have provided his family with real territorial power in East Anglia, which was viewed as being a worthwhile exchange. As Cannadine stated in his account of the Calthorpe's development of Edgebaston:

> Even though the rental from Culford was less than the combined income from Edgebaston and Gray's Inn, the attraction of the greater social and political influence which would accrue from consolidated holdings gave the Culford estate 'a value beyond a *mere* investment for money'.[8]

While the political value of agricultural land declined during the nineteenth century, political considerations still exercised considerable influence over aristocratic investment decisions. In addition to these non-monetary factors there was also an important financial rationale behind development of urban estates on 99 year leases, rather than simply selling the land. The financial gains from the development of landed estates were usually greater than the income that might be earned from the reinvestment of the proceeds from outright sales 'by a handsome margin'.[9] Between 1810 and 1888 £47 000 was invested by the Calthorpes in the making of roads at Edgebaston. This produced a rise in rentals from £5233 to £28 882. Even more dramatic rises in income were produced on other building estates, for example, the School of King Edward VI in Birmingham owned a smaller Birmingham estate than the Calthorpes, but benefited from a rise in ground rents from £11 000 to £27 000 over the much shorter period of 1863–1885. The Derby family's income from estates in Liverpool and Bury quadrupled between 1800 and 1837, while the Bishop of London's income from his Paddington estate grew from £5797 to £27 703 from 1833 to 1853. As Cannadine stated:

> The figures speak for themselves, and do so even more eloquently when it is recalled how much other families were investing in agricultural improvements in these years, and how little they were getting back.[10]

Such gains were usually 'windfall bonanzas', arising from the fortuitous ownership of land which happened to have development potential, rather than the result of any carefully thought-out policy to acquire such land. Traditional landowners might buy, sell or develop property to consolidate and improve relatively fixed estates, but did not undertake investment on the basis of purely financial considerations. As such, the aristocracy were not active participants in the urban property investment market, since they were not engaged in extensive purchases of property but concentrated, instead, on realizing the development potential of their existing holdings.

What was true of the aristocracy appears to have been equally true for other traditional landowners, such as the Crown, the Church, Oxford and Cambridge Colleges, public schools and City livery companies. The Ecclesiastical Commissioners undertook some redevelopment of the Church's urban estates from the late 1860s. In order not to be seen to be exploiting their tenants, many of whom were among the poorest members of London's working classes, this redevelopment was largely conducted on a philanthropic basis, with a target return to capital invested of 4% or less.[11] During this period the Commissioners did not undertake any extensive redeployment of resources into urban property, other than this redevelopment activity, and appear to have faced constraints similar to those experienced by aristocratic landlords.

The Oxford and Cambridge colleges were in a better position to develop and rationalize their property holdings during the late nineteenth century than had been the case in previous years, as a result of legislation introduced in the late 1850s which gave them powers to grant building leases of over 40 years' duration and to sell land.[12] Land sales during the late nineteenth century were generally undertaken in order to consolidate holdings, rather than as part of a strategy to move funds from agricultural land to urban property, or other assets.[13] Colleges made more extensive use of their powers to grant 99 year building leases, however, the proceeds of urban development resulting in an increase in college incomes from 1870 to 1913, despite a fall in rents from agricultural holdings. The behaviour of the colleges therefore conforms closely to that of the aristocracy, developing what urban land they happened to own but not redeploying their resources to increase holdings of such land.

Traditional landlords therefore appear to have pursued policies more akin to estate management of landholdings, the boundaries of which were broadly taken as given, rather than the switching of resources between securities, with the sole aim of maximizing profits. Legal restrictions and non-monetary factors such as tradition, prestige and political power appear to have been the main factors behind this relative immobility of funds. However, even during the early nineteenth century, a number of investors not encumbered by tradition or existing holdings did take an active role in the urban property market. These included a group of institutions which were to play a major role in commercial property investment during the following century, the insurance companies.

2.3 INSURANCE COMPANY INVESTMENT IN THE PROPERTY SECTOR 1800–1870

While there was little direct investment in property by insurance companies prior to 1870, other than buildings purchased for occupation, a considerable volume of insurance company funds were channelled to the property sector indirectly via mortgage lending. The first half of the nineteenth century witnessed a substantial switch of insurance company funds from government stock to mortgages. Mortgagees were initially mainly aristocrats, borrowing on the security of agricultural land. However, loans on urban property formed a growing proportion of mortgages as the nineteenth century progressed.

Substantial insurance company mortgage lending for residential development took place from the 1850s. This period saw an increase in the scale of urban development, the traditional system of constructing houses to order (which required only limited funds) being replaced by more extensive speculative development of estates, involving building ahead of demand.[14] The increased scale of development was associated with rail and other transport improvements which opened up new areas to residential occupation as proximity to the workplace became less important. In addition to their direct loans to speculative builders the insurance companies also contributed to the development process by lending to ground landlords. For example, the National Provident Institution made loans to Eton College in 1865 and 1869 to facilitate its development of the Chalcots estate in South Hampstead.[15]

Insurance company lending to urban developers was concentrated in large schemes, involving middle class or 'superior' working class dwellings. However, even speculative builders constructing cheaper houses appear to have found it relatively easy to raise capital. D.J. Olsen has characterized the supply of capital for the London speculative building industry during the nineteenth century as 'buoyant', with investors 'virtually forcing their money on the builders'.[16] Much of this finance, especially in the case of small and medium speculative builders who made up the vast bulk of the industry, was provided by solicitors on behalf of their investing clients. Building societies also made extensive loans to finance development projects during this period

In addition to mortgage funding, insurance companies occasionally undertook investment in building estates directly, either voluntarily or as a result of mortgage default by the originator of the development scheme. One of the earliest companies to become involved in this type of investment was Standard Life.

The Standard Life Assurance Co. was established in Edinburgh in 1825 and invested the bulk of its assets in loans on landed property from the 1840s to the 1870s. Direct investment in land was undertaken by Standard as early as the 1850s, for reasons that also loomed large in property investment thinking in the late twentieth century, namely the value of land as a hedge against inflation. At a meeting of the Board's Law Committee of 31st March 1852, Standard's manager, W.T. Thomson, argued that the increase in the supply of gold was likely to have inflationary results, eroding the real value of fixed interest securities such as mortgages. He suggested that the company 'should have some portion of their funds at least invested in such a manner as would allow them to derive benefit from the rise in value',[17] proposing land as the most suitable investment.

Thomson suggested several estates to the Committee, of which two, Leithenhopes and Lundin, were selected for further consideration. The estate at

Lundin, Fife, was purchased in June 1852 for £90 000. A committee of five directors was established to manage the estate, comprising nine farms, all but one of which were let. A tenant was quickly found for the remaining farm and leases were renegotiated as they fell in during the winter of 1852/3. The development of the Fife coast railways offered the prospect of residential development of the seaside estate as a resort, bringing it within reach of Edinburgh commuters. The railway's promoters realized its value and in September 1852 they asked Standard to subscribe £2000 of the share capital.

The Lundin Estate Committee postponed any decision regarding capital subscription. However, in January 1855 the Board was informed that the projected line from Leven to Kilconquhar would pass through the estate and after extensive negotiations with the East of Fife Railway it was agreed, in April 1856, to provide land for the line at what was considered a very reasonable price. In return the railway company agreed to fence the line, provide reasonable access for the estate's tenants and build a station at Lundin Links, which the Board intended to develop as a holiday resort with golfing facilities.[18] Standard also supplied capital to the railway company, via a loan of £8000, made on the personal security of its directors.

Building feus were drawn up on the estate and were available from May 1855, but though the estate was brought within reasonable commuting distance of Edinburgh by the railway no one came forward to build on the site. This did not dent the Board's enthusiasm for this type of investment; a continuous estate in Balcormo was also purchased, for £13 650. Upon the opening of the East of Fife Railway a new feuing plan was immediately prepared, the Board agreeing to spend £2000 on house building and undertake other work to encourage development.[19] The financial crisis of 1857 dealt a severe blow to the building industry however, and by early 1860 only two villas had been built on the site by Standard.

While the income produced by the estate was never outstandingly good, the value of the estate and adjacent land gradually rose, adding a capital component to the venture. By 1861 the company had spent a total of £126 875 on the project, while its estimated value was put at £153 360. Net yields to that date averaged just over 4% on the original purchase price, or 3.35% on the 1861 valuation.[19] The estimated value of the estate was not quite matched by capital gain to the company. The Board decided to sell the estate in July 1870 at a price set at £165 000. No purchaser was found and, nearly two years later, the estate was finally sold for £151 250; this raised the overall return on purchase cost over the 20 years it was held by the company to over 5% per annum.[20]

Another insurance company which purchased an estate for development during the mid-nineteenth century was the British Empire Mutual Life Assurance Co. In 1866 it acquired part of an estate in Camberwell, South London, on which an abortive attempt at development had occurred as early as 1836. The land was purchased for £14 765, which, at about £615 per acre, was a rather high price for the time.[21] It is not clear whether the company intended to simply lay out building plots and allow someone else to conduct the actual development or to undertake development of the site itself. However, a week after the contract to purchase was signed, the dramatic collapse of the discount house of Overend, Gurney & Co. led to a shortage of short-term credit which depressed the building industry and may have forced British Empire Life into providing more extensive finance for the estate's development than was originally intended.[22]

The following three years were spent laying out the estate, in conjunction with the Perpetual Investment and Building Society, which shared the costs of preparing the site with British Empire Life.[22] The first building agreements were signed in 1869; six years later only 150 houses had been built, or were in the course of construction, and the majority of the land remained under grass.[23] The long-term success or failure of this venture is not known. By 1874, the last date for which figures are available, the company had invested £46 665 in the project, the income from which (£1331 a year) offered a yield of only 2.85%.[24] It would be unfair, however, to assess the financial success of a long-term project such as this, the development of which was not completed until about 1896, on the basis of figures for only the first eight years.

These ventures provide rare examples of insurance companies voluntarily acquiring estates during the mid-nineteenth century. A number of other insurance offices became owners of estates via mortgage default on the part of their original developers. The experience of managing such estates acquainted them with the problems of active participation in the property market, while the element of adverse selection involved in their acquisition meant that the benefits of such involvement were not similarly emphasized. Their ownership did little, therefore, to commend property as an investment medium.

The Royal Exchange Assurance (REA) began to lend extensively to urban developers in the 1840s. In about 1842–1843 it was forced to take over an estate of leasehold houses in Clapton and Denmark Hill, on the security of which it had loaned £49 000. Additional funds were invested in repairing faulty construction of some of the houses and managing the estate, which yielded a fair return to REA for over 60 years.[25] In 1846 another London developer, George Wyatt, who was REA's largest debtor with a loan of £135 400 plus arrears of interest amounting to £3229, went bankrupt. As a result, REA acquired house property which had been valued at twice the sum loaned. Despite this considerable margin of security it was estimated that sales would produce an overall loss of £30 000 (including lost interest). This default led to a drastic reduction in REA's mortgage lending for urban development.[26]

The National Provident Institution (NPI) was established in 1835. The Institution's first major venture into the property market was initiated in 1856 by a loan of £50 000 to Colonel Greville, the owner of an estate in Milford Haven, for a scheme to develop the town as a major port. He envisaged the construction of an ocean terminal for the export of South Wales coal, in a location which Admiral Nelson had described as one of the finest natural harbours in the world.[27] A favourable report from NPI's surveyor and three directors who visited the town convinced the Board that this was an excellent proposition. However, the estate was never able to generate sufficient income to justify the initial and subsequent investment, and failure to meet mortgage payments led to ownership of the estate passing to NPI.[27] To improve the prospects of their investment NPI lent money on debenture to the Milford Dock & Railway Co., which was developing the port. NPI became closely involved with this company; in March 1868 it was resolved that three NPI directors would join its Board.[28]

However, by July 1882 the Milford Dock & Railway Co. had been taken into receivership, NPI's attempts to develop the estate by aiding the company having proved unsuccessful.[29] By 1877 it had been decided to dispose of the investment at the first opportunity, but it was not until 1920 that this proved possible.

The Legal & General Life Assurance Society, founded in 1836, also became involved in property development funding in the mid-1850s, investing heavily in the development of land in Birkenhead, near London, as a residential estate.[30] Default by the mortgagee led to the acquisition of the Birkenhead estate by Legal & General, which decided to continue with the plan of its previous owner to develop the site as residential building land. Building plots proved difficult to sell however, despite initial optimism, and the estate constituted an unproductive and time-consuming asset for the Society for a number of years.

These estates formed only part of larger programmes of mortgage lending to property developers by these three institutions. By the very nature of such lending only those investments which could not meet mortgage payments would pass into the ownership of the mortgagor; they might, therefore, be best regarded as an indication of the possible costs of mortgage lending for such projects rather than evidence of the risks of property development *per se*. It is not clear, however, that the institutions concerned took this view of what proved to be some of their largest and most troublesome single investments.

2.4 THE CHANGING NATURE OF THE PROPERTY INVESTMENT MARKET 1850–1914

While traditional investors continued to dominate the property market during the first half of the nineteenth century, from the 1850s the transformation of urban real estate from a social institution to a financial asset began in earnest. A number of factors contributed to this process. The second half of the century witnessed a substantial rise in the income of Britain's growing middle class, and the consequent growth of middle-class savings.[31] A class of investors therefore arose who were concerned almost exclusively with financial returns, rather than the non-monetary factors which had considerable influence over the investment patterns of the aristocracy.

THE CHANGING NATURE OF THE PROPERTY INVESTMENT MARKET

The investment opportunities available to the lower middle class channelled much of their savings towards the property sector. The businessman George Cross recalled the investment advice available to modestly successful tradesmen and professional men, such as his father, in the late Victorian period:

> They go to the bank manager, and he will advise them to buy stocks or shares through the bank ... They may try a solicitor. In nine cases out of ten, before the expansion of building societies, a lawyer would tell them to invest in mortgages ... they may turn to bricks and mortar, and consult an estate agent. They would then have a large selection of securities placed before them ...[32]

Even if the investor took advice from the bank, his/her savings might still end up in the property sector, via the stock market. One of the main obstacles to investment in real estate is the indivisibility of property assets, direct investment as part of a diversified portfolio being possible only for investors with very substantial funds. In order to overcome this problem various means of 'unitization', i.e. the division of property assets into a number of identical financial units, have been attempted. The first of these was the property company.

The ownership, maintenance and development of property by limited liability companies offered investors the possibility of acquiring a stake in the property sector without the need to commit substantial funds or deal with management and maintenance. Important factors behind the emergence of property companies during the second half of the nineteenth century included the growing acceptability of limited liability investment, the rising volume of middle-class savings referred to above, and high perceived returns on property investment as a result of the rise of the City as a commercial centre, urban growth and transport improvements, particularly the spread of the railways.

Table 2.1 shows new public property companies established prior to 1914, which were still in existence in 1950. The table excludes the subsidiaries of other companies whose ordinary shares were entirely held by the parent (though not property companies which had been taken over by other companies and continued as subsidiaries), companies which were originally founded to conduct a business other than property and companies investing mainly in overseas property. The table refers to public companies, though some were originally incorporated as private companies, and are included as of the date of their incorporation.

The first property company listed was founded, by Royal Charter, in 1845. The period from 1850 to 1870 saw a moderate growth in the number of new property companies, more rapid growth occurring in the 1880s and 1890s. This was followed by a decline during the 1903–1913 period as a result of adverse conditions in the property investment market and the political threat to property posed by the 1906 Liberal government.

Table 2.1 New public property companies 1845–1913

	Total	Annual average
1845–1849	1	0.2
1850–1859	2	0.2
1860–1869	7	0.7
1870–1879	7	0.7
1880–1889	14	1.4
1890–1902	20	1.5
1903–1913	9	0.8

Source: Thomas Skinner & Co. (1950) *Skinner's Property Share Annual 1950–51*, Thomas Skinner, London.

A further factor behind the growth of the property investment market during this period was the development of market intermediaries and a market press. The 1850s saw the establishment of a number of journals covering property transactions, including *The Freeholder* (1850), *The Journal of Auctions* (1853), *The Freehold Land Times* (1854) and *The Estates Gazette* (1858). Most of these publications were short-lived, with the exception of *The Estates Gazette*, which soon became the dominant journal of the property market. The development of a property press provided a vital source of market intelligence for an investment medium characterized by the lack of a centralized market, an extremely heterogeneous product, and the absence of any clear indicator of prevailing market conditions. As F.M.L. Thompson stated:

> On the whole there seems to have been a fairly widespread feeling that there was a need for a specialised advertising medium for real estate and for some attempt to provide regular property market reports. That this feeling arose when it did seems to imply that there was heightened activity in the land market, calling for alterations and improvements in its machinery.[33]

During the 1850s an attempt was also made to establish a centralized marketplace for land, which would be the property market's equivalent of the London Stock Exchange. The Estates Exchange was set up in 1857 by a group of London auctioneers. The Exchange maintained registers of property for sale, and property sold, so that market information covering the entire country could be examined in one office.[33] The ultimate goal of the Exchange's founders, to develop the institution into an actual market, with brokers and jobbers operating in a similar fashion to the Stock Exchange, was never fulfilled. However, the registers were continued, providing valuable information on prevailing yields and market turnover.[34]

The second half of the nineteenth century also saw far-reaching changes in the character of commercial property, which were both facilitated by the growing property investment market and, in turn, stimulated the development of that market. At the beginning of Chapter 3 five conditions are outlined, each of which are necessary, or at least of considerable importance, for a substantial volume of property investment activity to take place:

1. A substantial number of commercial property transactions, providing a steady stream of investment opportunities.
2. The growth of an efficient and integrated property investment market, covering the geographical area in which the investments are situated.
3. The development of a 'scarcity premium'[35] for investment property, the scarcity of prime property sites ensuring that investment property is likely to maintain, or increase, its value over time.
4. A financial climate in which commercial enterprises have considerable incentives to rent, rather than own, their business premises, or to dispose of some property rights to their premises in return for an immediate capital sum.
5. The emergence of a significant differential between the performance of property and alternative assets, property performing sufficiently better than other investment media to overcome its inherent disadvantages of lack of marketability, uncertainty of value, indivisibility and higher management costs.

Some, though by no means all, of these conditions were fulfilled during the late nineteenth century in London's most advanced commercial property market, that for City of London offices.

2.5 CITY OFFICE DEVELOPMENT AND THE PROPERTY MARKET

The development of the office block as a distinct building form began in the early nineteenth century; as late as 1793 the Treasury employed only 37 people, while the office functions of business were dealt with in coffee houses, counting houses and the homes of merchants.[36] However, it was during the eighteenth century that the first specially designed office and banking buildings had appeared in the City, including those of the East India Company (1726) and the Bank of England (1732).[36]

Whereas during the eighteenth century many businessmen could operate from their favourite coffee house, the need for a formal office grew significantly during the early nineteenth century. As Jon Lawrence has noted 'The growth of paper documentation, the need to be personally contactable, and the need to be near to public sources of commercial information such as the Post Office, reading rooms and coffee houses, made it increasingly essential to maintain a formal office at no great distance from the key centres of commercial activity'.[37] As a result the period from 1800 to 1850 saw a considerable increase in the proportion of City buildings used as offices, largely displacing the City's residential population.

Until the 1830s this process usually involved the adaption of existing premises, so that their previously residential upper floors could be let for commercial use.[38] London's first speculative office block was erected in Clements Lane by a Mr Voysey around 1823, though the number of speculative office developments only became at all significant in the 1840s,[39] gathering pace during the mid-Victorian period. According to estimates by John Lawrence, between 1817 and 1851 the aggregate assessed rental value of property in the City's financial district rose by 2.11% per annum in real terms; this rate of growth accelerated to 3.96% per annum from 1851 to 1871.[40] The pattern of City office development was mirrored by that of other major commercial centres. For example, during the early nineteenth century the growing demand for offices in Liverpool's central commercial district was met largely through the conversion of private houses and other buildings, the first purpose-built office blocks appearing in the 1830s and 1840s.[41]

The two decades after 1850 saw rapid commercialization of the City, which lost 60 000 people, almost half its resident population, as houses were replaced by offices and warehouses. Banks, insurance companies and other financial institutions sought prestigious offices, preferably in the heart of the City near the Bank of England, and the resulting pressure of demand on limited space resulted in a boom in City property values and rents.[42]

By the 1860s office development had become 'so considerable that almost all the eligible sites in the City have been converted to this purpose'.[43] The diminishing supply of available sites led to rising land values. In discussion of a paper read by Edward L'Anson to the Royal Institute of British Architects in 1864, a Mr J.J. Cole stated that he had recently entered into an agreement to sell some land in Throgmorton Street, purchased in 1854 at £4 per sq. ft., at a price equal to about £15 per sq. ft., confirming L'Anson's claim that land prices had trebled in some parts of the City during the previous few years.[44]

The rapid growth in office employment continued during the late nineteenth century. The mid-Victorian period had seen the advent of large-scale (though still largely unmechanized) data processing; by 1862 the Railway Clearing House employed 800 clerks, the Post Office Savings Bank had a clerical workforce of 300 in 1870,[45] and the Prudential (which pioneered the recruitment of women clerks) employed over 200 clerks by 1873.[46] A similar growth in office employment occurred in Britain's other major commercial centres; it has been estimated by B.G. Orchard that by 1870 9000 clerks were employed within 10 minutes' walk of Liverpool's Exchange.[47] This growth of large-scale clerical employment was facilitated by advances in communications and other office technology, such as the spread of the telegraph and the invention of the telephone, typewriter, arithmometer and stencil duplicator.

Investors soon sought to capitalize on these trends; 1864 saw the establishment of two public companies whose sole aim was investment in, and development of, City office property – City Offices Co. Ltd and the City of London Real Property Co. Ltd. These public companies operated alongside a number of smaller property companies established by groups of businessmen to develop specific sites, often partially for their own use, such as the Gresham Chambers Co. Ltd, also founded in 1864.[48]

City Offices was floated by the Mercantile Credit Association Ltd, in conjunction with the French bank Credit Mobilier Ltd.[49] The company's initial capital was set at £1 000 000, although it was only intended to call up £400 000 of this, with any further capital to be raised by new share issues. The company's prospectus stressed the potential for long-term capital appreciation offered by property in the vicinity of the Bank of England, stating:

> Of all securities or property that can come under the definition of 'first class', both for stability as well as increasing in value year by year, none can at all compare with freehold or long leasehold City of London property, especially that portion situated within a radius of 1,000 yards of the Bank of England; this fact is so well known, that the greatest competition is excited when any property of this description is offered in the market, and large as the rise in value of such property has been during the last few years, it is admitted with all conversant with the subject, that a still greater increase must continue to take place from the impossibility of supplying the suitable accommodation required by joint stock companies and commercial and professional firms.[49]

City Offices had arranged to purchase five blocks of office property, in Lombard Street, Leadenhall Street, Cornhill, Great Tower Street and Bishopsgate Street, on freeholds or long leaseholds, for £365 000. This worked out, according to the prospectus, at £9 per sq. ft. of floorspace. The company's plan of operations was to adapt those blocks that could be easily converted to multiple office use and rebuild the remainder. It was estimated that 'from calculations made, based upon moderate rentals, a return of from 15 to 20% on the outlay may be fairly relied on.'[49]

A highly geared capital structure was envisaged, as was to become the norm for property companies during the twentieth century: 'By the issue of debenture or otherwise, the paid up share capital will be kept at the lowest possible point.'[49] The reasons behind this capital structure were outlined in the prospectus for a £200 000 debenture bond issue by the company in August 1865:

> The advantage to shareholders from an issue of debentures such as the Directors propose is very great in as much as a property which pays 8% per annum on its cost; if three fourths of such cost be raised by debentures at 5%, and one fourth by share capital in the ordinary way, gives 5% on the debenture capital and 17% on the share capital.

The principle of gearing, i.e. the use of fixed-interest finance with a yield below that expected on the assets of a property company to magnify the yield on the ordinary shares, was, therefore, applied from the very beginnings of the property company sector.

The City of London Real Property Co. Ltd (CLRP) was founded by two brothers, James and John Innes, who were partners in a firm which imported rum into England and owned large sugar plantations in Jamaica. The abolition of slavery in the West Indies, the campaign for which they were said to have supported, adversely affected their Jamaican interests, leading them to switch their attention to City property.[50]

Unlike City Offices, which concentrated on property in the present heart of the City, CLRP's strategy was to anticipate the future movement of the City's commercial centre. Following the advice of a builder, William Verry, who believed that the City's future would involve expansion eastward, the Innes brothers purchased property in Mincing Lane (where they already held some property) and several more buildings in neighbouring Mark Lane.[51] Most were located in the produce markets and occupied by merchants and brokers connected with the London Corn Exchange and the London Commercial Sale Rooms. They accumulated a total of 12 properties, covering 1.5 acres, for £329 000. CLRP was established to hold these assets, with an authorized capital of £500 000.[51]

Conditions in the City property market were not to prove as buoyant as City Offices had anticipated; within a few years of its establishment the company was facing difficulties letting its accommodation. City Offices' half-yearly report for September 1867 noted that 'progress in letting has been less rapid than was hoped, still, considering the very depressed state of commerce during the past six months, this cannot occasion much surprise'.[52]

These difficulties continued for some years, the company's shares yielding only about 4% on paid-up capital, compared to the 15–20% estimated in the 1864 prospectus. In the late 1870s this increased to 6% but fell once again during 1879 and 1880. At the company's half-yearly meeting of 28th March 1881, the Chairman commented on the continuing depression in the City office market. The growing supply of office property had depressed letting conditions:

> ... it was well known that the supply of offices had wonderfully increased during the last two years. Those who were acquainted with the Drapers' Gardens Estate knew that the 5.5 acres of land which once formed gardens and long gravelled walks were now covered with offices, either tenanted or in the course of being occupied. Lime Street and other parts of the City had also been covered with new offices, and these alterations had created such an additional number of offices that the Board had some apprehension that the time might yet be a little distant ... when they would get what they considered the full value of their offices ... Their comfort had always been that their property was situated in the very best parts of the City of London, and that as trade and commerce in the City increased, there could be no doubt the demand for the offices would ultimately come all right.[53]

City Offices' problems persisted during the 1880s and 1890s, dividends seldom rising above 5%. At a Board meeting of 27th February 1899 the company's accounts included a revaluation of their properties. The freeholds were valued at £200 750 and the leaseholds at £235 761, compared to figures on a cost basis in the previous year's accounts of £234 481 and £483 993, respectively. City Offices' strategy of acquiring property in the most sought-after locations, to benefit from future capital appreciation, had not met with success; instead, the first 35 years of the company's history had seen falls, rather than rises, in the value of its assets.[54]

City Offices' disappointing performance was not due to any lack of demand for offices. The City's working population increased more than fourfold between 1866 and 1911.[55] However, the expansion in demand was met by a substantial increase in the supply of office accommodation. It has been estimated that 80% of the City's 1855 building stock had been replaced by 1905.[56] Furthermore, this rebuilding took place at considerably higher densities than the buildings they replaced. The floorspace density of street blocks in the City increased from about twice the gross area in 1840 to about four times (for those largely occupied by recent buildings) in the late Victorian period. It was estimated that during the Victorian period the City's overall floorspace increased by at least 50%.[57] Together with the considerable increase in the proportion of floorspace accounted for by offices, this resulted in a substantial increase in office supply.

The advent of the hydraulic lift also added considerably to the effective supply of desirable City floorspace. Prior to the 1860s the absence of lifts limited the attraction of upper floors to commercial tenants. In 1864 J.J. Cole noted that 'The rates at which offices in and near the Stock Exchange were let were about nine shillings per foot for the ground floor, seven shillings for the first floor, and five shillings for the second floor; the third floor at three shillings was not generally considered profitable.'[58]

The hydraulic lift was to fundamentally alter the price gradient of upper floors compared to their ground-floor counterparts. The City Offices Co. installed a lift in Palmerston Buildings as early as 1873, though they only became common after 1882, when the London Hydraulic Power Co. started operation.[59] By 1895 this company, which provided power for the vast majority of lifts in operation in the City at that time, supplied 221 passenger lifts in the EC postal district, plus 114 hydraulic goods lifts and cranes. This 'vertical transport revolution' provided a powerful impetus to the continued expansion of the City when the supply of non-office buildings for redevelopment was beginning to run short.[60]

The rapid spread of hydraulic lifts in major City office buildings significantly increased the rental value of offices on the upper storeys of buildings, relative to ground-floor values, as was shown in a recent study by Ralph Turvey.[61] The lift also opened up the possibility of much taller buildings (made technologically possible by the development of steel-framed construction techniques in America during the late nineteenth century), though the London Building Acts prevented the very substantial increase in office-building heights seen in the United States.

City Offices had undertaken little expansion beyond its initial portfolio during this period. Meanwhile, CLRP had pursued a policy of vigorous growth. The Innes brothers had amassed a further portfolio of properties, which were acquired by CLRP in a second share issue in 1881. Further expansion involved the acquisition of properties beyond the company's original City base, in addition to the redevelopment of some of their existing assets. In 1914, when CLRP and City Offices celebrated their 50th anniversaries, the balance sheet value of CLRP's assets had risen more than sevenfold, while the assets of City Offices had increased by less than one-third of their December 1864 figure. CLRP's strategy of acquiring properties to the east of the current economic centre of the City, in the hope of rising rents and capital appreciation as a result of the City's future expansion, had proved much more successful than City Office's strategy of purchasing properties on the most expensive sites.

Both City Offices' and CLRP's activities involved a large element of property development rather than pure investment. However, their histories do reveal a great deal about the investment profitability of City office property during the late Victorian period. Offices are less location-specific (within recognized commercial districts) than the High Street shops around which the property investment boom of the 1930s was based. The profitability of such investment appears to have depended on acquiring properties in anticipation of future demand, rather than where current demand was already high and prices had adjusted accordingly. During a period of very considerable expansion of the City's office stock, competition from newly-developed offices continually acted to depress rents, the most central sites offering only a relatively weak 'scarcity premium'.[62]

Yields in the region of 5% were no greater than those obtainable on conventional insurance company investments, and were certainly too low to compensate for the indivisibility, management costs and greater risks associated with office investment. Thus, the fifth and most important of the five conditions for the existence of an active property investment market, a sufficiently high return to investment (relative to other assets) to overcome property's inherent disadvantages as an investment medium, does not appear to have been met in the City of London during the late Victorian and Edwardian periods.

2.6 INSURANCE COMPANIES AND THE PROPERTY MARKET 1870–1918

The late nineteenth century saw the rapid growth of the insurance company sector. From 1870 to 1914 the value of life insurance company assets increased by about fivefold to over £500 million, equivalent to over 5% of national capital.[63] This period saw a drastic change in the pattern of insurance company investment, with a decline in the proportion of funds invested in property-based securities. From the mid-1870s yields on insurance company investments fell and companies faced considerable pressure to find higher yielding assets.

According to Board of Trade figures the overall yield on life assurance funds fell from 4.5% in the late 1860s and early 1870s to 4.2% in 1885, 4% in 1890 and 3.76% in 1900.[64] This was largely the result of a decline in the yield on consols, which reached a low-point of 2.5% in 1896–1898. Mortgage interest rates, which were closely linked to consol yields, also fell, while depression in the agricultural sector led to a reduction in the margin of security offered by agricultural land.

Falling yields threatened to reduce investment earnings below the level necessary to cover policy premiums, most life assurance companies basing premiums on an assumed rate of interest of 3% per annum.[65] Insurance companies therefore moved funds into higher-yielding assets. The main area of growth was investment in stock exchange securities, particularly foreign securities,[64] which increased from 7% to over 40% of total life assurance company assets between 1870 and 1914.

Cheap money led to a decrease in indirect insurance company investment in the property sector via mortgages and a rise in direct investment in property. The proportion of life assurance funds invested in mortgages, property and other assets is shown in Table 2.2. The table shows that despite a doubling in the percentage of insurance funds invested in land, property and ground rents between 1870 and 1913, the proportion invested in all property securities declined steadily from 1870 to 1890, fell sharply during the 1890s and continued to decrease during the 1900s. As mortgages include loans secured on assets other than real property, the table exaggerates the reduction in insurance company property-based assets a little, though the volume of non-property-based mortgages is not sufficient to significantly distort the figures.

Table 2.2 The percentage distribution of life assurance company assets 1870–1913

	1870	1880	1890	1900	1905	1913
Mortgages	49.8	48.1	42.7	29.0	26.6	22.6
Land, property and ground rents	4.5	5.6	8.1	9.9	10.2	9.2
Other	45.7	46.3	49.2	61.1	63.2	68.2

Source: B. Supple (1970) *The Royal Exchange Assurance,* Cambridge University Press, Cambridge, p.333.

The decline in mortgage lending was accompanied by a change in the distribution of mortgages between agricultural land, residential estates and commercial and industrial property. For example, Standard Life, which had hitherto concentrated its mortgage lending on agricultural estates, found that by the late

1860s it was unable to place sufficient funds in this sector. It therefore began to lend significantly to urban property developers, such lending being further increased following the onset of agricultural depression in the mid-1870s. However, doubts about the stability of urban rents and capital values prevented urban mortgages from becoming a major part of Standard's overall portfolio. Mortgage lending to commercial and industrial concerns was also regarded with more favour by Standard in this period, particularly after 1900, though while some individual transactions involved large sums, overall lending for such purposes never amounted to more than a small proportion of total mortgages. Rather than switching very substantial funds from agricultural to non-agricultural mortgages Standard moved into other securities, in common with most insurance companies.

Direct investment in property offered a means of placing money at substantially higher yields than could be obtained on consols or mortgages, and during this period the proportion of insurance company funds invested in land, property and ground rents more than doubled. However, such investment was made up largely of ground rents, property occupied by the insurance company in question and properties acquired as the result of mortgage default.

One reason for the lack of a more active property investment policy on the part of the insurance companies stemmed from actuarial theory regarding the criteria for insurance company investment. The most influential statement of this theory was presented in a paper given by A.H. Bailey to the Institute of Actuaries in 1862. Bailey set out five canons of investment policy. The first consideration of an investor should be the security of capital employed. Secondly, the highest available rate of interest should be earned subject to this capital security. Thirdly, a small proportion of funds should be kept in readily convertible securities for the payment of claims and for making loans. Fourthly, the bulk of funds might be invested in assets that were not readily convertible since these commanded a higher rate of interest due to their long-term nature. Finally, funds should be used, as far as practicable, to assist life assurance business.[66]

Such thinking favoured indirect investment in property via mortgages, which were secured on collateral with an estimated market value 50–100% in excess of the sum loaned, rather than direct investment, which offered no such margin and was subject to considerable variations in capital value. However, Bailey's ideas became increasingly obsolete during the late nineteenth century, as the growing size of insurance companies allowed greater scope for investment in higher-yielding securities, overall yields being protected by a policy of diversification. Such thoughts found formal expression prior to the First World War, in a paper presented by the Prudential's Actuary, George Ernest May, to the Institute of Actuaries in 1912. May stated that insurance companies could improve average yields by investing in a wider range of securities, both with regard to type of security and geographical area, accepting securities with higher risks but compensating higher yields.[67]

May's ideas were reflected in the Prudential's investment policy. At the company's 1912 AGM the chairman referred to the merits of a policy of investment diversification, stating:

> We believe that the greatest security and the smallest liability to temporary fluctuations are afforded by distributing our funds over as wide an area as possible, both in regard to the classes of security and to their geographical distribution. We have for years past based our practice on this principle.[68]

By the end of 1912 the Prudential's assets included properties with a book value of almost £4 million, and ground rents, feu duties and rent charges valued at over £5 million. However, land, property and ground rents formed only 10.8% of the Prudential's total assets, a figure only slightly higher than the average for all insurance companies.

Despite the Prudential's innovative investment policy, Bailey's canons remained widely accepted in the insurance world until the 1930s. Whether prevailing actuarial dogma would have been sufficient to prevent insurance companies from investing in property under favourable investment market conditions is uncertain. However, it appears that such conditions did not arise during this period. As section 2.5 has shown, investment in City offices does not appear to have offered sufficiently high returns to overcome the illiquidity, indivisibility, high transactions costs and other factors which disadvantage property compared to stock exchange securities. Other areas of the pre-1914 property investment market also offered little prospect for lucrative investment in developed property.

This period saw rapid suburban development around London and other large cities, offering the prospect of substantial capital gains for landowners as values escalated in the new suburbs. However, the profits from such activity accrued largely to the speculator (or established landowners who reaped windfall gains) rather than the investor in developed property. As H.J. Dyos stated 'The roughest calculations seemed to show that most of the money to be made out of this process went to the men who were first on the scene, who leased or bought land before its rise, and developed or sold it on the very top of the tide.'[69]

Furthermore, the major category of property involved in such development – houses and flats – entailed high management and maintenance costs, with a greater risk of non-payment, or arrears, of rent than was the case with commercial property. Yields on such investments were also very volatile, increasing by a considerable margin from 1902 to 1912 according to a study by Avner Offer.[70]

Insurance companies did participate in this market (in addition to mortgage lending to property developers) by purchasing the ground rents created when building leases were granted to developers. These provided a secure long-term income, with little management cost and the (distant) prospect of acquiring the properties themselves following the termination of the building lease. Despite offering a lower yield than other forms of investment property ground rents, together with similar securities such as rent charges and Scottish feu duties, formed about half of insurance company direct property holdings in 1870. This proportion fell to about 36% during the late 1890s and then rose slightly during the early years of this century, to about 39% by 1914.[71] Evidence indicates that insurance companies invested most heavily in this class of property securities, relative to rack-rented property, when the property market was depressed, switching to riskier but higher-yielding investment in buildings, rather than ground rents, at times of more buoyant property market conditions.

Some insurance offices invested substantial funds in property by acquiring the premises from which they conducted business while avoiding investment in property they did not wholly or partly occupy. Standard Life's holdings of land, property and ground rents formed 8.79% of total assets in 1900, almost all of which consisted of premises occupied by the company. Though such acquisitions were undertaken not solely for investment reasons, the decision to buy, rather than rent, was often arrived at by a careful assessment of the investment value of the property concerned.

Direct insurance company investment in property not occupied by the company in question, other than the purchase of ground rents, did not occur to any substantial extent prior to the First World War. There was, however, some investment of this type. Insurance companies obtained a certain amount of property involuntarily due to mortgage default. The Royal Exchange Assurance acquired shops and dwellings on Shaftesbury Avenue, a large agricultural estate, and land in London on which it proceeded to build two large blocks of flats, at a cost of £152 000, as a result of mortgage defaults.[72] As had been the case with building estates acquired in this way by insurance companies in the mid-nineteenth century, such assets were usually among the less attractive of their kind and did little to increase the confidence of insurance company managers in property as an investment medium.

Some insurance companies also acquired properties voluntarily during this period. The National Provident Institution (NPI) initiated a programme of property acquisitions in the 1890s. In 1895 a block of 25 flats in Bloomsbury, Great Russell Mansions, was purchased for £26 000.[73] During 1897 and 1898 NPI acquired a number of shop and flat properties in and around Sloane Street, in three deals, involving 26 flats, 11 shops and four houses/shops, at a total cost of about £85 000.[74] These were good quality properties and involved little management cost; the report on the purchase of one set of properties stated that the shops were all let on repairing leases and the flats were let 'to responsible tenants for a term of years'.[75] Further properties acquired in 1898 included shops and offices at London Wall, and an office in Fleet Street, which was managed by an agent for 3% of the gross annual rental. The total book value of NPI's property portfolio, excluding ground rents and the Milford Haven estate, was estimated at £188 115 in 1898, producing gross and net rents of £21 096 and £10 440, respectively. The net yield on book value of these holdings during 1898, 5.55%, compared favourably with yields on other insurance company securities, though the lack of rental figures for other years prevents a more thorough assessment of net yields on NPI's properties. NPI reversed its active property investment policy after 1899, purchasing no further properties (other than a few ground rents) until 1922.

The Clerical, Medical and General Life Assurance Society (Clerical Medical), also began investing in property in the 1890s, but on a much more limited scale than NPI. The Society, founded in 1824, began to purchase freehold ground rents in around 1898 and had also begun to invest in buildings by the early 1900s. Some of Clerical Medical's early property investment deals involved cooperation with potential tenants, using techniques which were to be further developed during the inter-war years. For example, in 1902 the Society considered a proposal by the London, City and Midland Bank, involving the purchase of a newly-erected building at Richmond, for a price equal to its construction cost, about £13 000. The bank would then lease the premises for a fixed term of 21 years at a rent amounting to 4% of this sum. The Finance Committee recommended the scheme, subject to the bank taking out a redemption assurance policy to provide a sinking fund against any possible depreciation of the property.[76] It is unclear whether the bank owned the property in question. If so, this proposal constitutes a very early example of the sale and leaseback arrangement, which was not systematically applied until the 1930s.

In 1906 Clerical Medical considered another proposal by a potential tenant, the caterers J. Lyons & Co. Ltd. It was suggested that the Society purchase a London hotel site, which had cost Lyons £100 000 and was offered to the Society for the same price on condition that it grant them a building lease of £4000 per annum for 99 years, thereby converting the security into a ground rent. Lyons undertook to develop a 'high-class' restaurant on the site, at a minimum cost of £30 000. They also promised to pay £16 000, or possibly more, to effect two or more capital redemption policies for a total of £100 000. When these policies matured they were to be used to repay Clerical Medical's purchase costs, in return for which the site would be reconveyed to Lyons when the lease expired.[77] After commissioning a report on the site the Society responded with the amended proposal that they purchase the site for £100 000 and receive back £20 000 for the purchase of the reversionary interest.[78]

From 1902 the property investment market experienced a slump, as was shown by Avner Offer, who estimated that the volume of property market turnover, and capital values for London dwellings, fell sharply from 1902 to 1912.[79] This was partly the result of rising interest rates; ground rents, and property let on long leases, constituted virtually fixed-interest securities, which were subject to depreciation in value during times of dearer money. Fears of anti-property legislation, and the general downturn in economic conditions, also served to depress the market.

These conditions do not appear to have adversely affected insurance company property income to any significant extent, as rents showed much greater stability than capital values,[79] and insurance companies had little need to dispose of their property holdings. Furthermore, ground rents represented a much more secure income stream than other classes of property, while property the insurance companies themselves occupied constituted an extremely secure source of income. However, the erosion in general market confidence, combined with an increase in interest rates, appears to have reduced the volume of insurance company investment in property compared to the previous decade, as is shown in Table 2.2.

The property taxation measures of the 1906 Liberal government presented a much more serious threat to insurance company property holdings. Its attempts to tax land values constituted the most direct challenge to the 'landed interest' in Britain this century; the aim being to transfer the 'unearned increment' in land values to the community. The unearned increment referred to both increases in land values as a result of land acquiring a potential for a change to more profitable use and increases in value due to higher demand for its existing use.[80] The radical attack on the landed interest had been gathering momentum since the 1880s and reached its zenith during the years 1906–1912; 400 Liberal and Labour MPs in the 1906 Parliament were signatories to a memorial which called for the taxation of land values, while the 'Land Value Group' and the 'Land Nationalization Society' had 280 and 130 parliamentary members, respectively.[81]

The first series of land taxes were introduced in Lloyd George's 1909 'People's Budget'. These included taxes on vacant land, mining royalties and ground rents. There was also a 20% capital gains tax on land transfers.[82] Furthermore, the budget provided for a valuation of all land in the UK, facilitating future taxation of land values themselves, one of the prime aims of the radical land campaign.

In 1909 the Housing and Town Planning Act was introduced – the first town planning Act to reach the statute books in Great Britain. This Act also marked the first attempt to introduce a levy on the 'betterment' of land, betterment being taxed at the rate of 50% on increases in the value of undeveloped land resulting from local authority town planning schemes.[83]

The Liberal land campaign of 1912–1914 marked a further attack on the landed interest. While focused in large part on rural land, it also involved the promise of security of tenure for urban leaseholders and a strengthening in their position *vis-à-vis* ground landlords.[84] Further proposals included a national tax on land values, which would supplement local rates.

The land taxes were abolished in the very different political climate following the First World War, while the betterment provisions of the 1909 Housing and Town Planning Act proved impossible to implement since the extent of betterment following implementation of a town planning scheme could not be measured. However, the uncertainty that this legislation created in the short term regarding its effects on property values and yields, together with fears of further anti-property measures, led to a reduction in property investment activity and a marked downturn in market conditions.

Fears regarding the effects of Lloyd George's property legislation prompted the Prudential to write down the value of its freehold ground rents by £250 000 in 1911.[85] The adverse affects of this legislation on property values was discussed by the Prudential's chairman, W. Edgar Horne, in his 1913 AGM speech:

> This class of security [investments in, and mortgages on, real estate] has suffered, not only from the rise in the prevailing rate of interest but also from the fear of oppressive legislation. The effect of the first cause can be, and has been, adequately met by us: the effect of the second it is impossible to foretell. There is, however, one consideration that cannot be too strongly urged, and that is that any legislation adversely affecting the value of land as a security not only strikes the landowner, but also directly affects the millions of industrial workers who have entrusted their savings to the Prudential Company.[86]

The First World War led to a virtual halt in property investment by insurance companies. Under the 'Gentleman's Agreement' between the insurance companies and the Treasury, new insurance company funds were invested in government securities, which grew from 1.1 to 35.0% of total life insurance company assets between 1913 and 1921.[87] The war also led to a reduction in income from existing properties, due to increased taxation.

The effects of the war on property market conditions were discussed in Clerical Medical's 1915 asset investigation. After noting that, according to balance-sheet values, the average yield on their freehold and leasehold ground rents was 4.2 and 4.7%, respectively, the report stated that due to falling security values as a result of the war yields of 5% on freehold and 6% on leasehold ground rents might be suitable under present conditions.[88] In 1917 Clerical Medical considered the purchase of an improved leasehold ground rent, although at a price which was low enough to yield 7.5% free of tax.[89]

While property values appear to have fallen significantly as a result of wartime conditions, the income produced by investment property was much less seriously affected. NPI's 1917 valuation report indicated that, if 'special cases' were removed, 'the remainder account for a total book value of £336,822, producing a net revenue of £15,808, yielding 4.7% net, equivalent to a gross yield of 6.25%'.[90]

The majority of NPI's properties were flats. These had experienced rising rents during 1917, outweighing an adverse movement in rents for the Institution's shop properties. A Board minute of 2nd October 1917 reinforced the view that the war had not adversely affected property income to a significant degree, stating that, with the exception of one ground rent, rents on the Institution's London properties appeared to be secured by a larger margin of rental value than was the case five years earlier.

The period from 1870 to 1918 had seen extensive involvement in the property market by the insurance companies with regard to low-yielding secure investments, mortgages and ground rents and the first steps towards a more active property investment policy on the part of a few companies. The cheap money era of the 1890s had generated some interest in property investment as insurance companies sought to maintain yields at a time of falling interest rates but found that the market offered few suitable investment opportunities. The next period of cheap money, the 1930s, was to see much greater direct involvement in the property investment market by the financial institutions, due to a transformation in the character of that market.

REFERENCES AND NOTES

1. H.J. Dyos (1961) *Victorian Suburb: A study of the growth of Camberwell*, Leicester University Press, Leicester, p. 87.
2. D.J. Olsen (1976) *The Growth of Victorian London*, Batsford, London, p. 127.
3. D.J. Olsen (1976) *The Growth of Victorian London*, Batsford, London, p. 154.
4. D.J. Olsen (1976) *The Growth of Victorian London*, Batsford, London, pp. 156–7.
5. D. Cannadine (1980) *Lords and Landlords: The Aristocracy and the Towns 1774–1967*, Leicester University Press, Leicester, p. 271.
6. D. Cannadine (1980) *Lords and Landlords: The Aristocracy and the Towns 1774–1967*, Leicester University Press, Leicester, p. 392.
7. F.M.L. Thompson (1963) *English Landed Society in the Nineteenth Century*, Routledge and Kegan Paul, London, p. 307.
8. D. Cannadine (1980) *Lords and Landlords: The Aristocracy and the Towns 1774–1967*, Leicester University Press, Leicester, p. 139. (The section in quotation marks is quoted from estate correspondence.)
9. H.J. Dyos (1961) *Victorian Suburb: A study of the growth of Camberwell*, Leicester University Press, Leicester, p. 88.
10. D. Cannadine (1980) *Lords and Landlords: The Aristocracy and the Towns 1774–1967*, Leicester University Press, Leicester, p. 222.
11. G.F.A. Best (1964) *Temporal Pillars*, Cambridge University Press, Cambridge, p. 494.
12. Information supplied by J. Dunbabin. Some colleges had obtained legal powers to grant building leases of over 40 years prior to the 1850s. Colleges also had some, very limited, powers to sell land prior to this time.
13. Oxbridge colleges were legally restricted to trustee securities in the range of investments they could undertake.
14. B. Supple (1970) *The Royal Exchange Assurance*, Cambridge University Press, Cambridge, p. 321.
15. D.J. Olsen (1976) *The Growth of Victorian London*, Batsford, London, p. 157.
16. D.J. Olson (1976) *The Growth of Victorian London*, Batsford, London, p. 155.
17. Standard (various years) *Private Minute Book A1*, p. 190.

REFERENCES AND NOTES

18. J.H. Treble and J. Butt (*c.* 1980) Unpublished typescript history of Standard Life, Ch. 4, p. 15.
19. J.H. Treble and J. Butt (*c.* 1980) Unpublished typescript history of Standard Life, Ch. 4, p. 16.
20. J.H. Treble and J. Butt (*c.* 1980) Unpublished typescript history of Standard Life, Ch. 4, p. 17.
21. H.J. Dyos (1961) *Victorian Suburb: A study of the growth of Camberwell*, Leicester University Press, Leicester, p. 119.
22. H.J. Dyos (1961) *Victorian Suburb: A study of the growth of Camberwell*, Leicester University Press, Leicester, p. 120.
23. H.J. Dyos (1961) *Victorian Suburb: A study of the growth of Camberwell*, Leicester University Press, Leicester, p. 121.
24. H.J. Dyos (1961) *Victorian Suburb: A study of the growth of Camberwell*, Leicester University Press, Leicester, p. 122.
25. B. Supple (1970) *The Royal Exchange Assurance*, Cambridge University Press, Cambridge, p. 329.
26. B. Supple (1970) *The Royal Exchange Assurance*, Cambridge University Press, Cambridge, p. 330.
27. N. Toulson (1985) *The Squirrel and the Clock: National Provident Institution 1835–1985*, H. Melland, London, p. 47.
28. NPI (1868) *Board Resolutions*, 27 Mar.
29. NPI (1882) *Board Resolutions*, 4 July.
30. Douglas Sun (1991) Legal & General Group PLC, in *The St James Press International Directory of Company Histories, Vol. III*, St James Press, London.
31. S. Pollard and D.W. Crossley (1968) *The Wealth of Britain 1085–1966*, Batsford, London, p. 212.
32. George Cross (1939) *Suffolk Punch: A Business Man's Autobiography*, Faber & Faber, London, p. 57.
33. F.M.L. Thompson (1968) The land market in the nineteenth century, in *Essays in Agrarian History, Vol. 2* (ed. W.E. Michinton), David & Charles, Newton Abbot, p. 40.
34. F.M.L. Thompson (1968) The land market in the nineteenth century, in *Essays in Agrarian History, Vol. 2* (ed. W.E. Michinton), David & Charles, Newton Abbot, p. 41.
35. See Chapter 3.
36. P. Cowen *et al.* (1969) *The Office: A facet of urban growth*, Heinemann, London, p. 25.
37. John Lawrence (1994) From counting house to office: the transformation of London's central financial district, 1693–1871. Unpublished paper, p. 15.
38. John Lawrence (1994) From counting house to office: the transformation of London's central financial district, 1693–1871. Unpublished paper, p. 14.
39. Edward L'Anson (1864) Some notice of office buildings in the City of London. *Royal Institute of British Architects, Transactions*, 25–6.
40. John Lawrence (1994) From counting house to office: the transformation of London's central financial district, 1693–1871. Unpublished paper, Table 4 and Figure 1.

41. D.K. Stenhouse (1984) Liverpool's office district, 1875–1905. *Historical Society of Lancashire and Cheshire*, **133**, 72.
42. St Quintin (*c.* 1981) *The History of St Quintin Chartered Surveyors 1831–1981*. Privately published, London, pp. 16–17.
43. Edward L'Anson (1864) Some notice of office buildings in the City of London. *Royal Institute of British Architects, Transactions*, 26.
44. Edward L'Anson (1864) Some notice of office buildings in the City of London. *Royal Institute of British Architects, Transactions*, 34.
45. M. Campbell-Kelly (1992) Large-scale data processing in the Prudential, 1850–1930. *Accounting, Business and Financial History*, **2**, 134.
46. M. Campbell-Kelly (1992) Large-scale data processing in the Prudential, 1850–1930. *Accounting, Business and Financial History*, **2**, 121.
47. D.K. Stenhouse (1984) Liverpool's office district, 1875–1905. *Historical Society of Lancashire and Cheshire*, **133**, 73.
48. John Lawrence (1994) From counting house to office: the transformation of London's central financial district, 1693–1871. Unpublished paper, p. 24.
49. City Offices (1864) Verbatim copy of prospectus, Board minutes, 7 Apr.
50. City of London Real Property Co. Ltd (*c.* 1964) *The City of London Real Property Co. Ltd: 1864–1964*. Privately published, London, p. 10.
51. City of London Real Property Co. Ltd (*c.* 1964) *The City of London Real Property Co. Ltd: 1864–1964*. Privately published, London, p. 15.
52. City Offices (1867) Half-yearly Report, Sept.
53. City Offices (1881) Report of Chairman's speech, Annual Report, 28 Mar.
54. The figures for the leasehold properties are also affected by a reduction in the time left before the termination of the leases.
55. R. Turvey (1994) City of London office rents: 1864–1914. Unpublished paper, p. 2.
56. P. Cowen *et al.* (1969) *The Office: A facet of urban growth*, Heinemann, London, p. 157.
57. C.H. Holden and W.G. Holford (1951) *The City of London: Record of Destruction and Survival*, London, pp.165–6; cited in R. Turvey (1994) City of London office rents: 1864–1914. Unpublished paper, p. 2.
58. Edward L'Anson (1864) Some notice of office buildings in the City of London. *Royal Institute of British Architects, Transactions*, pp. 34–5.
59. R. Turvey (1993–4) London lifts and hydraulic power. *The Newcomen Society, Transactions*, **65**, 147.
60. R. Turvey (1994) City of London office rents: 1864–1914. Unpublished paper, p. 4.
61. R. Turvey (1994) City of London office rents: 1864–1914. Unpublished paper, pp. 9–11.
62. See Chapter 3.
63. B. Supple (1970) *The Royal Exchange Assurance*, Cambridge University Press, Cambridge, p. 309.
64. B. Supple (1970) *The Royal Exchange Assurance*, Cambridge University Press, Cambridge, p. 331.
65. B.Supple (1970) *The Royal Exchange Assurance*, Cambridge University Press, Cambridge, p. 313.

REFERENCES AND NOTES

66. G. Clayton and W.T. Osborn (1965) *Insurance Company Investment*, Allen & Unwin, London, p. 62.
67. G.E. May (1912) The investment of life assurance funds. *Journal of the Institute of Actuaries*, **46**.
68. Prudential Assurance Company (1912) Annual Report.
69. D. Cannadine and D. Reeder (eds) (1982) *Exploring the Urban Past: Essays in urban history by H.J. Dyos*, Cambridge University Press, Cambridge, p. 157.
70. A. Offer (1981) *Property and Politics 1870–1914*, Cambridge University Press, Cambridge, pp. 268–70.
71. See section 9.3.
72. B. Supple (1970) *The Royal Exchange Assurance*, Cambridge University Press, Cambridge, p. 338.
73. NPI Assets Report: (1898 & 1936).
74. N. Toulson (1985) *The Squirrel and the Clock: National Provident Institution 1835–1985*, H. Melland, London, p. 49.
75. NPI (1898) *Board Resolutions*, 4 Mar.
76. NPI (1902) Asset Report, 15 Oct.
77. NPI (1906) Asset Report, 28 Feb.
78. This sum of money being equal to the value of a redemption policy which would repay the purchase cost at the end of the amended lease period.
79. A. Offer (1981) *Property and Politics 1870–1914*, Cambridge University Press, Cambridge, pp. 259–72.
80. E. Reade (1987) *British Town and Country Planning*, Open University Press, Milton Keynes, p. 36.
81. E. Reade (1987) *British Town and Country Planning*, Open University Press, Milton Keynes, p. 37.
82. A. Offer (1981) *Property and Politics 1870–1914*, Cambridge University Press, Cambridge, p. 326.
83. E. Reade (1987) *British Town and Country Planning*, Open University Press, Milton Keynes, pp. 39–40.
84. A. Offer (1981) *Property and Politics 1870–1914*, Cambridge University Press, Cambridge, p. 390.
85. Prudential (1911) Chairman's speech, Annual General Meeting.
86. Prudential (1913) Chairman's speech, Annual General Meeting.
87. J. Johnson and G.W. Murphy (1957) The growth of life assurance in U.K. since 1880. *Transactions of the Manchester Statistical Society*, **1956–57**, 49.
88. Clerical Medical (1915) Asset Investigation, 20 July.
89. Clerical Medical (1917) Finance Committee minutes, 21 Feb.
90. NPI (1917) Asset Report. (Yields are based on 1916 figures.)

3 | The growth of a national commercial property market, 1919–1939

3.1 INTRODUCTION

The inter-war years saw profound changes in the character and pattern of commercial property development in Britain. Motorized transport provided Britain's population with much greater mobility, allowing increased geographical concentration of economic activities and the buildings which housed them. Multiple retailing chains displaced local traders in Britain's High Streets, shopping parades sprang up alongside the new housing estates which mushroomed during this era of rapid suburban expansion and industrial estates were developed along London's arterial roads to accommodate the 'new', light, consumer-goods industries.

These trends were greatly facilitated by the emergence of a buoyant nationwide property investment market during these years. Although a limited investment market in urban property had emerged during the late nineteenth century, it was not until the inter-war period that property came to be viewed by a few of the most progressive financial institutions as a significant outlet for funds. It was during these years that many basic features of Britain's modern property investment market emerged, including the growth of market intermediaries covering the national property market, the 'securitization' of investment property and the development of funding links between financial institutions and property developers.

If institutional investors were to undertake a substantial volume of direct property investment, the property market had to meet the following conditions:

1. A substantial number of commercial property transactions, providing a steady stream of investment opportunities.
2. The growth of an efficient and integrated market, covering the geographical area in which the investments were situated.
3. The development of a 'scarcity premium' for particular properties, increasing the proportion of a property's value which results from its location and

THE GROWTH OF MARKET INTERMEDIARIES

therefore giving it a high value and marketability, relative to other property, which extends beyond the life of both the tenancy and the present building, capital appreciation for the site counteracting depreciation of the building.
4. The development of a financial climate in which commercial enterprises had considerable incentives to rent, rather than own, their business premises, or to dispose of some property rights to their premises in return for an immediate capital sum.
5. The emergence of a significant differential between the performance of property and alternative assets, the rate of return on investment property being sufficiently higher than that for other investment media to overcome its inherent disadvantages of lack of marketability, uncertainty of value, indivisibility and higher management costs.

The first four factors might be regarded as vital, or at least extremely important, pre-conditions, while the fifth is a necessary condition for investment to take place in a market where conditions 1–4 have been largely met. This chapter will show how the growth of the multiple retailers during the inter-war years, together with the associated development of nationally-based property market intermediaries, led to conditions 1–4 being fulfilled.

Condition 5 was met during the 1930s, largely as a consequence of the government's successful cheap-money policy. This is outlined in Chapter 4, which also includes a discussion of commercial property development and development funding, and the importance of more entrepreneurial attitudes on the part of institutional investors as a factor behind the rapid expansion of the property investment market during the 1930s.

3.2 THE GROWTH OF MARKET INTERMEDIARIES

Because of the highly localized nature of the nineteenth century property market, Victorian estate agents and solicitors had only needed to act as property market intermediaries over a relatively small geographical area. The multiple retailers, whose expansion (outlined in section 3.3) provided the main source of new commercial property development and investment during the inter-war years, operated on a national scale. Thus, a need arose for market intermediaries covering the entire country. A group of commercial estate agents based in the West End of London stepped into this void, providing a range of specialist property services for the expanding multiples.

The insurance companies had a similar need for property market intermediaries if a substantial volume of property investment was to take place. Property lacked a central market, unlike stock exchange securities, and was a far more complex asset than mortgages or other loans undertaken by insurance companies, where the desirability of the investment was a simple function of the rate of interest, the covenant of the borrower and the provision of an adequate margin of security. Before going on to look in detail at the evolution of commercial estate agency during this period it might be useful to briefly list the basic services which commercial estate agents provided in order to facilitate the user, investment and development markets for commercial property.

1. A centralized market, bringing together buyers, sellers, developers, investors and occupiers.
2. A repository of market information regarding prevailing rents and property values in a particular area, and prevailing yields on different classes of investment property.
3. A source of market intelligence, providing advice regarding the desirability of investment propositions, the most valuable shopping pitches in particular centres, the expected future course of the property market and the long-term factors influencing the market.
4. The active lobbying of independent local traders to dispose of their shops, making them aware of the sums that could be raised by selling properties in desirable locations to the expanding multiples.
5. Services covering the legal aspects of property investment and development, including such things as planning permission and landlord and tenant legislation.
6. A source of innovation in the property market. Innovative activities included devising new ways in which property transactions might be arranged, to suit the requirements of both the investor and the occupier or developer of the property in question, and putting together property investment and development propositions on a speculative basis, which were presented to clients on the initiative of the agency rather than being instigated by the client.
7. The diffusion of innovations and best practice techniques in property investment and development.
8. The marketing and advertising of property as a suitable outlet for institutional and private investment funds.

The booming trade in shop property in the 1920s and 1930s led to the rapid expansion of the commercial estate agency industry. The displacement of individual local traders by the multiples, a process which had begun in the late nineteenth century, reached its peak during the inter-war years. Major retailers such as Boots, Marks & Spencer, F.W. Woolworth and Montague Burton found that they could afford to pay rents far above those paid by local traders. In the case of shops with low profit margins, such as food retailers, the purchase price offered by a major multiple wishing to acquire a property was often more than the profit the trader could expect to make, if he retained the shop, for the rest of his days. The incentive to sell was therefore very strong.

A number of estate agents began to specialize in shop property, facilitating the transfer of ownership of Britain's High Street shops from the small local trader to the nationwide chain store. One of the first agents to specialize in this area was Hillier, Parker, founded in 1895 by William Hillier and Thomas

THE GROWTH OF MARKET INTERMEDIARIES

Parker. In 1921 Hillier, Parker merged with May & Rowden of 27 Maddox Street.[1] The merged agency established itself at that address, starting a trend towards the concentration of the industry in and around Maddox Street, which became the home of commercial estate agency between the wars.

In 1922 Hillier, Parker began to issue annual reports regarding conditions in the property market, including such information as the volume of annual sales and lettings they had arranged, building leases and grounds rent transactions, property valuations, and total business conducted by the practice during the year. These were quoted in the property columns of *The Times*, and *The Daily Telegraph*, and information regarding their annual volume of business also appeared in these newspapers via full- or half-page adverts, usually on the back page of an early January issue. These listed the towns and cities in which the practice had conducted business during the year, in alphabetical order, covering nearly every letter of the alphabet. By the beginning of 1924 properties shown in these adverts already revealed the nationwide market coverage of the agency, including towns and cities in the North, Scotland and Wales.[2]

The statistical information presented in the reports, covering business conducted by the practice during each year, provides an indication of changes in the overall buoyancy of the commercial property sector during the inter-war years. Hillier, Parker grew to be one of the two leading commercial estate agents during this period, and while these figures cannot be regarded as providing a comprehensive picture of conditions in the market for shop property, they do provide a unique guide to the changing fortunes of one of the most important agencies. The figures refer mainly to shops; a small volume of office and industrial property business was also conducted by the firm. Available data on business transacted by Hillier, Parker during the years 1922–1937 is shown in Table 3.1.

The lack of comprehensive information on business other than sales and rents on lettings makes analysis of other transactions difficult, though a trend towards a reduction in letting premiums is shown, alongside an increase in rentals paid on lettings. This was due to a change in the terms under which properties were acquired. During the 1920s property boom it was necessary to pay heavy premiums to secure the best High Street sites, while during the boom of the mid-1930s retailers preferred to buy such properties outright rather than pay large premiums on leaseholds.

Figures for sales and rents on lettings are available for each year from 1922 to 1937, allowing a more detailed analysis. In order to examine changes in the volume of these categories of business the figures in Table 3.1 were put into index form, and corrected for changes in retail prices, as is shown in Table 3.2 and Figure 3.1. During the years 1922–1928 business in both categories increased substantially, though sales grew much faster than rents, the index values for the two variables in 1928 being 279 and 146% of their 1922 values, respectively.

During the more unsettled years from 1929 to 1931 sales fell while rentals rose to their peak levels. This may reflect a tendency, noted in Hillier, Parker's reports, for retailers to rent rather than buy property at times of economic uncertainty, the balance between rental and sale business tilting in favour of sales upon a return to more stable conditions. In 1932 rents fell substantially while sales rose modestly; thereafter, the two categories of business both increased, rents lagging behind sales slightly. The close correlation between sales and rents may be largely due to the large volume of sale and leaseback transactions (discussed below) during these years, which generated both sale and letting business. Overall the figures show a boom in the practice's business during the inter-war years, the real value of sales and rents from lettings, set at 100 for 1922, growing to 332 and 287, respectively, in 1937, and reaching peak levels of 350 and 333 in 1936, before economic recession and the deteriorating international situation led to a fall in property market activity.

Table 3.1 Business conducted by Hillier, Parker, May & Rowden, 1922–1937

Year	Sales (£)	Letting rents (£)	Letting premiums (£)	Building leases and ground rents (£)	Properties valued (£)	All business (£)
1922	1 800 000	175 000	150 000	N/A	N/A	N/A
1923	2 549 346	194 063	110 132	35 144	N/A	N/A
1924	2 570 366	225 420	144 215	39 713	1 000 000	N/A
1925	3 608 756	259 316	N/A	N/A	N/A	9 000 000
1926	3 541 500	230 040	N/A	21 800	N/A	10 000 000
1927	3 776 000	257 000	215 214	30 395	N/A	10 000 000
1928	4 566 698	233 087	130 040	60 500	5 705 219	12 500 000
1929	3 998 207	301 945	N/A	30 949	N/A	N/A
1930	2 995 709	389 767	N/A	N/A	N/A	N/A
1931	2 396 567	480 000	N/A	N/A	N/A	N/A
1932	2 463 000	216 472	N/A	N/A	11 500 000	N/A
1933	3 962 805	238 372	N/A	N/A	N/A	N/A
1934	3 986 000	340 000	N/A	36 000	4 930 000	N/A
1935	4 153 896	393 500	57 107	38 833	4 550 000	N/A
1936	5 065 151	468 310	93 385	100 465	N/A	N/A
1937	5 078 400	425 996	79 320	59 099	8 769 057	N/A

Source: Hillier, Parker's reports on their activities for the years shown, as quoted in *The Times* and *The Daily Telegraph*. Some figures are approximate and as the statistical information quoted varies in detail from year to year some figures are missing for some years. N/A, not available.

The other large commercial estate agency during the 1930s, Healey & Baker, also experienced rapid growth over this period. Healey & Baker had been founded in 1820, but concentrated on residential property until 1927, when a West End office was opened, managed by Arthur Hemens and Aubrey Orchard-Lisle. Initial successes in the shop property market following the opening of this office led the agency to concentrate on commercial property, the residential side of the practice eventually being closed down.

THE GROWTH OF MARKET INTERMEDIARIES

Table 3.2 Indices of sales and lettings by Hillier, Parker, May & Rowden, corrected for changes in retail prices: 1922–1937

Year	Sales	Letting rents
1922	100	100
1923	148	116
1924	149	134
1925	207	153
1926	208	139
1927	229	160
1928	279	146
1929	247	192
1930	192	257
1931	164	339
1932	174	157
1933	288	178
1934	287	252
1935	295	287
1936	350	333
1937	332	287

Sources: As for Table 3.1. Retail price index is from C.H. Feinstein (1972) *Statistical Tables of National Income, Expenditure and Output of the U.K., 1855–1965,* Cambridge University Press, Cambridge, Table 65.

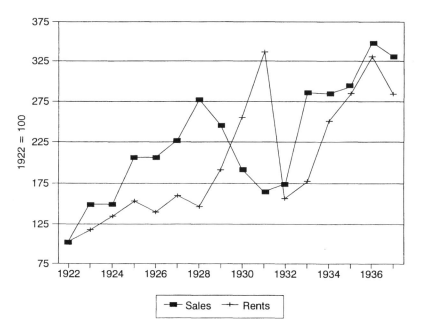

Figure 3.1 Sales and lettings arranged by Hillier, Parker, May & Rowden, corrected for changes in retail prices: 1922–1937. Source: Table 3.2.

During the 1930s the firm grew rapidly, its staff increasing from 15 in 1930 to over 100 by 1939. Strong ties were developed between the agency and major multiple retailers, including Montague Burton, J. Sainsbury, Marks & Spencer, F.W. Woolworth, Price's Fifty Shilling Tailors, J. Lyons & Co, Boots, Finlays, Barrats, Allied Suppliers and Tesco.[3] These links enabled them to build up a substantial volume of business as intermediaries in the property development sector; Healey & Baker handled over 150 different shop developments between 1934 and 1939. The presence of a large multiple trader could determine the success or failure of a development, as is discussed in section 3.3. The practice's links with these traders therefore made association with Healey & Baker extremely useful for shop developers, as once they had persuaded Woolworth's or Sainsbury's to locate in a new shopping parade other retailers would follow. Aubrey Orchard-Lisle recalled that 'We used to dictate our own terms ... we'd have a parade of shops and we were letting agents and one of our conditions was when the development was completed and let, we would then have the sole agencies to sell the completed investment to an institution ... so we had a jackpot at the end of every scheme.'[4] Such was the power of the market intermediary.

The way in which these agents were able to cover the national market for shop property, providing the services listed at the beginning of this section, can be illustrated by the career of one influential inter-war agent, Edward Erdman. Erdman's initial career ambition was to become a lawyer, but insufficient funds led him to apply instead for a job as an office boy, and in 1923 he was engaged in this capacity by the estate agents and auctioneers Gordon, Thomas & Co., thus entering the property world almost by accident.

In 1924 Erdman moved to the practice of Douglas Kershaw & Co. He had applied for a post as a junior negotiator but was engaged instead as a canvasser for the West End Department. This involved visiting shops to hunt for business. Erdman recalls that he adopted a standard approach upon entering a store. He would walk in boldly, dressed in bowler hat and stiff collar, and ask to see the chairman. When asked if he had an appointment he would reply that he hadn't, but that his mission was special and personal. Having gained access to the manager his usual opening statement was 'I have a point blank question to put to you – are you prepared to sell this building if you secure a satisfactory price'.[5] This approach usually secured an average of at least six properties a day. If a positive reply was obtained an instruction form was later written up, to be followed up by one of the firm's negotiators.

In addition to approaching property owners in person, estate agents also engaged in extensive canvassing by mail. The method of doing this was already well established prior to the First World War, and as George Cross, who worked as a London estate agent for some years, recalled, involved sending a letter along the following lines:

THE GROWTH OF MARKET INTERMEDIARIES

Dear Sir,
> We have an applicant, whose name and address we shall be pleased to disclose at once, who is requiring a business such as yours, and is prepared to pay its full value. We shall be glad to learn if you are willing to sell, and if so, perhaps you will kindly forward us a few particulars, which will be treated in the strictest confidence.[6]

Such speculative approaches led many commercial property owners, who had not hitherto intended to sell their businesses, to seriously consider the financial benefits of doing so, thereby increasing the volume of property entering the market. Doing business in this way was made possible by the unregulated nature of commercial estate agency in the inter-war period. The industry then had few of the codes and regulations regarding professional conduct which have since been evolved, and operated in an atmosphere of fierce competition, in which soliciting business by direct personal call and touting for instructions on premises where other agent's boards were on exhibition were regarded as normal practice.[7] As Erdman recalls 'Agents regarded each other as adversaries and not as brother agents. Transactions were small and fees were correspondingly limited; consequently, one of the driving forces was the necessity to carry out several transactions week by week.'[8]

After working for a number of commercial estate agencies, Erdman went into business on his own account in 1934 as a result of a chance conversation with a satisfied client, Walter Sawyer, who offered to lend him the necessary capital to establish himself in business – £1500.[9] He established an office at 35 Maddox Street; by this time the area around Maddox Street had already become the market place for commercial property, as is shown in Figure 3.2. Other agents not shown were also located nearby; including Douglas Kershaw & Co. in Princes Street, White Druce & Brown in Hanover Square, Garret, White & Brown in Hanover Street and Knight, Frank & Rutley (who were not yet dealing in commercial property) in Hanover Square. Rather than having to travel throughout the country in search of suitable sites for new shops, retailers had only to visit a small area of London to gather information on available premises throughout the country. The national market for one category of property – shops – had gained a degree of centralization that it had never hitherto enjoyed.

A system was devised by Erdman to keep track of which clients might be interested in particular properties. Street plans were prepared, showing all the shops in particular London suburbs. This was later extended to provincial towns and cities. When a town was initially examined by the practice a notebook map of the shopping centre, showing the name and trade of every shop, was drawn, which was later converted into a detailed street map. Traders in the town were approached in writing to ascertain whether they would be interested in selling their premises, and from the replies and subsequent negotiations the firm gained an impression of prevailing rents. By working in this fashion the practice was able to cover the provincial markets without the need to establish provincial offices. A register was also established, with the names of multiple retailers without a branch in a particular area, so that they could be contacted once a suitable property came on the market, and a further register was kept showing the premiums that each company was prepared to pay to secure a lease.[10]

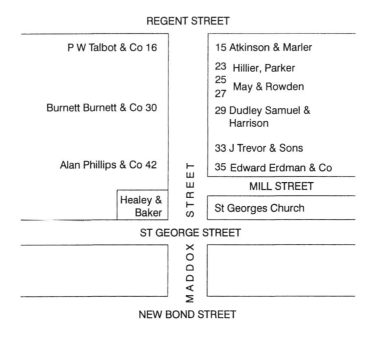

Figure 3.2 Maddox Street in the 1930s. Source: provided by Mr Edward Erdman.

The drawing up of 'key plans', showing every shop in a shopping centre with both the name of the trader and the type of business, as well as an indication of the relative size of the shop, was introduced by Healey & Baker in 1928. These plans were based on Ordnance Survey maps and were gradually enhanced until Healey & Baker had compiled a complete record of every shopping centre of any consequence, together with accompanying records of the frontage for every store shown on the maps.[11] At one point during the 1930s the Prudential offered Healey & Baker £5000 for a complete set of these maps, but the offer was rejected, as the information contained in them was considered too valuable for the practice to lose. The technique of using key plans was copied by a number of other agencies in the 1930s; by 1938 they were being used by at least three major firms. Staff moved frequently between agencies at this time, as is shown by the above example of Edward Erdman's early career, and analytical techniques such as the key plan may have been diffused among different agencies due to staff applying techniques acquired at one practice to another following a move there.

THE GROWTH OF MARKET INTERMEDIARIES

Key plans were particularly important for retailers, as location is all important in this sector, a factor that was appreciated by leading inter-war traders who had very clear ideas regarding their locational requirements. The proximity of leading multiples, the most important being Marks & Spencer, Woolworth's and Montague Burton, was the most important factor influencing the desirability of a shopping pitch and, by improving market efficiency, key plans, and associated analytical techniques, may have substantially accelerated the process of concentration of the multiples into 'High Street' areas, where turnover was maximized and rents and property values underwent substantial appreciation.

The transition of ownership from the individual retailer to the multiple store often involved the property speculator. The way in which speculators used the commercial estate agents as a source of information on prevailing rents reveals the importance of these agencies as repositories of market information. Speculators would approach agents and ask what rent could be earned for a shop in a particular street. The agent would supply this information for no charge and the speculator would buy the store at a price calculated to yield 10% at the rent quoted. He would then return to the agent and ask him to rent the property to a multiple retailer, at the rent he had earlier stated. Once the property had been re-let a mortgage was secured for two-thirds of the purchase price.

The property could then either be sold to an investor, kept by the speculator or sold directly to the multiple concerned.[12] As shop property let to well-known multiples might command an initial yield of 5%, the speculator could make up to 100% capital appreciation on the purchase price of the property, while the use of mortgage finance could boost this appreciation to 300% of the speculator's own long-term capital investment (assuming the store was successfully re-let to a major multiple). In addition to providing information on prevailing rents to speculators, estate agents also constituted an important repository of information on rents and capital values for retailers and long-term investors in commercial property.

Commercial estate agents often arranged funding for property investors and speculators, both via mortgages (using their links with the financial institutions and solicitors acting as trustees for clients) and leaseback arrangements with insurance companies. They also acted as an intermediary between developers and local authorities, dealing with matters such as planning permission and providing information on planning regulations for clients.

Their final major function in the property market was to promote property as an investment medium. The leading West End agents engaged in substantial newspaper advertising, which served to publicize particular properties, the agencies conducting the advertising, and property as an investment asset. Advertisements ranged from single lines in the property classified sections of newspapers such as *The Times* and *The Daily Telegraph* to full-page advertisements, often on the back pages of these newspapers.

3.3 THE 'MULTIPLE REVOLUTION' AND THE PROPERTY INVESTMENT MARKET

For a significant amount of investment in an asset to take place, a substantial volume of market activity is required, with new units of the asset entering the market, thus providing a steady stream of investment opportunities. The most important development in the urban property market which led to this condition being met during the years 1919–1939 was the spread of the multiple retailers.

The growth of the multiples had been significant since the 1880s, but it was during the inter-war period that they came to dominate urban retailing. Some of Britain's best known retailers, such as Marks & Spencer, Montague Burton and Woolworth's, became household names during these years, while local retailers almost disappeared from the High Street, being replaced by such names as Sainsbury's, Price's Fifty Shilling Tailors, J. Lyons & Co., Boots, Finlays, Liptons, Meakers, Barratts, Home and Colonial, Tesco's, Scotch Wool & Hosiery Stores, Great Universal Stores, Dolcis, Trueform, J.H. Dewhurst, Maypole Dairy and British Home Stores. In 1920 multiple chains accounted for between 7 and 10% of all retail sales; by 1939 this had increased to 18–19.5%.[13]

The average yearly net increase in the number of branches of multiple traders with 25 or more branches, over the period 1876–1950, is shown in Table 3.3. Measuring the expansion of multiple retailing in this way underestimates the acceleration in the growth of this sector during the inter-war years however, as multiple retailers were acquiring property adjacent to their existing branches in order to increase the size of those branches, in addition to adding to their branch network.[14] This was particularly true of the 'variety chain stores', such as Woolworth's, Marks & Spencer and British Home Stores, which experienced very rapid growth during this period; by 1938 the average employment per variety chain store branch was over seven times the average for all other multiples.[15]

The other forms of large-scale retailing, Co-operative Societies and department stores, also experienced substantial growth in turnover,[16] their growth, together with that of the multiples, occurring at the expense of the independent local trader. By 1939 large-scale retailers accounted for 33–36% of all retail sales.[17] However, their influence on town centre High Street trading was much greater; by the end of the 1930s there were very few good High Street shopping pitches which were not occupied by multiple retailers, department stores or Co-ops.[18]

There were a number of reasons behind the rapid growth of multiple retailing during the inter-war years. Most of these represented a continuation of trends which had become apparent before 1914, such as the growth of nationally-advertised, branded goods; changing patterns of demand; economies of scale with regard to the expansion of average shop sizes,[19] and other scale economies, particularly in purchasing and transport, that could be gained from large-scale organization in the retailing industry. Motor transport – the car for the middle classes and the motor bus for the working classes – also contributed to this process by increasing the accessibility of town centres for shopping. A further important factor behind the pace of these trends, however, which has not been emphasized in previous accounts of the growth of large-scale retailing in Britain, was the development of a commercial property investment market which placed a premium value on High Street property occupied by well-known multiples and allowed the multiples to use their property assets as a major source of expansion finance.

Table 3.3 Average annual net increase in number of branches of multiple retailers with 25 or more branches: 1876–1950

Year	Net increase
1876–1880	74
1881–1885	167
1886–1890	308
1891–1895	510
1896–1900	648
1901–1905	626
1906–1910	815
1911–1915	505
1916–1920	323
1921–1925	791
1926–1930	1207
1931–1935	788
1936–1939	1120

Source: J.B. Jefferys (1954) *Retail Trading in Britain 1900–1950*, Cambridge University Press, Cambridge, pp.22 and 61.

The influence of the 'multiple revolution' on property values was noted by Hillier, Parker in its first annual report on the property market, covering the year 1921, which stated 'The tendency throughout the country continues for trade to concentrate in the recognised centres, and in these spots values are consequently increasing'.[20] Ten years later Hillier, Parker's report for 1931 discussed future trends in shop rents and values in the light of the spread of the multiples, stating:

> ... our considered opinion is that the rental value of the shop property in the leading cash positions serving the masses will not only be maintained but show a steady advance. More and more retail trade becomes a question of turnover, and rent plays a comparatively small part of the total overhead charge. When turnover can be maintained or increased, rental values will do likewise, and even assuming that there is a decreased spending power of the public, this will be more than counterbalanced by the tendency for customers to concentrate in the leading retail centres.[21]

As a result of these trends, shop values appreciated substantially during the inter-war years. By the mid-1930s the best central shop sites were worth more than the shops major retailers developed on them. An analysis of 10 store developments undertaken by Woolworth's, British Home Stores and Marks & Spencer during 1936–1937 shows that in all cases shop sites exceeded store development costs; for the largest developments site costs could exceed development costs by 100%.[22] However, this concentration of trade in High Street areas may have had an adverse effect on rents and values for shops in less central shopping districts, which were made increasingly peripheral by these trends. As the report of estate agents John D. Wood and Co., for the year 1936, stated 'The "multiple" concerns can gradually command a monopoly of the likeliest shopping positions, and, as those concerns crowd out the old-established resident shopkeepers, the rental value of less favoured positions declines'.[23]

In addition to the concentration of retailing activity in prime High Street areas, the very spread of the multiples increased the attractiveness of the properties they occupied to investors. Occupation by a multiple retailer was viewed by investors as an important factor in enhancing the value of a shop. There were two reasons for this. Firstly, rental income was seen as being more secure if it was backed by the resources of a large business organization. Secondly, occupation by, or proximity to, one of the leading multiples, particularly Marks & Spencer, Woolworth's, Montague Burton or Boots, enhanced the value of a shop site since these stores were able to draw an increased number of people to the immediate area in which they located. As a result retailers were very particular with regard to location, for example, Woolworth's, Marks & Spencer and Burton's would often locate alongside, or close to, each other, sometimes engaging jointly in store development projects.

The effect of occupation by a multiple retailer on shop values can be illustrated by the following example, given by R.M. Lester in a contemporary textbook for property investors.[24] Lester stated that while a well-located shop occupied by a local trader might be purchased to yield 5%, the same shop let to a leading multiple would yield only 4½% and, if it was in possession of a multiple of the stature of Woolworth's, the yield would be as low as 4%. This had an important effect on property values:

> ... assume shop A to be let to Woolworth's at £800 per annum net, and that the market price obtainable is £20,000 freehold, namely 25 years' purchase. Shop B is exactly similar in every respect, but is let to Mr X, a private trader, at £550 per annum. The market price of his shop is found to be £11,000, i.e. 20 years' purchase.[25]

Even if Woolworth's paid the same rent for the shop as the private trader the higher years purchase[26] figure by which it was valued as an investment meant that occupation by Woolworth's would result in a capital appreciation of 25%.

The expansion of multiple retailing, the associated growth of nationwide property market intermediaries and the appreciation of property values arising from the concentration of retail activity in High Street shopping pitches, thus fulfilled conditions 1–3 listed at the beginning of this chapter. Condition 4 (the development of a financial climate in which commercial enterprises had consid-

THE 'MULTIPLE REVOLUTION' AND THE PROPERTY INVESTMENT MARKET

erable incentives to rent, rather than own, their premises, or to dispose of some property rights to their premises in return for an immediate capital sum) was fulfilled as the result of the financial advantages which multiple retailers could accrue from disposing of property rights in their stores to long-term investors.

In his classic study of the growth of retailing in Britain during the century after 1850 J.B. Jefferys noted that:

> The trend in the inter-war years towards main-street shopping sites with large and costly shop premises gave the multiple shop retailers a great advantage over the small-scale retailer in that the former were able to raise the capital for the development of such sites with greater ease.[27]

However, it was the opportunities for corporate finance which resulted from that concentration of retailing activity which gave the multiple retailers their most important financial advantage, enabling them not only to out-bid independent traders for sites and develop large, expensive premises, but also to achieve extremely rapid rates of growth using property-based finance. Some of the most successful and rapidly expanding multiples during this period capitalized on these trends, using property assets as a key element of corporate finance. The integration of retailing and property development and investment gave the multiples access to methods of finance which made fund-raising both relatively easy and inexpensive. The advantages of combining these two areas of enterprise worked in both directions as the major multiples had the ability to increase property values and rents in an area by the very act of locating there.

Some retailers did not fully exploit the opportunities their property assets offered them; for example, one of the main attractions of J. Sears & Co. to Charles Clore, when he launched his successful take-over bid for the company in 1953, lay in the large portfolio of unexploited property assets assembled by the company during the inter-war period. Those retailers which were most successful at using property to fund corporate expansion included a high proportion of firms founded by Jewish businessmen, who also formed the majority of successful property developers in the two decades after 1945.[28] Many of these people, such as Montague Burton, Simon Marks, Isaac Wolfson of Great Universal Stores, Harry Salmon of J. Lyons & Co. and Jack Cohen of Tesco's, were regarded in the property world as important property entrepreneurs in their own right.

Some growing multiple retailers, especially those in the food and other 'convenience' trades, which had a large proportion of outlets in suburban centres and other locations away from the central High Street areas favoured by 'comparison' shopping traders, preferred to rent, rather than buy, their stores. They were able to exploit their ability to raise property values in areas in which they located by bargaining with shop developers for preferential terms. During this period developers such as Edward Lotery were engaged in extensive development of shopping parades in the new suburban districts. To ensure their success it was necessary to attract a well-known multiple which would take the best located unit in the parade. In return for locating there, and thus increasing the desirability of the remaining shops for other traders and the overall value of the development, expanding retailers such as Woolworth's and Tesco's were able to secure substantial concessions from the developer, such as a very low initial rent, a five-year rent-free period or free shop-fittings.[29]

Those multiples which concentrated on town centre High Street sites, and preferred to own the freehold, or a very long leasehold, on their premises, were able to achieve even greater benefits from their ability to raise property values. The way in which inter-war multiple retailers used property holdings to finance expansion can be illustrated by examining two of the most prominent multiples, Montague Burton and Marks & Spencer.

Montague Burton opened his first tailor's shop in Chesterfield in 1904, having migrated to Britain from Lithuania about four years earlier. His business experienced rapid growth, based on a strategy of developing a large turnover with moderate profit margins on each item sold and the integration of manufacture and distribution. The number of Burton's branches rose from 14 on the eve of the First World War to 37 in 1919, and further rapid growth occurred during the 1920s. In 1929 Burton's was launched as a public company, Montague Burton, the Tailor of Taste, Ltd, with an authorized capital of £4 million. During the 1930s the number of Burton branches continued to grow rapidly, from 364 in 1928 to 641 in 1939, and Burton's, which had also greatly expanded the clothing manufacture side of the business, became one of the 10 largest employers in Britain.[30]

Securing adequate capital for the company's expansion plans represented a continual problem, and on several occasions the firm was forced to halt expansion temporarily because of a shortage of funds.[31] This problem was largely overcome by the use of the firm's property assets to raise new finance. Prior to the early 1920s Burton's did not purchase their shop premises, preferring to rent. The firm's fixed capital, mainly consisting of shop fixtures, proved too low to act as security for all borrowings and while new capital requirements were largely financed via mortgages the company relied on a considerable overdraft with Parr's Bank, which stood at over £83 000 in 1920 compared to assets of less than £22 000 and profits of just over £27 000.[32] Parr's Bank became alarmed at the firm's rapid turnover growth, with which its fixed capital failed to keep pace, and, after failures to meet a repayment deadline, the bank refused further financial support.[33] This experience may have led Montague Burton to pay more attention to the property side of his business. During the following years he amassed one of the largest retailing property empires in the country.

By 1931 Burton's real estate holdings were valued at £6 million, compared to an annual sales turnover of £3 million.[34] Finance raised on these properties, together with a high level of retained profits,[35] was used to fund the firm's continued rapid expansion from the mid-1920s onwards. Prior to the 1930s, the main mechanism for raising finance via these assets was their use as security on mortgages and bank overdrafts. Borrowing on overdraft, on the security of the company's properties, was a very important source of development capital prior to Burton's becoming a public company, such borrowing amounting to an average of 27.3% of the firm's total liabilities between 1920 and 1929.[36]

The amount that could be raised by overdraft, or mortgage finance, was limited, however, as the sum for which property was accepted as security for loans was calculated at only two-thirds of its market value. During the 1930s the company made use of a financing technique which realized the full purchase and development cost of stores – the sale and leaseback. Branches were sold to investors at their cost price and simultaneously leased back by Burton's on 999 year leases, at a rent yielding between 4 and 5% of the sum raised by selling the property. For example, Clerical Medical was offered a package of eight freehold Burton branches in September 1934 at a yield of 5%. The initial proposal was rejected, but two of the properties offered were later purchased.[37] A much more substantial deal was done with Eagle Star, who bought 14 Burton's branches, out of 17 initially offered, in 1935.[38] The stores were sold at a price calculated to cover initial purchase and rebuilding costs, plus expenditure on shop fronts, fixtures, architect's fees and any other costs involved in getting them to the state at which they were ready to open, covering Burton's entire outlay on each property.

Leaseback finance was adopted by several other retailers by the end of the 1930s, including Great Universal Stores, Woolworth's, Meakers, Town Tailors and British Home Stores. British Home Stores and Woolworth's also used a related financing technique. This involved the acquisition of a site by a financial institution on behalf of a retailer, the provision of funding by the financial institution for the development of a store on the site and the letting of the completed building to the retailer on a long lease. A number of such agreements were made between British Home Stores and Clerical Medical from 1934.[39] British Home Stores and Woolworth's also agreed several deals of this type with the Prudential during the mid-1930s.[40]

Leaseback arrangements provided the financial institutions with a fixed interest security, with a very stable income stream and a yield in excess of that offered by gilts. The reversion of the property at the end of the lease was of little importance in the case of 99 year leases, due to the very long time period involved, and could be virtually ignored if the lease term was 999 years. Leaseback deals therefore left Burton's with what they regarded as, in this respect, effectively a freehold interest in the property, which might, in future years, be used as security for further finance, as soon as the fixed rent had fallen substantially below the true value of the property.

The only disadvantage of this form of finance was a degree of loss of freedom of action with regard to the future development of leased-back stores. Extensions, and other structural changes to properties sold on this basis, required the permission of the financial institutions which had purchased them. This gave the institutions a good deal of bargaining power when the retailer wished to extend the property, which might be used to obtain a higher rent if rental values had risen substantially since the initial lease was signed. This made leaseback deals unsuitable for companies such as Marks & Spencer, which frequently redeveloped their stores, as is discussed below.

Despite this imperfection, leaseback finance enabled Burton's, and other retailers which made use of it, to grow at a much more rapid rate than was possible with mortgage, overdraft or debenture finance. This was due to certain multiplier effects which were inherent in both mortgage and leaseback finance, but were much more powerful in the latter. These can be illustrated by the following hypothetical example. A retailer has mortgageable property of value P. The sum that can be raised on the property by mortgage or debenture is equal to PX, where X is the amount financial institutions are prepared to lend, as a percentage of the estimated market value of the security offered. It is assumed that the retailer invests the sum borrowed entirely in mortgageable property. The sum raised by doing so is $(PX)X$. If this process is repeated until no further finance can be raised the total sum raised by the series of transactions, Y, is given by the following equation, which I will refer to as the capital security multiplier.

$$Y = [X/(1 - X)]P$$

During the inter-war years it was customary for insurance companies to loan a maximum of only two-thirds of the estimated market value of property assets used as security for mortgages. The total sum raised according to the above equation would therefore be 200% of initial capital, P, for mortgage or debenture finance. However, with sale and leaseback finance, X is equal to 100%, as shops were sold to insurance companies for their full development cost. The multiplier therefore becomes infinite, there being no limit to the amount of finance that could be raised.

This model ignores other sources of finance not secured on property, such as retained profits, and ordinary and preference share issues. It also ignores the length of time taken to complete each round of transactions. In order to take account of such factors it is necessary to develop a more comprehensive model of corporate growth. In discussing corporate growth models, Robbin Marris stated that:

> The theory of the growth of the firm begins with a consideration of the underlying dynamic constraints, or better, restraints, which either limit the maximum growth rate or permit faster growth only at the expense of other desiderata such as profits, dividends and stock-market values. The effects of these restraints are usually formulated in a steady-state system, in which the firm grows at a constant rate over time with constant values of associated 'state' variables.[41]

In the model below, as in the earlier capital security multiplier model, all the company's assets are assumed to consist of mortgageable property.[42] TA_Y is the

THE 'MULTIPLE REVOLUTION' AND THE PROPERTY INVESTMENT MARKET

total value of the firm's assets in year Y, and is given by the equation $TA_Y = TA_{Y-1} + IC_{Y-1}$, where IC_Y represents new capital invested in year Y. MA_Y represents the firm's mortgageable assets in year Y, and is given by the equation $MA_Y = IC_{Y-1}$, since all other assets have already been committed as security for loans. CR_Y represents capital raised in year Y, and is given by the equation: $CR_Y = MA_Y\alpha$, where α is the value at which mortgageable assets are accepted as loan security, as a percentage of their market value.[43]

E_Y represents investment from retained earnings, or ordinary or preference share issue, in year Y, and is given by the equation $E_Y = TA_Y\beta$.[44] Investment from such sources is assumed to be set at a fixed percentage of the firm's total assets in a given year, β indicating that percentage. New capital invested in a given year, IC_Y, is therefore equal to capital raised from borrowing secured by property, CR_Y, plus investment from other sources, E_Y, and is given by the equation $IC_Y = CR_Y + E_Y$. F_Y represents the growth rate of the firm in terms of additions to capital as a percentage of total capital.

Table 3.4 shows the asset growth of a firm, according to this model, over 16 years, growth being entirely funded by mortgage or debenture finance. The firm's initial capital is set at £100 000, though the rate of growth would be the same for an initial capital with any positive value. It is assumed that the time taken between deciding to expand the number of shops and developing the shops to a state at which they can be used as collateral for fund-raising on debenture or mortgage is one year. It is further assumed that all mortgages are granted for a period in excess of that covered in Table 3.4, repayment of principal and all interest payments being taken out of earnings not earmarked for capital investment, and therefore not included in the model. With mortgage or debenture funding as the only source of capital, $\alpha = 2/3$ (as properties are only accepted as security at two-thirds of their market value for such finance) and $\beta = 0$.

As Table 3.4 shows, the firm's rate of growth with mortgage finance alone falls dramatically during the first few years, to below 1% by year nine. Eventually it approaches zero. Table 3.5 shows a second worked example, again using mortgage finance, but with finance from other sources, E, set at 5% of total assets; $\alpha = 2/3$ and $\beta = 0.05$. Growth is initially rapid, as in Table 3.4, but rather than declining to zero over time it settles at a 'steady state' of 12.3%.

The steady state rate of growth, $L(\alpha,\beta)$, is given by the following equation, the mathematical derivation of which is outlined in the Appendix to this chapter:

$$L(\alpha,\beta) = \{(\alpha + \beta - 1) + [(1 - \alpha - \beta)^2 + 4\beta)]^{\frac{1}{2}}\}/2$$

The effects of the income stream, E, are magnified in the firm's expansion, as assets purchased with this income are used for mortgage finance in subsequent years. Thus, the long-term growth provided by the income stream is greater than the value of that income. With an income flow, E, equal to 10 or 15% of total assets a similar, but somewhat reduced, magnification occurs, expansion settling down to an eventual rate of 22.04 and 30.63%, respectively. Marris argued that 'under assumptions of certainty and full comprehension the value of a growth rate is independent of the method of finance'. This is because 'in steady state the growth rate of debt must equal the growth rate of assets; if that were not the case, the ratio of debt to assets would be continually changing, implying, *inter alia*, that the rate of profit net of interest itself must be changing. Steady state must imply a stable leverage ratio.'[45] Such an argument might be true for share issue and straightforward borrowing, both of which would be included in variable E in this model, but does not take account of the multiplier effects associated with borrowing secured on a firm's properties or leaseback finance.

Table 3.4 Corporate growth with mortgage finance

Year	TA	MA	CR	IC	F(%)
1	100 000	100 000	66 667	66 667	66.67
2	166 667	66 667	44 444	44 444	26.67
3	211 111	44 444	29 630	29 630	14.04
4	240 741	29 630	19 753	19 753	8.21
5	260 494	19 753	13 169	13 169	5.06
6	273 663	13 169	8 779	8 779	3.21
7	282 442	8 779	5 853	5 853	2.07
8	288 294	5 853	3 902	3 902	1.35
9	292 196	3 902	2 601	2 601	0.89
10	294 798	2 601	1 734	1 734	0.59
11	296 532	1 734	1 156	1 156	0.39
12	297 688	1 156	771	771	0.26
13	298 459	771	514	514	0.17
14	298 972	514	343	343	0.11
15	299 315	343	228	228	0.08
16	299 543	228	152	152	0.05

Source: See Appendix 3.1.

The difference between the long-term growth rates achieved with mortgage and leaseback finance is substantial. For example, with income from earnings and share issues at 5% of total assets the use of leaseback, rather than mortgage financing, results in an approximate doubling in the steady state rate of expansion from 12.3 to 25%. If 10 or 15% of the firm's current asset value is invested each year this results, under sale and leaseback, in expansion rates of 37.01 and 46.95%, respectively. The difference in the rate of growth made possible by the two types of funding becomes greater when the investment of funds not connected with property is reduced, with $E = 1\%$ mortgage financing resulting in a growth rate of 2.84% per annum, while the figure for leaseback finance is 10.51%.

These results are sensitive to the time taken to develop properties and arrange leaseback transactions; if this is less than a year a substantial long-term rate of growth could be achieved using leaseback finance alone, with no other income. If it is assumed that the length of time taken by each phase of expansion was about a year in most cases, this has a number of interesting implications which were not evident in the earlier capital security multiplier model. These are:

1. Income from sources other than leaseback transactions would be necessary to produce significant long-term growth.
2. The switch from mortgage finance to sale and leaseback allows much more rapid growth with a given level of investment from earnings or share issue.
3. Both mortgage and leaseback finance magnify growth resulting from an injection of capital, as the assets produced by that capital are used to raise further finance in subsequent years.
4. This magnification is smaller the greater the value of funds invested.
5. Sale and leaseback provides a much greater magnification of new capital than does mortgage finance, though the differential between growth rates produced by the two forms of finance falls as the amount of funds not connected with property rises.

Table 3.5 Corporate growth with mortgage finance and 5% cash inflow

Year	TA	MA	CR	E(a)	IC(b)	F(%)
1	100 000	100 000	66 667	5 000	71 667	71.67
2	171 667	71 667	47 778	8 583	56 361	32.83
3	228 028	56 361	37 574	11 401	48 975	21.48
4	277 003	48 975	32 650	13 850	46 500	16.79
5	323 504	46 500	31 000	16 175	47 175	14.58
6	370 679	47 175	31 450	18 534	49 984	13.48
7	420 664	49 984	33 323	21 033	54 356	12.92
8	475 020	54 356	36 237	23 751	59 988	12.63
9	535 008	59 988	39 992	26 750	66 743	12.48
10	601 750	66 743	44 495	30 088	74 583	12.39
11	676 333	74 583	49 722	33 817	83 538	12.35
12	759 871	83 538	55 692	37 994	93 686	12.33
13	853 557	93 686	62 457	42 678	105 135	12.32
14	958 692	105 135	70 090	47 935	118 025	12.31
15	1 076 717	118 025	78 683	53 836	132 519	12.31
16	1 209 236	132 519	88 346	60 462	148 808	12.31
17	1 358 044	148 808	99 205	67 902	167 107	12.31
18	1 525 151	167 107	111 405	76 258	187 662	12.30

Source: See Appendix 3.1.
(a) = βTA; (b) = $CR + E$

Financing growth through earnings has the effect of depressing dividends, while the yields on both the ordinary and preference shares of multiple retailers were usually in excess of those on leaseback transactions, making new share issues relatively unattractive as a source of income for expansion. Other major sources of finance were directly connected to available collateral security.[46]

Overdraft, debenture, mortgage debenture and mortgage finance were about as costly as leaseback transactions in terms of interest rates, but unlike leaseback finance they entailed the repayment of the principal at some future date. Furthermore, they provided a ratio of income raised to collateral security of 67%, or less, compared to the 100% ratio offered by leaseback finance. Sale and leaseback therefore provided an extremely attractive funding mechanism, as it made very high growth rates practicable, at a moderate cost, without repayment of principal and significantly reduced the amount of capital from more expensive sources such as share issues that was needed to achieve a given rate of growth.

Montague Burton's financial innovations also included the establishment of a system of subsidiary companies to hold the Burton's premises. These were owned by Burton or his nominees, and allowed Burton's to minimize tax liabilities via a complicated system by which one of these companies (Key Estates Ltd) would acquire properties using finance lent by Montague Burton Ltd to the other (Henry Holding Ltd) and then re-loaned to Key Estates, which also received rent for the properties from Burton's.[47]

There are a number of similarities between the early history of Montague Burton Ltd and Marks & Spencer. Marks & Spencer was founded over 20 years before Burton's but both firms were started by Jewish immigrants, Michael Marks migrating from Slonim, Russia. Marks & Spencer's early growth strategy, like Burton's, was based on the principle of high turnover with low profit margins on each item sold. However, unlike Burton's, which concentrated on one line of goods, men's clothing, Marks & Spencer pioneered the variety chain store form of retailing, which was characterized by shops of very large size, sometimes covering several floors, the sale of a wide variety of goods, with clearly marked low prices, and a trend towards self-service methods of retailing.[48] This difference in sales policy had important implications for the store development policies pursued by the two companies, placing greater constraints on the types of property-based funding mechanisms that could be used by Marks & Spencer, as is discussed below.

The period after the First World War saw the first substantial property purchases by Marks & Spencer; as late as 1918 it had owned only three freeholds, and five leaseholds, other branches being let on tenancy agreements.[49] Store acquisition was associated with a fundamental change in the company's long-term strategy. In 1924 the chairman, Simon Marks, visited the United States, and was greatly impressed by the character of the new chain stores that were being developed there.[50] Marks introduced several innovations to the company as a result of his visit, one of the most important of which was the replacement of existing shops with new, larger 'super-stores', of the type that were already being developed in Britain by Woolworth's. Superstore development offered Marks & Spencer considerable opportunities to improve the efficiency, appearance and scale of their store network, but necessitated heavy capital expenditure.

THE 'MULTIPLE REVOLUTION' AND THE PROPERTY INVESTMENT MARKET

One way in which funds were raised for new development was via mortgages on the company's properties.[51] Retained profits were another important source of capital at this time, accounting for an average of 60% of total profits, after tax, between 1920 and 1925.[52] Loans and overdraft credit were also used extensively, bank loans and overdrafts amounting to £150 000 by 1924. By then it had been decided that in order to expand as quickly as was intended it would be necessary to launch Marks & Spencer as a public company. Initial negotiations with the British Foreign and Colonial Corporation, regarding the underwriting of a public issue to fund new store development, were unsuccessful. Two years later, however, the company succeeded in launching a public issue, with the help of the Industrial Finance and Investment Corporation Ltd, which was associated with the Prudential.[53] This marked the start of a long association between the Prudential and Marks & Spencer; following the public issue two nominees of the Prudential were appointed as directors of the company. Other retailers also developed strong financial links with insurance companies during this period. The major catering multiple, J. Lyons & Co., financed its inter-war expansion largely through immediately mortgaging newly acquired branches. All Lyons' mortgage business was conducted with a single insurance company, the Alliance Assurance Co. Like the relationship between Marks & Spencer and the Prudential, this involved an interlocking of directorates; two of Lyons' directors, H. Salmon and B. Salmon, were also directors of the Alliance.[54]

In April 1926 Hillier, Parker conducted a valuation of Marks & Spencer's properties. The freeholds were valued at £364 650, and the leaseholds at £223 420.[55] Most leasehold properties were held on relatively short leases, typically for 21 or 14 years. The valuations of the individual stores stressed proximity to other large multiples as a factor which contributed to the value of the properties, for example:

> 9/10 Newborough, Scarborough ... This property adjoins Messrs Montague Burton's premises and is close to Messrs. Boots Cash Chemists and occupying a good position in this principal shopping thoroughfare of the town ...

Following its launch as a public company, Marks & Spencer's policy of superstore development entailed changes in real estate policy. In the late 1920s and 1930s, as the company's property holdings grew, the proportion of leaseholds, particularly short leaseholds, declined. The reasons for this were outlined by Simon Marks in his AGM speech of 11th June 1928:

> It is almost impossible to lease stores at anything like an economic rent. Further than that, a large investment has to be made in reconstructing or rebuilding a property to suit our special requirements, and so far as the company is concerned it is better to make improvements in its own properties rather than in properties leased for a short term of years.[56]

Other retailers also changed their premises policy during this period, seeking either freehold ownership or lengthening the lease terms they considered acceptable. For example, evidence from the Prudential's minutes indicates that Woolworth's (a great deal of whose premises they acquired, both through direct dealings with the company and purchases from third parties) had moved from a policy of renting properties on 42 year leases in the 1920s to 60, and in some cases 99, year leases by the late 1930s.[57]

Store development entailed heavy capital costs and, despite the raising of funds by mortgaging newly acquired property, capital constraints placed a limit on Marks & Spencer's expansion, as was noted by Rees in his history of the company.[58] It was not until after 1929 that new branch openings were on a scale comparable with that before the First World War.

By 1928 the company's growth necessitated raising further finance; approximately £1 million was raised via overdraft facilities, an increase in authorized and issued ordinary share capital, additional mortgages and other borrowing arrangements.[59] In 1930 the company once more augmented its capital, with the issue of £1 million of mortgage debentures at 6%, and £1 million of preference shares to yield 7%. The debenture stock was to be secured by a trust deed in favour of the Prudential, as trustees, creating a first specific mortgage on the company's existing and future properties, and a floating charge on the remaining assets. Although a demand for this stock subsequently arose, the issue was of no initial attraction to the stock exchange, the Prudential's role as underwriter being crucial to its success.[60] It was used to replace almost £1 000 000 which had been borrowed on mortgage. Mortgages subsequently formed only a small proportion of funds raised on the security of property, mortgage debentures, or loans secured by mortgage debentures, providing the bulk of this type of finance. The debenture agreement gave Marks & Spencer the power to issue further debentures, providing that the amount of debenture and mortgage finance arranged was not allowed to exceed 60% of the value of the company's properties.

During the early 1930s the firm continued its programme of rapid expansion, profits rising each year from £167 243 in 1928 to £670 117 in 1932, despite the depression. In 1933 and 1934 Marks & Spencer took advantage of falling interest rates and a buoyant stock market with the arrangement of further substantial loans secured on mortgage debenture stock and an increase in the company's ordinary capital.

The expansion of Marks & Spencer during the years after 1934 was funded by retained profits and debenture finance, liabilities secured on the company's properties forming an increasing proportion of total liabilities in the late 1930s, as is shown in Table 3.6. Debenture finance was provided by the Prudential, under an arrangement which bore some similarity to the funding links developed between property companies and insurance companies discussed in section 4.3. Such finance made it unnecessary to go to the stock exchange to fund the company's development programme.[61]

THE 'MULTIPLE REVOLUTION' AND THE PROPERTY INVESTMENT MARKET

A programme of extending existing stores had been instigated in the 1920s; extensions, rather than the opening of new branches, formed a growing proportion of Marks & Spencer's store developments during the inter-war period, becoming the dominant form of development activity during the late 1930s.[62] The pace of store modification and extension was such that some branches were rebuilt several times during the inter-war years. This shift from extending the branch network to increasing the size of individual stores was also true, to a lesser extent, for multiple retailers in general during these years, though it was particularly notable for variety goods retailers.

Table 3.6 An outline of Marks & Spencer's growth: 1926–1939

Year	Rate of growth of balance sheet assets (£)	Freeholds as percentage of total property (£)	Property assets as percentage of total assets (£)	Property liabilities [a] as percentage of total liabilities (£)
1926		62.01		
1927			75.08	14.19
1928	18.80		80.59	12.88
1929	58.31	71.36	80.10	17.17
1930	43.13	74.04	79.44	33.25
1931	39.40	69.46	68.57	27.59
1932	8.37	65.38	76.84	24.55
1933	13.62	62.48	74.67	26.32
1934	21.03	61.89	74.20	29.74
1935	21.45	63.41	71.70	24.51
1936	16.98	65.50	71.68	29.07
1937	20.06	64.01	68.10	33.57
1938	4.57	62.87	71.52	32.00
1939	10.08	63.07	71.91	33.07
Average	22.98	65.46	74.18	25.99
Average 1930–1939	19.87	65.21	72.86	29.37

[a] Mortgages, mortgage debentures and loans secured on mortgage debentures.

Source: Derived from Marks & Spencer's published accounts and April 1926 property valuation.

During the 10 years to March 1937 Marks & Spencer had invested over £8 million, of which over £7.25 million was spent on the firm's properties. The company used its property portfolio as security for mortgage and mortgage debenture funding, but did not make extensive use of leaseback finance. This was due to Marks & Spencer's policy of frequent store redevelopment, arising from the variety goods nature of its business and a wish to develop superstores along American lines, which would have been more difficult had stores been sold to financial institutions. Redevelopment of a property leased to an investor was only possible after negotiation with the institution each time redevelopment was proposed, the institution's veto over development putting it in a strong position to demand a higher rent for the redeveloped property. Such considerations resulted in leaseback finance being relatively unattractive for Marks & Spencer, which compensated for the lower multiplier effects inherent in mortgage and mortgage debenture finance by reinvesting a large proportion of profits and using sources of finance not connected with property more extensively than Burton's.

However, Marks & Spencer did use a related finance technique, which involved the purchase of sites on which the company wished to erect stores by an insurance company, which would then grant Marks & Spencer a 99 year building lease. The Prudential acquired a number of sites for Marks & Spencer on these terms.[63] In 1937, when Marks & Spencer began its ambitious plan to develop the Pantheon, Oxford Street, a variety store on a department store scale, the cost required them to go further down the road of ceding property rights to the financial institution in return for the necessary finance. In July 1937 the Prudential approved a plan involving its purchasing the 27 000 sq. ft. site on which the Pantheon was to be built (recently acquired by Marks & Spencer for £175 000) and paying £70 000 to fund the development of the new superstore. On completion Marks & Spencer agreed to take a full-repairing 99 year lease on the property, providing a 4.25% yield on the Prudential's total outlay of £245 000.[64]

Both Montague Burton Ltd and Marks & Spencer made use of property as a means of raising finance to the maximum extent possible given their different growth strategies. The relative success of each strategy is difficult to measure, due to the absence of regular market valuations of their property assets. Marks & Spencer's strategy produced a high and growing profits:assets ratio; pre-tax profits averaged 12.7% of balance sheet assets over the period 1927–1938; over the period 1932–1938 this rose to 14.6%.[65] However, the continued redevelopment and expansion of shop premises entailed heavy capital expenditure which limited capital appreciation for the stores. A March 1955 valuation of Marks & Spencer's property portfolio put the value of their property assets at £35 297 191. This was over twice the properties' original cost, though subsequent rebuilding and extensions, to the value of £16 million, had more or less cancelled out this appreciation.[66]

Montague Burton's more property-orientated policy produced a much lower profits:assets ratio, which averaged only 4.5%, for post-tax profits, from 1930 to 1939. However, it also resulted in the development of a massive and appreciating property portfolio which proved to be of immense value to the company during the following decades. In February 1946 Montague Burton stated that the market value of properties owned by Burton's and its 100% subsidiaries was about £20 million, compared to a cost price of £11 million.[67] An independent valuation of Burton's branch premises, conducted in January 1947,[68] broadly confirmed these figures, the group's properties (excluding factories, stores let to other retailers and other non-branch premises)[69] being valued at £15 781 566.

Montague Burton's policy therefore led to the growth of an asset base which appreciated significantly in value during the medium term, while Marks & Spencer's policy of regular and systematic store redevelopment sacrificed such medium-term capital appreciation but produced a much higher level of profits in relation to the firm's assets. The success of both policies is evident from the balance sheets of the two companies, Marks & Spencer achieving an annual average rate of growth of almost 23% per annum from 1928 to 1939, while Montague Burton Ltd grew at an average rate of over 25% between 1920 and 1938.

Property-based finance facilitated the rapid transformation of Britain's High Streets into a series of branches of nationally-based retailing companies, and as such accelerated the modernization and improved the efficiency of British retailing. At a time when the British capital market was widely viewed as an impediment to the growth of companies of the size of Marks & Spencer and Montague Burton at the beginning of the inter-war period, the institutional property investment market allowed retailers to raise funds not solely on their reputation as business concerns, but on the strength of their capital assets. These methods of finance were unavailable to most other areas of industry; raising money on the strength of factory property via the institutional property investment market was almost impossible at this time, with regard to both mortgage and leaseback finance.[70] The role of alleged flaws in Britain's capital market in inhibiting the modernization of industry during this period is beyond the scope of this study, though the example of multiple retailing does illustrate that in one area of the British economy abundant and inexpensive finance, together with other favourable conditions, did lead to rapid growth and structural change.

APPENDIX: THE MATHEMATICAL DERIVATION OF THE STEADY STATE OF CORPORATE GROWTH EQUATION

By definition:

$$TA_Y = TA_{Y-1} + IC_{Y-1} \tag{3.1}[71]$$

$$IC_Y = CR_Y + E_Y \tag{3.2}$$

$$= \alpha MA_Y + \beta TA_Y$$

$$MA_Y = IC_{Y-1} \tag{3.3}$$

In steady state growth:

$$F_Y = F_{Y-1} = IC_Y/TA_Y = IC_{Y-1}/TA_{Y-1} \tag{3.4}$$

From Equation 3.4:

$$F_Y = IC_Y/TA_Y \quad (3.5)$$

Substituting from Equations 3.2 and 3.3:

$$F_Y = \frac{\alpha IC_{Y-1} + \beta TA_Y}{TA_Y} \quad (3.6)$$

Substituting from Equation 3.1:

$$F_Y = \frac{\alpha IC_{Y-1}}{TA_{Y-1} + IC_{Y-1}} + \beta \quad (3.7)$$

Dividing the first term on the right-hand side by TA_{Y-1}:

$$F_Y = \frac{\alpha(IC_{Y-1}/TA_{Y-1})}{1 + (IC_{Y-1}/TA_{Y-1})} + \beta \quad (3.8)$$

Substituting from Equation 3.4:

$$F_Y = \frac{\alpha F_Y}{1 + F_Y} + \beta \quad (3.9)$$

$$\backslash F_Y + (F_Y)^2 - \alpha F_Y - \beta(1 + F_Y) = 0 \quad (3.10)$$
$$\backslash (F_Y)^2 + F_Y(1 - \alpha - \beta) - \beta = 0 \quad (3.11)$$

Note: For $ax^2 + bx + c = 0$, $x = [-b \pm \sqrt{(b^2 - 4ac)}]/2a$

$$\backslash L(\alpha,\beta) = \{(\alpha + \beta - 1) \pm \sqrt{[(1 - \alpha - \beta)^2 + 4\beta]}\}/2 \quad (3.12)$$

Given that the steady state rate of growth is known to be greater than zero with a positive value of β, the positive root of this equation applies. The steady state of growth equation is therefore the positive root of this last equation, as given below.

$$L(\alpha,\beta) = \{(\alpha + \beta - 1) + \sqrt{[(1 - \alpha - \beta)^2 + 4\beta]}\}/2$$

REFERENCES AND NOTES

1. The practice was known thereafter as Hillier, Parker, May & Rowden, but is referred to in this text using the abbreviated title Hillier, Parker.
2. *The Times* (1924) 1 Jan.
3. Healey & Baker (c. 1970) *Healey & Baker: 1820–1970*. Privately published, p. 10.
4. Source: Typescript of interview by Kerrie Walkingshaw with Mervyn and Aubrey Orchard-Lisle (1987) 1 July.
5. Edward L. Erdman (1982) *People & Property*, Batsford, London, p. 6.
6. G. Cross (1939) *Suffolk Punch: A Business Man's Autobiography*, Faber & Faber, London, p. 201.
7. G. Cross (1939) *Suffolk Punch: A Business Man's Autobiography*, Faber & Faber, London, p. 5.

REFERENCES AND NOTES

8. Edward L. Erdman (1990) *Edward Erdman: Surveyors – A Brief History of the Practice.* Unfinished draft copy, Ch. One, p. 9 (held by the author).
9. Edward L. Erdman (1990) *Edward Erdman: Surveyors – A Brief History of the Practice.* Unfinished draft copy, Ch. One, p. 2 (held by the author).
10. Edward L. Erdman (1990) *Edward Erdman: Surveyors – A Brief History of the Practice.* Unfinished draft copy, Ch. One, p. 8 (held by the author).
11. Healey & Baker (*c.* 1970) *Healey & Baker: 1820–1970.* Privately published, p. 9.
12. Source: Interview with Mr Edward L. Erdman (1990) 28 Aug.
13. Source: J.B. Jefferys (1954) *Retail Trading in Britain 1900–1950*, Cambridge University Press, Cambridge, p. 51. (These figures do not include sales by department stores and Co-operative Societies.)
14. G. Rees (1969) *St Michael: A History of Marks & Spencer*, Pan, London, p. 57.
15. J.B. Jefferys (1954) *Retail Trading in Britain 1900–1950*, Cambridge University Press, Cambridge, p. 63.
16. J.B. Jefferys (1954) *Retail Trading in Britain 1900–1950*, Cambridge University Press, Cambridge, pp. 58–61.
17. J.B. Jefferys (1954) *Retail Trading in Britain 1900–1950*, Cambridge University Press, Cambridge, p. 74.
18. J.B. Jefferys (1954) *Retail Trading in Britain 1900–1950*, Cambridge University Press, Cambridge, p. 90.
19. J.B. Jefferys (1954) *Retail Trading in Britain 1900–1950*, Cambridge University Press, Cambridge, p. 82.
20. Hillier, Parker, Report for the year 1921 (1922) *The Daily Telegraph*, 4 Jan., 2.
21. Hillier, Parker, report for the year 1931 (1932) *The Times*, 4 Jan., 21.
22. Prudential (1936–37) Board minutes.
23. John D. Wood & Co. (1937) *Report for the Year 1936*. Privately published, p. 9.
24. R.M. Lester (1937) *Property Investment*, Pitman, London, p. 15.
25. R.M. Lester (1937) *Property Investment*, Pitman, London, p. 16.
26. 'Years purchase' is the number the yield on an asset has to be multiplied by to arrive at the capital value – the inverse of the yield.
27. R.M. Lester (1937) *Property Investment*, Pitman, London, p. 89.
28. See Chapter 5.
29. O. Marriott (1967) *The Property Boom*, Hamish Hamilton, London, p. 18.
30. K. Honeyman (1993) Montague Burton Ltd: The creators of well-dressed men, in *Leeds City Business, 1893–1993: Essays Celebrating the Centenary* (eds J. Chartres and K. Honeyman), Leeds University Press, Leeds, p. 186.
31. Aubrey Orchard-Lisle (1983) Recollection on Sir Montague Burton. Unpublished paper, supplied to the author by Mr Stanley H. Burton.
32. K. Honeyman (1993) Montague Burton Ltd: The creators of well-dressed men, in *Leeds City Business, 1893–1993: Essays Celebrating the Centenary* (eds J. Chartres and K. Honeyman), Leeds University Press, Leeds, p. 192.

33. K. Honeyman (1993) Montague Burton Ltd: The creators of well-dressed men, in *Leeds City Business, 1893–1993: Essays Celebrating the Centenary* (eds J. Chartres and K. Honeyman), Leeds University Press, Leeds, p. 193.
34. E.B. Behrens (1931) Critical survey of Montague Burton Ltd. West Yorkshire Archive Service, Montague Burton Papers, Box 114, Sept., p. 1.
35. K. Honeyman (1993) Montague Burton Ltd: The creators of well-dressed men, in *Leeds City Business, 1893–1993: Essays Celebrating the Centenary* (eds J. Chartres and K. Honeyman), Leeds University Press, Leeds, p. 194.
36. E.M. Sigsworth (1990) *Montague Burton: The Tailor of Taste*, Manchester University Press, Manchester, p. 74.
37. Clerical Medical (1934) Investigation into the Society's assets.
38. Eagle Star (undated ledger) Report on general fund properties.
39. Clerical, Medical and General Life Assurance Society (1934) Board minutes, 17 Jan.
40. Prudential (1935–36) Board minutes.
41. R. Marris (1971) An introduction to theories of corporate growth, in *The Corporate Economy: Growth, Competition, and Innovative Potential* (eds R. Marris and A. Wood), Macmillan, London, p. 5.
42. For a typical large inter-war multiple retailer the proportion of mortgageable property to total balance sheet assets would be about 70–75%.
43. $\alpha = 2/3$ for mortgage finance and 1 for sale and leaseback.
44. β = investment from company earnings, share issue or other sources not involving property assets as security, as a percentage of total assets.
45. R. Marris (1971) An introduction to theories of corporate growth, in *The Corporate Economy: Growth, Competition, and Innovative Potential* (eds R. Marris and A. Wood), Macmillan, London, pp. 21 and 23.
46. With regard to overdrafts, collateral security might not be required for relatively small advances if the firm had a strong balance sheet or a good relationship with the bank in question, but was usually required by the 1930s for large overdrafts amounting to a substantial proportion of a firm's liabilities. See W.A. Thomas (1978) *The Finance of British Industry, 1918–1976*, Methuen, London, p. 56.
47. See K. Honeyman (1993) Montague Burton Ltd: The creators of well-dressed men, in *Leeds City Business, 1893–1993: Essays Celebrating the Centenary* (eds J. Chartres and K. Honeyman), Leeds University Press, Leeds, pp. 194–5, for a fuller description of Burton's property companies and other tax avoidance measures.
48. J.B. Jefferys (1954) *Retail Trading in Britain 1900–1950*, Cambridge University Press, Cambridge, p. 70.
49. G. Rees (1969) *St Michael: A History of Marks & Spencer*, Pan, London, p. 59.
50. Michael Marks died in 1907 and during the following years the firm was managed by William Chapman and Bernard Steel, who represented the

REFERENCES AND NOTES

interests of the Spencer and Marks families, respectively. This partnership did not prove a success and, after a prolonged power struggle between the representatives of the two families, Simon Marks, Michael Marks' son, replaced Chapman as chairman in 1916.

51. Marks & Spencer, Memo of Interview Ledger, D54.
52. G. Rees (1969) *St Michael: A History of Marks & Spencer*, Pan, London, p. 63.
53. G. Rees (1969) *St Michael: A History of Marks & Spencer*, Pan, London, p. 70.
54. D.J. Richardson (1970) The history of the catering industry, with special reference to the development of J. Lyons and Co. Ltd., to 1939. PhD thesis, University of Kent, p. 348.
55. Hillier, Parker, May & Rowden (April 1926) Report and valuation of Marks & Spencer's properties. (These figures exclude trade fittings, furniture, etc.)
56. Marks & Spencer (1928) Annual Report, 11 June.
57. Prudential (1919–39) Board minutes.
58. G. Rees (1969) *St Michael: A History of Marks & Spencer*, Pan, London, p. 59.
59. G. Rees (1969) *St Michael: A History of Marks & Spencer*, Pan, London, p. 71.
60. G. Rees (1969) *St Michael: A History of Marks & Spencer*, Pan, London, p. 72.
61. G. Rees (1969) *St Michael: A History of Marks & Spencer*, Pan, London, p. 77.
62. G. Rees (1969) *St Michael: A History of Marks & Spencer*, Pan, London, p. 81.
63. Prudential (1933) Board minutes, 2 Nov.
64. Prudential (1937) Board minutes, 8 July. (Marks & Spencer also undertook to pay interest on the project's costs prior to signing the lease.)
65. Source: Marks & Spencer (1927–38) Annual Reports.
66. Marks & Spencer (1955) AGM Chairman's Speech, 9 June.
67. Letter from Montague Burton to Major Waddington of Myers & Co. (5 Feb. 1946) West Yorkshire Archive Service, Montague Burton Papers, Box 110.
68. Valuation of Montague Burton branches by Healey & Baker (7 Jan. 1947) West Yorkshire Archive Service, Montague Burton Papers, Box 135.
69. An earlier independent valuation, conducted in October 1946, had estimated the rental value of the group's properties which were available for letting, or might be available before the beginning of 1950, at £155 158, implying a market value in the region of £3 million which, together with the values of the firm's factory properties, warehouses and sites, must be added to the value of the branch premises to arrive at the value for the total portfolio.
70. See section 4.3.
71. *Note*: For stocks y = value at beginning of year.

4 Property investment, development and the capital market between the wars

4.1 THE PATTERN OF INSTITUTIONAL PROPERTY INVESTMENT 1919–1939

The most important group of institutional investors in urban property during the inter-war years were the insurance companies. From 1922 to 1937 the funds of UK-based life assurance companies more than doubled from £804 million to £1655 million. Net[1] investment of new insurance company funds in land, property and ground rents is shown in Table 4.1, which also shows the volume of money invested in another property-based security – mortgages.

The table shows a very low level of investment in real property during the 1920s, with negative figures for 1924 and 1925.[2] Significantly higher levels of net property investment were recorded in 1930 and 1931 than for any previous year shown, despite these being years of depression in the property market. During the mid-1930s a boom in insurance company property investment occurred, the proportion of new insurance company funds directed to property rising to a peak of over 11% in 1935. The contrast between insurance company property investment in the 1920s and the 1930s is striking; the average proportion of net new funds invested in this sector from 1923 to 1929 amounted to only 0.86%, compared to 7.38% for the years 1930–1937.

The pattern of insurance company property investment during the inter-war period did not follow the cycle of the property market, which was broadly one of relatively prosperous conditions in the 1920s, depression from 1929 to 1932 and boom thereafter, falling off from 1937. There were certainly opportunities for institutional investment in property during the 1920s, as is shown by the example of Clerical Medical.

Clerical Medical was one the few insurance companies to undertake a substantial volume of property investment during the 1920s. The Society's

Finance Committee reviewed what was then a very small property portfolio, in 1921, and found that their properties had appreciated significantly in value, their estimated market value being at least £40 000 above book cost[3] of £62 054. Clerical Medical's property acquisitions during the 1920s consisted almost exclusively of office property, including some office blocks let to several tenants which were managed by their agents, Messrs Goddard & Smith, for a fee of 3-5% of gross rental income. In 1926 Clerical Medical estimated the market value of its freehold properties at £150 000, a margin of £16 500 over their book value, and stated that their sale value would probably exceed £150 000. As with the previous valuation, it was shown that in addition to high rental income the properties were proving to be appreciating assets.[4]

Table 4.1 Net (inflow–outflow) new insurance company funds and net investment in mortgages and property 1922–1937

	Net mortgages	Net property [a]	Net total funds	Mortgages as percentage of total	Property as percentage of total
1923	149 608	342 804	46 718 573	0.32	0.73
1924	3 505 676	−1 170 262	31 008 485	11.31	−3.77
1925	9 171 684	−77 661	64 490 051	14.22	−0.12
1926	9 581 490	1 361 997	57 717 260	16.60	2.36
1927	11 921 621	1 060 264	60 447 254	19.72	1.75
1928	5 944 704	1 551 294	56 383 098	10.54	2.75
1929	16 411 818	1 364 275	58 673 495	27.97	2.33
1930	11 429 524	1 928 880	39 114 766	29.22	4.93
1931	2 978 502	2 934 653	45 650 169	6.52	6.43
1932	13 182 292	2 488 289	45 179 782	29.18	5.51
1933	−16 325 059	4 338 964	60 711 667	−26.89	7.15
1934	2 760 222	4 534 850	65 118 757	4.24	6.96
1935	3 601 845	7 371 761	66 900 431	5.38	11.02
1936	14 691 371	8 376 129	77 746 775	18.90	10.77
1937	8 646 339	4 730 667	75 560 977	11.44	6.26

Source: Board of Trade (1923–1938) *Annual Report On Life And Other Long-term Assurance Business,* HMSO, London. These figures include both the life and non-life funds of life assurance companies.

[a] Includes land, property, ground rents and office furniture.

The Society does not appear to have found difficulty in obtaining properties on favourable yields during the 1920s, the proportion of total funds invested in property, excluding ground rents, rising from 2% in 1920 to 7% in 1931.[5] The 1930s saw a more general move towards investment in property on the part of the insurance companies, in contrast to a very low level of investment in this sector during the 1920s. This was largely due to the government's successful policy of cheap money, introduced following Britain's departure from the gold standard in 1931 in order to stimulate economic recovery.

Section 10.4 provides figures for initial yields on consols, ordinary shares and various classes of prime investment property from 1920 to 1938. After a period of stable consol yields during the 1920s the yield on consols fell significantly in 1932, and continued to decline during the following years, to a low-point of 2.9% in 1935–1936. This caused serious problems for the insurance companies, as many policies were calculated at an assumed rate of interest of 3%. The fall in the consol yield, and the associated fall in yields for other fixed-interest assets such as mortgages, led the insurance companies to look for higher yielding, but relatively safe, investments. Investment property constituted a virtually fixed-interest asset at this time, being subject to long leases at fixed rents, but offered a yield that was over 2% higher than that for consols. Furthermore, the yield differential between prime shops and gilts widened somewhat during the cheap money period, from an average of 1.9% from 1921 to 1929 to 2.5% from 1930 to 1938.

Increased yield differentials between property and gilts were reflected in higher total returns to investment in property during the 1930s, as is outlined in section 10.6. From 1921 to 1929 the average rate of return on primary shops was 5.2%, compared to 6.3% for consols. During 1930–1938 the position was reversed however, with average rates of return on shops and consols of 6.9 and 6.7%, respectively. Equities out-performed both asset classes by a wide margin during the 1920s but produced an average return of only 6.6% from 1930 to 1938.

The unsettled nature of the equity market, and general economic conditions, during the early 1930s was a further factor behind the increase in property investment activity. Hillier, Parker's report on business for the year 1932 noted that despite the depression considerable interest had been shown in property by investors during the previous two years:

> The increase in interest in property investments experienced in 1931 – by many considered extraordinary and therefore much commented upon – has continued throughout the past year at a rate surpassing anything in our previous experience. The reason for the rapid rise in demand is found in events like the departure of this country from the gold standard, the uncertainties of industrial investment, and the introduction of a new fiscal policy.[6]

The same trends were noted in an *Estates Gazette* article of September 1932, which stated:

> The departure from the gold standard and the conversion of the Great War Loan have worked a vast change in the economic situation, and practically every 'safe' security in the true sense of the word, has reached a price which still further reduces low yields. In these circumstances it is natural enough that investors, whether restrained by legal bonds or free to do as they please, should turn to a security which may almost invariably

THE PATTERN OF INSTITUTIONAL PROPERTY INVESTMENT 1919-1939

be depended upon to yield a substantially higher return than Government obligations, with practically equal safety.[7]

During the 1930s two books offering advice on property investment were published, each of which stressed the relative stability of real property as an investment medium.[8] The more substantial of the two, by R.M. Lester, began:

> The rapid fluctuations and uncertainties of the Stock Exchange have caused many investors to turn their attention to the property market during recent years. The security of capital, higher yield of interest, and freedom from disturbance in the enjoyment of such investments, are the outstanding attractions.[9]

Cheap money, and the consequent low yields on the largest categories of insurance company assets, gilts, mortgages and debentures, put considerable pressure on insurance companies to find higher yielding outlets for new funds. Such problems were discussed in Legal & General's actuarial report on the quinquennial investigation into the Society's assets for the five years to 31st December 1936, which noted that:

> The rapid expansion of the Society's funds (from £25,438,023 to £39,724,653) during the quinquennium has coincided with a particularly difficult period for investment. In the first year, 1932, the conversion of the 5% War Loan to a 3.5% basis set up new conditions in the investment market involving a substantial fall in the yield on investment of new money. The position was aggravated by the industrial depression and consequent release of large funds for which the only outlet has been investment in British Government Securities. Throughout the quinquennium the latter have yielded barely 3% gross. It has been in these conditions that the Society has had to invest over £14,000,000 of new funds and has had to follow the course taken by other investment institutions to increase its holdings of gilt edged securities. The change in the character of its investments is well illustrated by contrasting the position at the beginning with that at the end of the quinquennium, as shown in the following table:

	31/12/31 Amount	%	31/12/36 Amount	%
Investments in:				
(1) Mortgages, loans etc.	17,971,491	75	19,601,620	52
(2) Stock Exchange Investments	6,071,129	25	18,067,471	48
Total	24,024,620		37,669,091	

Included in the Stock Exchange Investments there was at the outset of the quinquennium the sum of £1,270,670 or 5.3% of the total funds invested in British Government securities while at the end of the period this had increased to £5,478,847, or 14.5% of the total funds. This position may be contrasted with that at 31/12/1913, the last year of account preceding the War when apart from the statutory deposit of £20,000, the Society held only £90,722 in British Government securities, representing less than 1% of total funds.

The change in the character of its investments has naturally been accompanied by a reduction in the net interest yield, shown in the annual account for the quinquennium as follows.

Year	1932	£4.12.1%
	1933	£4.7.4%
	1934	£4.4.10%
	1935	£4.2.11%
	1936	£4.-.9%.[10]

Low gilt yields, and the lack of sufficient new mortgage business to absorb the same proportion of the Society's rapidly growing funds as had been possible in the 1920s, led Legal & General to undertake some limited direct investment in property from 1934. In 1930 land, property and ground rents had accounted for only 2.2% of the Society's total assets, being composed entirely of properties occupied by the Society, buildings acquired as the result of mortgage foreclosure and ground rents. By 1938 the proportion of Legal & General's funds invested in property had more than doubled, to 4.9%.

Legal & General's foray into the property market was dwarfed, however, by the property investment activities of Britain's largest life assurance company, the Prudential. The Prudential had already invested substantial sums in property prior to the First World War, as noted in Chapter 2. Some additional property investment was undertaken in the 1920s, including the commencement of the flat development programme discussed in section 4.2, though the extent of direct investment in this sector was increased substantially during the 1930s. At the Prudential's Annual General Meeting in March 1935 the chairman, W. Edgar Horne, stated:

> The main problem at the moment is of course to find suitable investments giving an adequate yield. As I mentioned last March, the fall in the rate of interest is a most serious matter for all investors, and in 1934 it has been more difficult than ever before to obtain a satisfactory yield ...
>
> During the past 12 months we have invested substantial sums in the property market, on which the return is somewhat higher than that obtainable from stock exchange securities.[11]

During the 1935 financial year alone the Prudential invested over £3.5 million in property, achieving a yield of 5.25%,[12] at a time when gilts were yielding only 2.9%.

During the 1930s Clerical Medical continued its active property investment policy, property assets, excluding ground rents, increasing from 7% of total assets in 1931 to 13% in 1938. Whereas office purchases had constituted the bulk of the Society's acquisitions in the 1920s, Clerical's property investments during the 1930s were concentrated in shops. Sale and leaseback transactions with property companies and developers, such as Town Investments Ltd and Edward Lotery, and retailers, such as Montague Burton Ltd, were the source of many of these acquisitions.

4.2 PROPERTY COMPANIES AND THE PROPERTY DEVELOPMENT MARKET

The 1930s saw a substantial expansion of the property company sector. Table 4.2 provides estimates of the number of public property companies established from the 1870s to the 1930s. It is a continuation of, and is based on the same sources as, Table 2.1.

Table 4.2 New public property companies 1870–1938

	Number	Annual average
1870–1879	7	0.7
1880–1889	14	1.4
1890–1902	20	1.5
1903–1913	9	0.8
1914–1932	26	1.4
1933–1938	40	6.7

Source: Thomas Skinner & Co. (1950) *Skinner's Property Share Annual 1950–51*, Thomas Skinner, London.

Unlike the upturn in insurance company property investment, which began in 1930 while the property development market was still depressed, the property company boom did not commence until the onset of cheap money in the latter half of 1932. During 1933 14 new public property companies were launched, and in 1934 there were a further 11. For the period 1919–1932 the number of new public property companies had averaged less than two, while only one company was launched in 1932, immediately prior to the boom. A similar trend is shown when figures for all new property companies (both public and private) are examined; Jordan's register of company formation gives a total figure for new property companies formed in 1933 of 813, compared with an average of only 380 during the previous six years. By 1935, at the height of the property boom, the number of new companies had risen to 1322.[13] Figures compiled by the Midland Bank, for the value of new capital issues by property companies, indicate an upturn in property company capital issues in the second half of 1932. They also show substantial growth in the property company sector relative to the stock market as a whole. During 1930-1931 new issues by property companies amounted to an average of only 1.9% of all new company issues. However, there was a substantial acceleration during the second half of 1932 after the War Loan conversion announcement which marked the advent of cheap money, and from 1933 to 1935 property company new issues averaged 6.9% of all new issues.[14]

Shop investment and development companies dominated the new property companies established during the 1930s. A notable example of this type of enterprise was the Covent Garden group. The Covent Garden Properties Company arose out of the sale of the Covent Garden Estate, owned by the Duke of Bedford, to Sir Joseph Beecham, after a dispute between the Duke and Mr (later Sir) Malleby-Deeley, a Member of Parliament and well-known land speculator. In 1924, several years after the death of Sir Joseph, his son, Thomas Beecham, together with the financier Philip Hill, arranged to float the estate (which was causing his family serious financial problems regarding mortgage repayment) together with the family pill business, as the Beecham Estate and Pill Corporation.

The company's early years were dominated by the pharmaceutical side of the business, and proposals, of which nothing came, to relocate the Covent Garden market and redevelop the present site. In November 1927, however, the company purchased the Bootle, Walton and Kirkdale estates, via Philip Hill, from Lord Derby. The deal involved the sale of the company's pharmaceutical interests to Hill, making it entirely a property company, a move which was reflected in a change of name to the Covent Garden Properties Co. Ltd.

The estates purchased from Lord Derby were broken up and sold off to give the company funds for reinvestment in commercial property. Some very prominent figures in the property world were recruited as directors, including Claude Goddard, of estate agents Goddard & Smith, and Sir Edward Mountain, the chairman of Eagle Star. Despite this, progress during the late 1920s and early 1930s was slow. Ordinary share dividends stood at 7.5%, compared to 10% prior to the sale of the pill business, and some of the proceeds from the sale of Lord Derby's estates had to be used to maintain dividends at even that level. An improvement in the company's fortunes during the 1930s was to arise due to a change in capital structure, improved conditions in the property market and, most importantly, cheap money.

In his book on the Covent Garden companies, *Cabbages and Things*, J.M. Keyworth, who worked as secretary to Philip Hill, explains the importance of a revised capital structure as follows:

> Based upon the proportion of assets backed by the various classes of Debenture Stock and share capital, for each £10,000 invested the interest and dividend costs were:-
>
> | | |
> |---|---:|
> | 500 Debentures at 5% | £250 |
> | 2200 7% Preference shares | 154 |
> | 1200 8% Preferred shares | 96 |
> | 1600 Ordinary shares, say 6.25% | 100 |
> | | £600 |

Thus to achieve a dividend of 6.25% on the ordinary shares would require a return on the investments averaging 6% after paying all the costs of financing and running the company. A dividend of 7.5% would require a return on assets of 6.2% ... it was not possible to obtain first-class investments to provide the return required by this capitalisation.[15]

However, if the capital structure was changed, with a higher proportion of debentures and preference shares, and these securities carried dividends which reflected present low interest rates, the following figures could be obtained for the same £10 000 of property:[16]

£5000 Debentures at 3.75%	£187
2500 5% Preference shares	125
2500 Ordinary, say 7.5%	188
	£500

With the above capital structure a dividend of 7.5% could be obtained from a net yield of 5% and a higher yield would magnify the dividend; if the yield was 6% the dividend would be 11.5%.[17] A new company, Second Covent Garden Property Co. Ltd, was formed with this capital structure in July 1933. Second Covent Garden concentrated on the purchase of shop properties, often let to leading multiples such as Woolworth's, Montague Burton and British Home Stores, which were regarded as extremely reliable tenants. As Philip Hill stated at the company's first General Meeting, 'In most instances the status of our tenants is such that their covenant for payment of rent, apart from the property itself, is worth as much as we paid for the property.'[18]

In addition to the shop investments two large offices were acquired by the new company, at a total cost of £1 600 000, via sale and leaseback arrangements. Their rents amounted to £75 000, giving an initial gross yield of only 4.7%. Despite their low yield, cheap money made them attractive investments, much of the purchase costs being raised by the issue of 3.5% debenture stock.

The links between cheap money, the upturn in the property market and the property company boom were discussed in *The Economist* in November 1933, it being stated that since the beginning of the year securities with a nominal value of approximately £9.5 million had been publicly or privately issued by property companies. The article noted:

> At first sight, it may appear surprising that this 'boomlet' in property issues should have coincided with a downward trend in rents, particularly noticeable in the London and suburban areas. Actually both phenomena have at least one cause in common – the cheapness of money. Rents are tending to fall because new buildings can be erected on cheaper money at decreasing costs ... Simultaneously, the rise in general security values, which has lowered all investment yields, has turned the attention of investors to the debentures and shares of property companies, which are regarded as affording safety of capital and steadiness of income.[19]

In addition to boosting investment demand for property, cheap money also added to the supply of such property, as was noted in the above extract from *The Economist*. For example, in the case of shops, cheap money lowered the interest element in development costs, reducing the rent needed to cover these costs when development was funded by sale and leaseback, without lowering yields on leaseback transactions. The stimulus that cheap money gave to retail building can be seen from Figure 4.1, which shows annual percentage changes in retail sales and the volume of retail building during the years 1930–1938. The advent of cheap money led to a sharp increase in building activity, which stabilized at a high level in the late 1930s, while the upward trend in retail sales was not evident until 1934 and was much more gradual. The sharp rise in building in 1933 may represent construction which had been deferred in previous years due to the depression, and was undertaken before the rise in retail sales to take advantage of low interest rates and consequent low development costs.

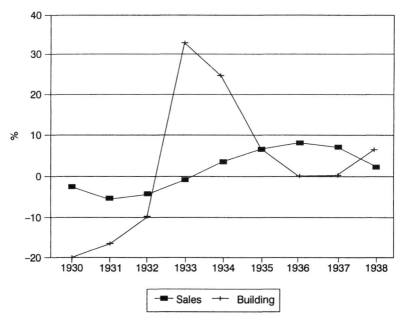

Figure 4.1 Annual percentage changes in the volume of retail sales and retail construction 1930–1938, in real (1930) prices. Sources: retail sales – Bank of England Statistical Summary (1932–1939); volume of retail building activity – H.W. Richardson and D.H. Aldcroft (1968) *Building in the British Economy Between the Wars*, Allen & Unwin, London, p. 62 (figures are taken from a series by C.H. Feinstein). Both series have been corrected for changes in retail prices. The values for retail building activity were estimated as a proportion of residential building activity for the 1930s rather than being independently determined and are therefore subject to a certain margin of error, although discussion of the state of the retail property market in the Estates columns of *The Times* and the *Daily Telegraph* also indicate a boom in retail building activity in the years after 1932 of a magnitude greater than that indicated by changes in the level of retail sales, broadly confirming the trends shown by the retail building series.

The inter-war years saw a considerable boom in shop development. In addition to the 'multiple revolution' on Britain's High Streets, discussed above, this period also saw a transformation of the character of local retail centres. Prior to the First World War isolated shops sprang up all over residential areas, usually involving the conversion of a private dwelling to supplement, or replace, the family's existing source of income. However, during the inter-war years isolated shops were frowned upon by private estate developers. In their place the development of small parades of half a dozen or so shops serving each estate became common. These were generally larger than the pre-war local shops, providing full-time employment to the proprietor, plus possibly one or more assistants, rather than merely supplementing household income.[20] As Jackson noted 'A common rule of thumb was that a small shopping centre required a minimum of 200 houses in support and should contain at least the six "essential trades" – butcher, baker, grocer, greengrocer, dairyman and newsagent/tobacconist/confectioner.'[21]

It has been argued that these sub-centres, one serving each estate, were economically inefficient since they led to an unnecessarily atomistic pattern of shop provision for communities which might have been better served by larger suburban shopping centres. However, for the estate developer they proved attractive, since they offered the opportunity of capitalizing on the potential retail business created by the estate.[22] Local councils sometimes tried to prevent the development of such estate shopping centres to protect existing shopkeepers in the district. However, their limited planning powers made such attempts much less effective than was the case after 1945.

The growth of motor transport encouraged the development of larger suburban shopping centres, along major arterial roads, while the expansion of the tube and rail network to the new London suburbs created opportunities for the establishment of suburban centres in their immediate vicinity. These served a much greater potential market than the local centres, often attracting the major multiples and providing very lucrative opportunities for developers. While the redevelopment of High Street shops was typically undertaken on a piecemeal basis, often by multiple retailers replacing the cramped premises of local traders by larger, more spacious stores, suburban and local shopping centres offered much greater opportunities for the speculative developer.

The importance of the multiple retailer to the speculative shop developer is illustrated in the autobiography of the businessman George Cross.[23] Cross had monitored Edgware as a potential future area of suburban expansion since 1910. In 1919 he purchased the 54 acre Edgware Manor estate (later adding a further 16 acres) for £175 per acre. This had almost a mile of existing road frontage and was close to the site of the possible new tube station, for which parliamentary powers had already been obtained (though the final decision to develop it had not yet been taken).[24]

A couple of years after purchasing the land Cross received a notice to treat from the London Electric Railway Co. for some acres of land, for lines, sidings and, most importantly, a station. He noted that 'The day I received that notice I knew I had made a huge fortune.[25] The tube separated a small portion of Cross's land from the rest. This, and another strip of land at the other side of the proposed station, were earmarked by Cross for shops. Shop development was potentially much more profitable than housing; Cross recalled that 'Average sites for private houses in the new London suburbs sell at from five to ten pounds a foot, whereas good shop frontage will command ten times as much'.[26]

Although Cross initially found it difficult to sell his residential plots at Edgware the shop sites he had designated on either side of the proposed tube station proved very successful. He was offered £21 per sq. ft. frontage for a one acre shopping pitch, which would have recouped almost all of the purchase price for the disposal of little more than one of the 70 acres he owned. However, on the advice of his brother, a local estate agent, he decided instead to develop the shops himself and let them to tenants. This proved a wise decision; Cross estimated that by the late 1930s the parts of this land closest to the station might fetch £400 per sq. ft. of frontage, or more (as a result of arbitration £550 a foot had been paid for shop frontage adjoining the station).[27]

Cross described the multiple traders he tried to attract to his shopping development as being 'very much like sheep in the way they follow one another, but, unlike sheep, they do not go astray'.[28] In order to herd the sheep in his direction Cross approached Sainsbury's, which was already a leading grocery multiple in the South East. J.B. Sainsbury informed him that:

> He was interested in Edgware, but he rather fancied the main road, so he said. My shop plans were no good to him; they always built their own premises on land they purchased, and would never rent. If I cared to offer him a 24 ft site at a nominal price, he might consider it. I knew that if only I could say Sainsbury's had bought a plot other traders would follow suit, and I asked him what he considered a nominal price.
>
> 'Oh, a couple of hundred pounds, or say ten pounds a foot,' he replied.
>
> 'But,' I objected, 'that would be giving it away.'
>
> 'Well, wise people have done that before now,' was his rejoinder.
>
> It did not take me long to make up my mind.
>
> 'You can have it at your own figure if you will let me put up a board stating it has been sold to you,' I said.[29]

Cross's next target was W.H. Smith. This company had a well-organized property acquisition policy with a separate Premises Department headed by a Mr Bayliss, who Cross described as a clever qualified architect and highly-skilled businessman.[30] Cross recalled that 'He wanted the largest shop in the

PROPERTY COMPANIES AND THE PROPERTY DEVELOPMENT MARKET

central position of the parade, built to an excessive depth, with special fittings for the trade, at a price which would leave me with precious little margin. Renting on any terms he would not hear of.'[30] Even with Sainsbury's already committed to his development, Cross considered W.H. Smith too good to pass up and was 'pleased to accept a price which in the end showed me a little return'.[30]

Within a short time of signing up Sainsbury's and W.H. Smith, Cross was able to attract another three traders (at least two of which were multiples) 'on plans only', without a single brick having been laid.[31] By the end of 1924 he had developed eight shops (not including Sainsbury's, who undertook their own development), seven of which were let before they were completed. Their success created a demand for others, and eventually Cross developed a total of 21 shops on his estate. Towards the end of the development Cross refused to sell the shops, rather than let them, whatever price he was offered.

Even on the basis of a land cost of £20 per sq. ft. of frontage (representing the cost of the land for shop development, rather than the much lower price he paid for it), Cross made a profit of £6119 on total costs for the 14 shops he sold. However, to this 8.6% return on his investment must be added the value of seven shops which he let rather than sold.[32] While Cross did not assign any value to these in the account he published of his development activities they would certainly take the total return well into double figures.

This success led Cross to buy shops and shop frontage in many of London's suburban districts. He 'did not seek for anything "ripe" but tried to look five to ten years ahead.'[33] Cross stated that while some of these shop speculations were not very remunerative few, if any, produced a loss.[34] His developments sometimes involved buying land in anticipation of transport or building projects which would raise the volume of retail trade in the area. At Wrythe Lane, Carshalton, Cross, together with his brother, bought shop frontage adjoining a country lane, in anticipation of the London County Council's (LCC) development of a 10 000 house estate. The LCC built their own rival shopping development, but most big multiples 'did not like this form of municipal trading',[33] and the LCC scheme was not very successful. Cross eventually developed 60 shops on the site, mainly let to multiple traders.

Multiple retailers formed one of the most important agents of shop development during this period. In addition to developing their own stores some undertook more comprehensive development. For example, Sainsbury's built a number of suburban shopping parades under the auspices of its development company, Cheyne Investments Ltd, at Kenton, Wembley, Ruislip, Haywards Heath and Amersham.[35] Sainsbury's would take the central shop, often attracting other multiples to the surrounding stores as a result of their presence.

A similar strategy was adopted by Burton's, though rather than developing shopping parades the company concentrated on large, consolidated, High Street sites. This typically involved the acquisition of a prominent street-corner site, the demolition of existing buildings and comprehensive redevelopment, with additional shops being built on either side of the Burton's store and the first floor of the premises developed as a billiard hall, offices or shops. The scale of its developments was such that Burton's became one of the largest shop developers in Britain; at the beginning of 1939 the company's Architects Department employed a total of 62 people, a number which the firm thought to be greater than that of any other organization, except the Office of Works.[36]

As early as the 1920s it had been decided to let property developed but not occupied by the company on short leases, of five or seven years. An important reason behind the development of additional property, and its letting on short leases, was Montague Burton's belief that the property of the largest retailers, in 'key' High Street locations, occupied a monopoly position which would lead to substantial capital appreciation over time, with a possible doubling of shop values, in real terms, every 20 years. The mechanism which he perceived to be behind such increases had nothing to do with general price inflation but was the result of the growing scarcity of prime High Street locations. He reasoned that as the best shopping pitches were limited in supply, as the economy grew they would become scarce relative to other goods and services, the supply of which could be more easily increased. Capital values of prime shop sites would therefore appreciate.

This belief formed the basis of his property development strategy, the hallmarks of which were developing stores to a high standard of construction, developing property adjacent to branch premises as an investment, funding this development via sale and leaseback transactions with a rent that was fixed for a very long period, preferably 999 years, and letting property not occupied by the firm on very short leases, of five to seven years, in order to benefit from rising rents.[37] Burton was in an ideal position as a developer, as the prestige of his retailing chain raised property values in areas where he located; by the very act of selecting a property for development he therefore increased the profitability, and reduced the risk, of the project. Such a process of changing the 'rules' of the development 'game' is an integral part of the property development process, as noted recently by Carol E. Heim:

> Property developers often do not take risks as given, instead seeking to transform the 'rules of the game' to their advantage. Ex ante risk may be high, but ex post risk may in fact be *lower*, in cases of successful property development. Far from simply maximising subject to constraints, property developers are active economic agents seeking to change those constraints.[38]

The inter-war years saw considerable improvements in office design, made possible by the introduction of steel-framed building technology. Developed in the United States and introduced to Britain just prior to the First World War, steel-frame construction provided the model for office building during the inter-war years. By freeing external walls of their load-bearing function it led to a new architectural style, marked by regular fenestration, enhanced daylight and flexible internal planning.[39]

However, the demand for new office space during the inter-war years proved insufficient to generate a substantial volume of speculative office development, the few really substantial speculative office blocks which were built during the era, such as London's first 'skyscraper', Bush House, often proving slow to let.[40] Those property companies which had been formed in the late nineteenth century to develop and manage City office property found conditions in the inter-war years less buoyant than during the five decades before 1914. City of London Real Property Co. undertook considerable expansion in the City area, but found itself severely hit by a depression in City office rents during the late 1920s and early 1930s; the company's rental income fell from a peak of £632 000 in 1925 to £366 000 in 1934.[41] City Offices Ltd also experienced the consequences of these depressed conditions. However, despite this, the company's rental income rose, from a virtually static portfolio, from £86 688 in 1925 to £97 525 in 1934, partly due to its policy of granting relatively long leases, and agreeing to rent reductions for tenants in difficulties.

The 1930s saw a very substantial boom in flat development in Central London. Cheap money and relatively low building costs in the wake of the depression stimulated supply, while the growing army of middle-class white-collar office workers provided the demand, cheap money facilitating their purchases by reducing mortgage costs.[42] This boom, together with the much greater boom in conventional house-building, are strictly beyond the scope of this study due to their residential nature. However, flat development during this period must be considered alongside that of commercial property, since the sector was one in which property investment companies, and the financial institutions, took an active role.

Residential property investment had traditionally been the preserve of the small investor, the Church and philanthropic trusts. However, unlike conventional houses, flats were seen as an attractive long-term investment medium by many property companies and financial institutions. As Hamnett and Randolph noted:

> While the involvement of large corporate developers partially reflected the size of many of the developments and the necessity for ready access to secure and plentiful finance, it also reflected the profitable and trouble-free nature of the block of flats as a safe investment commodity. The type of tenant for whom the flat market was orientated – the salaried, non-manual households – the level of rents obtainable, and the compact and orderly nature of the developments which allowed for efficiency in management, all contributed to the appeal of flat development for corporate investors.[43]

Some of the leading flat development companies of the era relied heavily on institutional finance. For example, the Bell Property Trust, which erected at least 2700 'up-market' flats in and around London during 1933–1939,[43] financed their activities via mortgages with the Royal Liver Friendly Society and Eagle Star, together with bank loans.[44]

One insurance company, the Prudential, embarked on its own flat development programme during the inter-war years. The Prudential's first flat development, involving the construction of 28 flats in Kensington, London, was completed in 1925. Between then and March 1930 developments included 52 flats in Brysanton Court, Marylebone, a further 48 flats in Kensington and 81 flats in two developments in Portman Square. In 1930 the Prudential began constructing three large blocks of flats at Hurlingham, with a total of 208 suites. These were all luxury flats, with inclusive rents ranging from £750 to £1300 per annum. Their amenities included central heating, refrigerators and jewel safes in bedrooms.[45]

These developments appear to have been very profitable. In January 1931 a board meeting approved the purchase of a freehold site of 3.3 acres in Belsize Park for £40 000. It was proposed to erect a block of 145 flats, together with 18 shops and 15 garages, at an estimated cost of £203 000. The expected net income from this development, £16 500, would yield 8.125% on development costs.[46] A further flat development project, considered by the Board in January 1932, offered an expected yield on capital expenditure of 7.25%.[47]

The results of the Prudential's flat development policy appear to have been a source of considerable pride to the company, their performance featuring prominently in the speeches of the company's chairman, W. Edgar Horne, at Annual General Meetings. At the AGM of March 1937 Horne noted that 'Two years ago I referred to the various large blocks of flats which we had erected, and told you that, with a total rent roll of approximately £250,000 per annum, only 7% of our flats were unlet. Today the rent roll is over £300,000 and the unlets amount to less than 4%.'[48] By the end of 1938 the proportion of unlet flat property held by the Prudential had fallen to 2.2%.[49]

Some insurance companies did venture into investment in lower-income housing during the inter-war years, providing the investment was offered on sufficiently attractive terms to make up for the inherent disadvantages of low-income residential property (rent control, high management and maintenance costs, and the risk of tenant default) or the deal had some special feature which reduced those disadvantages. For example, in 1935 Clerical Medical agreed to purchase a number of small houses and flats which had just been constructed in Slough, on a basis calculated to yield over 7%.[50] The yield, which was somewhat higher than that on commercial properties, reflected the greater costs and risks associated with this class of investment. Three months later the scheme was re-submitted in a modified form. In order to avoid the payment of stamp duty it was suggested that the Society should purchase the land and pay the cost of the houses by instalments during building. The Society purchased several sections of the estate on this basis, investing a total of £223 000 by July 1936.

Legal & General acquired some residential property via two deals conducted with building societies. In 1934 a number of small houses on the Lords Estate,

Oldham, were purchased from the Huddersfield Building Society, as part of a deal involving the cancellation of a mortgage guarantee policy. Sixty-five houses were originally purchased, subsequent acquisitions as part of the same deal increasing the number to 133 by the end of 1938. These, and 21 houses in and around London acquired in a similar deal in 1935 with the National Building Society, were managed, free of charge, by the respective building societies. The income from both these investments during the following years was considered to be disappointing by Legal & General. The net yield on purchase cost for the Lords Estate properties averaged about 4% during the 1930s, being depressed by heavy repair costs and low tenant income as a result of depression in the Oldham area.[51] Despite more prosperous conditions in London, heavy maintenance costs also depressed income from the properties acquired from the National Building Society.

In 1937 Legal & General purchased four blocks of flats in Putney and a block of 42 flats at 35/37 Grosvenor Square, at a total cost of £253 000. These catered for higher-income earners than their earlier purchases; the Grosvenor Square property included several maids' rooms and nine shops. Legal & General's residential property acquisitions were among the least successful of its property investments during the 1930s, the building society houses falling foul of the classic problems associated with investment in working-class housing. Flat property, of the type purchased by Legal & General in 1937 and built by the Prudential from 1925, was generally a much more profitable outlet for investment funds during this period, aiming at the decontrolled sector of the housing market and catering for affluent tenants.

4.3 PROPERTY DEVELOPMENT FINANCE

The property development boom of the 1930s was largely funded by the financial institutions. Insurance companies and banks had a strong bias towards lending backed by collateral security at this time, and property (or rather those classes of property which had a high estimated resale value) provided excellent collateral for loans. Businessmen in property-rich sectors, such as retailing and property development, therefore found it much easier to raise the funds necessary to expand their operations than was the case in most other industries.

The ease with which funds could be raised on real estate is illustrated by the example of the developer Percy Bilton, founder of the major property development company which bears his name. Bilton began his property career by buying a small piece of land in Mitcham, Surrey, on which he intended to develop 22 semi-detached houses. The cost of this land was £1200, on which he paid a deposit of £120. The property was purchased in his father-in-law's name, Bilton agreeing on paper to purchase the land from his father-in-law for £6000 (though the cash did not change hands).[52] With the price he paid for the land recorded at £6000 Bilton then placed the deeds of the property in the bank and asked for an overdraft of £4000 using the property as security. As his references, based on his previous oil business, were good the bank granted him this overdraft (which only appeared to amount to two-thirds of the property's value), providing Bilton with the finance for his development scheme.

While banks were an important source of short-term finance for the property development industry the major source of long-term finance was the insurance company sector. Insurance companies provided very substantial funds to property developers via mortgage lending. Mortgages averaged 11.91% of annual net insurance company investment during the years 1922–1937, while direct property investment averaged only 4.33%. Mortgage lending constituted a favoured means of placing insurance company funds in an asset that yielded substantially more than gilts and was relatively safe, the security for loans typically having an estimated value 50% higher than the sum loaned.

The various arrangements by which insurance company funds were channelled to property development can be illustrated by reference to one of the most active property-investing insurance companies, Clerical Medical. Clerical's mortgages constituted a larger proportion of total assets than was typical for life offices, amounting to 19% of assets in 1920, 30% in 1931 and 28% in 1938.[53] In 1920 the Board decided to move substantial funds out of government securities, which were experiencing large fluctuations in value, and reinvest them in mortgages.[54] During 1922 two substantial mortgages, of £550 000 and £80 000, were granted to property companies.

In March 1923 the Society entered into a financing agreement with a syndicate of property developers who had negotiated six 99 year building leases from the Duke of Bedford, one lease for each of six office blocks that they intended to develop. The funding arrangement was such that the degree of risk to the Society was minimized. The first portion of the advance would not be made until a substantial proportion of the first block was let and would be fixed at such a figure that interest on the loan would not exceed 50% of the net rental. The money would be spent on completing the second block and the same procedure would be followed for each subsequent block. The arrangement therefore provided that income would be produced by every building the Society accepted as security, at the time of the loan, and there was no danger of Clerical Medical having to take possession of uncompleted buildings.[55]

The establishment of a funding link between Clerical Medical and a newly-formed property company, Brixton Estate Ltd, took place in 1924. Brixton Estate, which was established to develop a six acre industrial estate in South London, was one of several specialist industrial property development companies which flourished during the inter-war years, providing modern single-storey factories along London's arterial roads for the new, light, consumer goods industries. A number of these developers relied on insurance companies for the bulk of their finance; most of Brixton Estate's development finance during the inter-war years was provided by Clerical Medical, while the National Provident Institution developed a similar relationship with another industrial estate developer, Allnatt Ltd, who were the major developers of London's giant Park Royal industrial estate, together with a number of smaller estates in West London. By the end of the inter-war period Brixton and Allnatt had become the largest single mortgage clients of these two insurance companies.

While insurance companies were sometimes prepared to provide extensive finance to industrial property developers, they were generally unwilling to grant mortgages on isolated factories during this period, or to buy such factories as direct investments. This was due to two main reasons:

1. The lack of marketability of industrial property. Industrial buildings were seen as being so specific in design as to be suitable only for the firms for which they were built, or other companies in the same industry. Should such a building become vacant it would, therefore, prove very difficult to re-let.
2. The lack of any 'scarcity premium' associated with industrial property, unlike the shop and, to a lesser extent, the office market, where site values could be expected to appreciate over time, counteracting any depreciation for buildings. Isolated factories were far less location-specific than other forms of investment property, reducing the likelihood of capital appreciation for sites.

The first condition listed above was much less true of factories developed during the inter-war years, particularly for the light, consumer goods industries, than had been the case prior to the First World War. The multi-storey factory of the Victorian era was giving way to 'the single storey block with saw-tooth layers of north-light roofing concealed at the front by offices',[56] though this change was slow to be appreciated by institutional investors. Even entrepreneurs such as Montague Burton, who had well-established links with the property investing institutions, had little success in arranging leaseback deals on factory premises. A letter from Healey & Baker to Montague Burton, dated December 1938, stated that factory property:

> ... is not regarded with favour in the investment market, although we believe that modern factory construction is today on far more 'unit lines' than the buildings of even fifteen or twenty years ago. One modern factory is machinery apart and as a building more nearly true to type than in the past, so that the old objection of factory premises being only suitable for one trade is to a certain extent in a limited way overcome. Nevertheless, there is still the objection to industrial property to be overcome and this does at the moment seem to be a stumbling block.[57]

Why were insurance companies prepared to lend extensively to industrial property development companies on the security of their factory properties while otherwise avoiding lending on this class of security? The answer lies in the emergence of the industrial estate as a new form of factory accommodation. Industrial estates had begun to appear before the First World War, but became much more widespread during the inter-war years. They were usually situated within easy reach of London, on sites that were chosen for their excellent road and rail links. Tenants were almost exclusively light-industrial concerns, in industries such as car manufacture, food processing and electrical goods, occupying shed-type factories that could be easily re-let to other tenants and often undertaking a significant volume of trading with other companies on the same estate. These factors gave the industrial estate a measure of 'scarcity premium' which was lacking for other types of industrial property, site values on such estates appreciating due to their good transport links and the other advantages of an industrial estate location.[58]

This capital appreciation overcame the problems of lack of resale value which weakened the suitability of other industrial property as loan security, making it possible for industrial estate developers such as Allnatt and Brixton Estate to borrow extensively from financial institutions on the strength of their property assets. These loans allowed them to provide factory accommodation on a rental basis, rather than solely via outright sales, in turn reducing the immediate capital costs of factory occupation for their tenants.

During the inter-war period the proportion of Clerical Medical's mortgages that were conducted with property companies rose substantially. In 1924 the Society held mortgages granted to property companies with a book value of £305 264, 23.3% of its total mortgage portfolio. By 1930 its holdings of property company mortgages had risen to £1 131 383, 35.5% of total mortgage holdings, and by 1940 property company mortgages had risen further to £2 218 096, 43.9% of all mortgages.[59] In addition to providing an outlet for a substantial proportion of Clerical's funds at an attractive rate of interest, the Society's inter-war property company funding arrangements paved the way for much closer links with some of these companies, involving substantial equity participation, following the Second World War.

NPI also provided extensive mortgage finance to the property company sector during the inter-war years. Table 4.3 shows NPI's mortgages with: (1) property companies; and (2) all borrowers, during this period. The figures are somewhat unreliable prior to 1932, as they represent mortgages granted in earlier years which had not been fully repaid by that year, and refer to the original amount borrowed rather than the sum outstanding. They do, however, provide a rough guide to the Society's mortgage lending in the 1920s. Figures from 1932 refer to the amount outstanding on loans at the end of each year. The table shows a substantial increase in the proportion of mortgages granted to property companies, from 14.69% in 1920 to 48.56% in 1932 and 58.85% in 1938. A decline in total funds invested in mortgages after 1932 is also indicated. Declining mortgage business was discussed in a major review of investment policy, undertaken by NPI in the last months of 1938. It was noted that 'good class' mortgages were becoming more and more difficult to obtain. Properties which were not 'good class' included mortgages on isolated factory premises and flats, and it was recommended that these be avoided.[60] Clerical Medical was also careful regarding the kind of properties which could be accepted as security for mortgages. A 1938 Board minute noted that having received enquiries for mortgages on cinema properties the Directors had decided that as a general principle these should not be entertained; however, loans of amounts considered adequately secured upon the site values of cinema properties would be considered on their individual merits.[61]

Table 4.3 The distribution of NPI's mortgages between property companies and other borrowers: 1920–1938

	Total mortgages (£)	(1) as percentage of (2) [a]
1920 (1)	95 000	14.69
(2)	646 795	
1925 (1)	627 734	39.86
(2)	1 574 729	
1929 (1)	824 734	37.11
(2)	2 222 629	
1932 (1)	1 280 777	48.56
(2)	2 637 581	
1935 (1)	1 055 058	49.23
(2)	2 143 091	
1938 (1)	1 169 782	58.85
(2)	1 987 237	

Source: NPI (1933–1939) Asset Reports.

[a] (1) Refers to mortgages to property companies and (2) refers to all mortgages.

Mortgage funding enabled many companies in sectors such as retailing and property, for which mortgageable property formed a high proportion of assets, to raise substantial finance at a net rate of interest that was usually only about 1% in excess of the yield on consols.[62] At a time when the security of capital invested assumed paramount importance in insurance company investment decision-making, mortgages provided a means by which capital could be invested in industry, secured on the value of marketable assets rather than the expected profitability of the mortgagee. While mortgage finance constituted a substantial channelling of insurance company funds to productive investment it was, however, highly selective in the industries which received such finance due to the restrictions concerning what was regarded as acceptable security. This was confined to a limited range of property – offices, shops and some industrial property located on industrial estates. Like direct property investment, mortgage funding gave significant advantages to a narrow range of enterprises with regard to raising finance for expansion, which may have been a significant cause of the very high rate of growth of these sectors relative to the economy as a whole during this period.

Insurance company property development finance was not limited to mortgage funding. Several insurance companies began to use the leaseback mechanism to provide development finance from the early 1930s. For example, during 1932 Clerical Medical purchased two properties from Town Investments Ltd on a leaseback basis, the vendors being granted full repairing leases for 99 years. A year later Clerical entered into a leaseback arrangement with one of the most important shop developers of the 1930s, Edward Lotery. In March 1933 the Board approved a proposal by Lotery that they buy the freeholds of seven shops in, or near, London from him and lease them back for 42 years, at a rent equal to 6% of the purchase price.[63] Two months later it was agreed to purchase another shop on similar terms[64] and another substantial leaseback transaction was conducted between Lotery and Clerical Medical in 1938. Such deals were not very far removed from mortgage funding, a fixed income stream being purchased for a capital sum. The main differences between the two funding mechanisms were:

1. Greater property rights were given to the institution under sale and leaseback than under mortgage funding, the lessor's permission being necessary if the lessee wished to make structural alterations to the property. The retailer also lost the ability to cease interest payments by repaying the capital sum, as was possible under mortgage finance.
2. At the end of the lease period (typically 99 or 999 years) the institution had the power to re-let the property at a rent which reflected prevailing market conditions.
3. The sum raised was usually equal to the full construction costs of the property, rather than two thirds as was the case for mortgages.

The third difference between the two funding mechanisms had far-reaching effects for the growth possibilities of firms whose major fixed asset constituted mortgageable property, as was outlined in section 3.3.

In addition to mortgage and leaseback transactions, Clerical Medical sometimes used other methods of property development finance. For example, in January 1934 it participated, together with a number of other insurance companies, in a scheme to finance the Arlington Property Co., which was currently being established by the merchant bank Messrs Schroder, to erect a block of flats costing about £225 000. The insurance offices were to acquire the freeholds of 18 and 20 Arlington Street for £110 000 and take building leases on the sites of 17 and 19 at a total ground rent of £5500 per annum. A building lease was then to be issued to the Arlington Property Co., for the whole of the land, at a ground rent of £10 300 per annum.

The Arlington Property Co. was to have a share capital of about £130 000 and issue 5% debentures for about £160 000, which the insurance offices were offered at par. Schroder's were to guarantee the interest and principal on the debentures for two years, or until the completion of the building, whichever was shorter. The Society approved participation to the extent of up to one-fifth of the whole. Shortly after it emerged that construction costs would exceed estimated costs by £20 000 and an increase in the debenture issue, and debenture take up, was agreed to cover this. In May 1936 the Society was offered 15% of the capital of the scheme, which had been subscribed by Messrs Schroder, and some or all of an additional loan of £40 000, incurred due to building costs having further increased, both of which were declined.[65]

Clerical Medical participated in the foundation of another property company, Western Ground Rents, in 1938. Western Ground Rents was established to acquire the Mountjoy Estate from Lord Bute. Clerical Medical and another insurance company, Equity & Law, each took 45% of the ordinary shares and half the preference capital of the company. The stake of both insurance companies in Western Ground Rents was enlarged in 1946, when it became a public company.[66] This arrangement represents one of the few instances prior to the Second World War of significant insurance company equity involvement in a property company. Another example is that of Sackville Estates, a property company formed in 1880 and registered in 1893. The firm had a family connection with the Provincial Insurance Co., which acquired a substantial equity stake in the company in March 1930.[67]

There was also at least one instance of an insurance company taking an equity stake in a property development via the establishment of a joint development company prior to the Second World War. In May 1926 the Prudential's board agreed to subscribe for the whole of the debentures, half the preference shares and 55% of the ordinary shares of Mayfair Hotel Ltd, a company established to undertake development of the Mayfair Hotel in London. The rest of the ordinary and preference shares were to be held by Gordon Hotels Ltd, who were also to manage the hotel.[68] Joint development companies of this type were to become very popular during the 1950s, as is discussed in Chapter 6.

The Prudential provided development finance to a number of companies during the 1930s by buying the sites on which they were to undertake development, and then granting them building leases. In addition to such arrangements with Marks & Spencer, discussed above, deals of this type were conducted with the Equity Investment Co. and Odeon Theatres Ltd. The rapidly expanding Odeon cinema chain made extensive use of sales of ground rents to insurance companies as a means of raising finance; in addition to the Prudential's purchases, at least 19 such deals were conducted with Eagle Star (whose Chairman, Sir Edward Mountain, was a director of Odeon Theatres) during 1935–1938.[69]

Sometimes the Prudential would go further, providing development finance in addition to the land. For example, in April 1935 a deal was agreed involving Woolworth's acquiring the leaseholds of two Clapham properties in which the Prudential had a freehold interest, purchasing a further property and selling the combined interests to the Prudential.[70] The Prudential would then finance the development of the consolidated site, which would be let to Woolworth's on a 99 year lease, providing the Prudential with an estimated yield on the proposed expenditure of 4.64%.[71]

In some deals with major retailers the Prudential adopted an even greater development role; for example, in 1936 its Board approved two deals, involving both the purchase of sites on behalf of Woolworth's and the development of stores to their specifications. These developments, let to Woolworth's on 60 year leases, provided estimated yields of 4.9 and 4.78%, respectively,[72] the extremely strong covenent of Woolworth's allowing the company to arrange for its stores to be developed at a very low annual cost in relation to their capital value. A few months later the Prudential approved a similar deal – involving both site purchase and development – with British Home Stores, yielding 5.25%.[73]

The Prudential even very occasionally purchased sites for speculative commercial property development (in addition to its flat development programme). For example, in June 1936 its board approved the purchase of 75 High Street, Godalming. This was to be redeveloped as shops with living accommodation or showrooms above. The estimated net yield on the completed development was over 6%.[74] Thus, even prior to the Second World War, at least one major insurance company was becoming directly involved in the property development market, in addition to providing development finance.

4.4 ENTREPRENEURSHIP IN THE INTER-WAR PROPERTY INVESTMENT MARKET

Only a small number of insurance companies became active in the property market during the inter-war years. However, these included some of the largest and most influential offices, notably the Prudential, Norwich Union, Commercial Union, Eagle Star and Clerical Medical. The investment policies of these companies were usually dominated by particular individuals, such as Sir Andrew Rowell of Clerical Medical, or Sir Edward Mountain of Eagle Star. These people generally had a greater understanding of property as an investment medium than was common in the actuarial world and were able to successfully put forward an active property investment policy, sometimes in the face of considerable opposition from some Board members who favoured more traditional investments. Insurance company investment policy was still influenced by the actuarial theory of the late nineteenth century, developed by A.H. Bailey and others,[75] which emphasized security and stability of investment capital values as the overwhelming criterion for investment, and saw assets such as equities and property as carrying too high a risk to be appropriate for insurance company portfolios. By the 1930s such thinking was already being displaced, to some extent, by ideas which gave property investment greater acceptability, though such a willingness to broaden the range of assets held, in order to increase yields, was only slowly gaining acceptance.

The entry of the insurance companies into the property investment market coincided with the expansion of insurance company investment departments and the growth of a group of professional investment managers, who were generally better established and more influential in the companies which entered the property market than in those that did not. As such, the move to property investment in the 1930s was a product not only of cheap money but of greater professionalism and entrepreneurship by some insurance companies. This consisted of a willingness to venture into new areas of investment, to use techniques such as the sale and leaseback which were not yet widely exploited by institutional investors, and to deal with the new property entrepreneurs whose business interests and social backgrounds were often very different from those of insurance company directors. More entrepreneurial attitudes on the part of insurance company investment managers had a similar effect on the commercial property market to that of building societies, which liberalized the conditions under which they would grant mortgages, in encouraging the residential building boom of the 1930s, as discussed in a recent article by Jane Humphries.[76]

There were, however, important limits to the adoption of entrepreneurial behaviour by the property investing insurance companies. Insurance offices estimated the value of their assets, usually on a quinquennial or triennial basis, and sometimes had their properties independently valued when doing so. Such valuations almost always showed capital appreciation for their property portfolios. However, despite this evidence of rising property values insurance companies made little effort to maximize this potential for capital appreciation. Leases were commonly granted for long terms, such as 99 or even 999 years, with no prospect of rent review, constituting a virtually fixed-interest security. An insistence on shorter leases, or periodic reviews of rent within leases, would have secured some of the future capital appreciation of their properties for the insurance companies, though such policies were not systematically pursued by most insurance offices until the mid-1950s.

Leading retailers and commercial estate agents expected High Street property to appreciate in value, due to the spread of the multiples and the consequent concentration of retailing activity in 'prime' shopping pitches, as discussed in section 3.3. A few insurance companies were also beginning to think in terms of capitalizing on these trends. The National Provident Institution, which did not participate in sale and leaseback deals involving long leases during this period, considered, in their 1938 review of investment policy, that property should only be purchased when leased for a period of less than 30 years, stating that a property on a 99 year lease is for all practical purposes a fixed-interest security offering no possibility of rising rents.

The Prudential had been pursuing a policy of seeking short leases for some years, 21 year lease terms being typical for the properties it purchased during the 1920s and early 1930s.[77] It also usually let properties which had come to the end of their current leases on new leases of about 21 years. From the mid-1930s the Prudential did engage in some leaseback transactions involving 99 year leases, possibly due to the difficulty of investing its massive funds at an acceptable yield during this period of cheap money, though it was still usually able to obtain short leases from retailers other than the most well-known multiples. However, most other insurance companies that were active in the property market were happy to accept properties on very long leases, paying little heed to the loss of potential capital appreciation which this entailed.

The best example of a long-term fund which pursued a capital growth maximizing strategy in its property investment activities is provided by the pension fund of estate agents Hillier, Parker. The Hillier, Parker, May & Rowden Superannuation Fund was founded in 1922, constituting one of the earliest private staff pension funds to be established in the UK. The fund's investment strategy was based around accruing the benefits of appreciating High Street shop property values for the practice's staff, investing solely in property, or property-related securities, at a time when most pension funds were afraid to stray outside the safe but low-yielding areas of government securities and mortgages.

Initially, contributions were invested in the shares of property companies which were associated with Hillier, Parker, though by 1928 the Fund's assets had grown sufficiently to allow it to invest directly in property. The Fund's investment policy was marked by a number of characteristics which set it apart from that pursued by most contemporary institutional property investors. Properties were usually purchased near the end of their lease term, with rents that were substantially below prevailing values. Following the termination of the lease the rent could be renegotiated at a substantially higher level. Tenants had security of tenure under the 1927 Landlord and Tenant Act, though the offer of a substantial cash payment to the tenant was often sufficient to persuade him to vacate the property. It could then be let to a multiple retailer, giving the investment a higher capital value in relation to its rental income, as a result of the lower yields that could be obtained for shops let to the multiples. The property might then be sold or kept as an investment.

Shops were always re-let on short leases, usually for 21 or 28 years with breaks every seven years, in order that the fund could benefit from any future capital appreciation. For example, in May 1936 it was reported at the meeting of the fund's Committee of Management that the tenant of a shop in Wealdstone had agreed to vacate the premises in return for the payment of £500. The premises were then re-let to Charles Phillips & Co. on a 21 year lease with seven yearly breaks.[78] Properties were also purchased with vacant possession, let to multiple retailers on short leases and either kept as investments or sold.

Another feature which distinguished the Fund from most contemporary institutional investors was the use of its assets as security for bank loans, which would then be used to purchase further properties. Money was originally borrowed by Hillier, Parker, which in turn loaned it to the Fund, though in June 1938 its rules were amended to allow borrowing directly, on the security of its assets, any sum up to one-third of the value of the Fund.[79] The use of this loan finance allowed the Fund to expand much more rapidly than would otherwise have been possible.

The Fund's annual report for the year ended 31st March 1950 contained a valuation of its assets. Its properties had an estimated market value of £107 927, an appreciation of over 26% compared to their cost price (after the amortization of leaseholds).[80] In addition to these unrealized capital gains, sales of properties and property company shares over the 28 years to 31st March 1950 had resulted in a surplus over costs of £17 299.[81]

Towards the end of the 1930s institutional investors do appear to have become wary of leaseback deals involving very long leases of 999 years without rent review, as is illustrated by a series of letters between Montague Burton and the company's estate agents, Healey & Baker. In June 1938 a letter was sent to Healey & Baker with a list of 12 shops and a factory that Burton's wished to sell, the sale being conditional on their being granted 999 year leases on a 4% basis. Particulars were only to be released to the Norwich Union, who had already agreed to purchase a shop at Hull on similar terms. Norwich Union rejected the property package. Healey & Baker's letter stated that they were in touch with several other finance houses who might be interested in the properties. Burton replied that the firm would be happy for the proposal to be offered to a pension fund, which Healey & Baker had mentioned previously, but it must not be offered to any insurance company, for the following reasons:

> As in the case of insurance companies, there is a kind of freemasonry amongst them, and if an application is made to one, a good many others seem to get to know – due probably to the following reasons:
>
> (a) There is an interlocking of Directorates.
>
> (b) There is an Insurance Institute where probably a good many of the junior and senior members of the staff will meet.[82]

This letter also indicates the importance of leaseback finance to the company, stating 'Our future development will depend upon whether we are able to dispose of properties as completed'.[82] Such a reliance on leaseback finance to bring about target rates of growth, due partly to the dynamics of such finance (outlined in section 3.3), placed the insurance companies in a very strong bargaining position *vis-à-vis* the retailers.

Healey & Baker's reply concurred with Burton's opinions regarding the 'freemasonry' of the property investing insurance companies:

the views you express with regard to the freemasonry between the insurance companies are we think sound and are certainly always before us in connection with business to be handled for you and other clients.

This is a point to which we attach considerable importance.

Obviously, whilst the amount available by insurance companies for investment is large, it is our constant aim to keep in touch with private people such as the Royal Pension Fund for Nurses and the Collegiate Authorities who proved purchasers at Tottenham and Hayes respectively.[83]

A further letter from Healey & Baker stated that the required 999 year lease period was an important factor in preventing some investors, who would be happy with a 99 year lease, from taking the property. While such a change would make very little difference, from an actuarial point of view, it was stated that:

in the purchase of property by a college, institution or other body which has an indeterminable period of existence and presumably goes on for all time, then 99 years is a term which will in 30 or 40 years' time possibly give them some possibility of increase or reversion. Furthermore, whilst it may seem very remote to mention the point, the question of inflation does arise, although not to such a marked extent as the reversionary aspect.[84]

Burton's stated the following by way of reply:

We ourselves look upon 999 years as a freehold, from the point of view of security of tenure. Again, a property on a 99 year lease is not as good a marketable proposition from an investor's point of view; further it is not as good a proposition from an Estates Management point of view, for, assuming that in 20 years' time there is an appreciation in the value of real estate, say to the extent of one third, then with a 999 year lease, we should derive the full benefit from such appreciation.[85]

Burton evidently believed that property was an asset capable of substantial long-term capital appreciation, and that any arrangements entered into by his company should retain as much of this capital appreciation element as possible. He wished only to dispose of very limited property rights to his premises, in such a way that virtually the whole of any future appreciation would accrue to his company. However, Burton was forced to modify his policy somewhat in order to secure the desired finance. He later agreed to offer 99 year leases for the leasehold properties and 999 year leases on the freeholds, stating that he would sell the whole package on 99 year leases if a deal could be done with one purchaser.

If the collective power of the insurance companies was as great as Burton believed, they had it in their power to build an equity element into investment property by insisting on clauses in leases allowing for the periodic adjustment of rents to market levels. That they did not do so suggests a degree of institutional inertia and the largely passive acceptance of property investment packages put together by retailers, with the help of the commercial estate agents; both groups, rather than the insurance companies, appearing to be the main repositories of entrepreneurial drive in the inter-war property investment market. As a result, the multiples were able to amass property empires whose values soared in the post-war years, while most insurance companies were left with only a series of virtually fixed-interest securities and enjoyed little of this wealth that they had helped to create.

REFERENCES AND NOTES

1. Net = purchases − sales.
2. The figures underestimate net investment in property, by perhaps 1%, due to the writing-down of property book values. This was undertaken to account for leasehold amortization for leasehold properties and as an allowance for depreciation for freeholds. Properties were also occasionally written-down in value if their market value was believed to be below book value, while the writing-up of properties with estimated market values in excess of book values rarely occurred, such properties being used as a 'hidden reserve' in insurance company accounts.
3. Book cost was calculated on a historical cost basis, being the purchase cost plus subsequent capital expenditure.
4. Clerical Medical (1926) *Investigation Into The Society's Assets*.
5. Source: Clerical Medical (1942) Note regarding the Society's investment policy, 7 Oct.
6. Hillier, Parker, Report for the year 1932 (1932) *The Times*, 31 Dec., p. 4.
7. *Estates Gazette* (1932) 24 Sept.
8. R.B. Sunnucks (1935) *Investment in Property*, 2nd edn, Banbury Publishing Co., London (first edition published in 1933); R.M. Lester (1937) *Property Investment*, Pitman, London.
9. R.M. Lester (1937) *Property Investment*, Pitman, London, p. vii.
10. Legal & General (1937) Actuarial report on the quinquennial investigation into the Society's assets for the five years to 31st December 1936.
11. Prudential (1935) Annual Report, 14 March.
12. Prudential (1936) Annual Report, 12 March.
13. E. Nevin (1955) *The Mechanism of Cheap Money*, University of Wales Press, Cardiff, p. 279.
14. E. Nevin (1955) *The Mechanism of Cheap Money*, University of Wales Press, Cardiff, p. 280.

15. E. Nevin (1955) *The Mechanism of Cheap Money*, University of Wales Press, Cardiff, pp. 150–1. *Note*: The above example does not include the effects of tax, as tax was then paid at a standard rate on profits and recovered at the same rate from dividends paid; therefore with a full distribution of profits Income Tax can be ignored for the purposes of this example.
16. E. Nevin (1955) *The Mechanism of Cheap Money*, University of Wales Press, Cardiff, p. 151.
17. E. Nevin (1955) *The Mechanism of Cheap Money*, University of Wales Press, Cardiff, p. 152.
18. E. Nevin (1955) *The Mechanism of Cheap Money*, University of Wales Press, Cardiff, p. 153.
19. Real property shares (1933) *The Economist*, **cxvii**, 25 Nov., 1022.
20. W. Burns (1959) *British Shopping Centres*, Leonard Hill, London, pp. 6–7.
21. A.A. Jackson (1991) *Semi-detached London: Suburban Development, Life and Transport 1900–39*, 2nd edn, Allen & Unwin, Didcot, p. 91.
22. W. Burns (1959) *British Shopping Centres*, Leonard Hill, London, p. 25.
23. G. Cross (1939) *Suffolk Punch: A Business Man's Autobiography*, Faber & Faber, London.
24. G. Cross (1939) *Suffolk Punch: A Business Man's Autobiography*, Faber & Faber, London, pp. 360–2.
25. G. Cross (1939) *Suffolk Punch: A Business Man's Autobiography*, Faber & Faber, London, p. 364.
26. G. Cross (1939) *Suffolk Punch: A Business Man's Autobiography*, Faber & Faber, London, p. 368.
27. G. Cross (1939) *Suffolk Punch: A Business Man's Autobiography*, Faber & Faber, London, pp. 369–70.
28. G. Cross (1939) *Suffolk Punch: A Business Man's Autobiography*, Faber & Faber, London, p. 372.
29. G. Cross (1939) *Suffolk Punch: A Business Man's Autobiography*, Faber & Faber, London, p. 374.
30. G. Cross (1939) *Suffolk Punch: A Business Man's Autobiography*, Faber & Faber, London, p. 375.
31. G. Cross (1939) *Suffolk Punch: A Business Man's Autobiography*, Faber & Faber, London, p. 376.
32. G. Cross (1939) *Suffolk Punch: A Business Man's Autobiography*, Faber & Faber, London, pp. 378–9.
33. G. Cross (1939) *Suffolk Punch: A Business Man's Autobiography*, Faber & Faber, London, p. 393.
34. G. Cross (1939) *Suffolk Punch: A Business Man's Autobiography*, Faber & Faber, London, p. 394.
35. B. Williams (1994) *The Best Butter in the World: A History of Sainsbury's*, Ebury, London, p. 78.
36. Letter from M. Burton to R.J. Pearson (21 Aug. 1939) and Pearson's reply (23 Aug. 1939) Montague Burton Papers, Box 130.

37. The archival records of Montague Burton Ltd show that the policies of letting shops on short leases, and constructing property which was surplus to the firm's requirements, were undertaken for investment reasons and not to facilitate the future expansion of the Burton's stores.
38. Carol E. Heim (1990) The Treasurer as developer–capitalist? British new town building in the 1950s. *Journal of Economic History*, **50**, 15.
39. S.J. Murphy (1984) *Continuity and Change: Building in the City of London 1834–1984*, Corporation of London, London, p. 66.
40. O. Marriott (1967) *The Property Boom*, Hamish Hamilton, London, pp. 20–1.
41. City of London Real Property Co. Ltd (c. 1964) *The City of London Real Property Co. Ltd: 1864–1964*. Privately published, London, p. 23.
42. See O. Marriott, *The Property Boom*, Hamish Hamilton, London, pp. 21–3; C. Hamnett and B. Randolph (1988) *Cities, Housing and Profits: Flat Break-up and the Decline of Private Renting*, Hutchinson, London, Ch. 2.
43. C. Hamnett and B. Randolph (1988) *Cities, Housing and Profits: Flat Break-up and the Decline of Private Renting*, Hutchinson, London, p. 23.
44. O. Marriott (1967) *The Property Boom*, Hamish Hamilton, London, p. 22.
45. *The Times* (1930) 25 March, 13.
46. Prudential (1931) Board minutes, 6 Jan.
47. Prudential (1932) Board minutes, 7 Jan.
48. Prudential (1937) Annual Report, 11 March.
49. Prudential (1939) Annual Report, 9 March.
50. Clerical Medical (1935) Board minutes, 5 June.
51. Legal & General (1985) Notes on the first 50 years of the Estate Department, p. 2 (written by an employee of the company).
52. Edward L. Erdman (1982) *People & Property*, Batsford, London, p. 152.
53. Clerical Medical (1942) Note regarding the Society's investment policy, 7 Oct.
54. Clerical Medical (1920) Finance Committee minutes, 15 Dec.
55. Clerical Medical (1923) Finance Committee minutes, 30 May.
56. C.G. Powell (1980) *An Economic History of the British Building Industry 1815–1979*, Methuen, London, p. 89.
57. Letter from Healey & Baker (29 Dec. 1938) Montague Burton papers, Box 129.
58. The appreciation of land values on London's industrial estates was noted in Hillier, Parker's report on business conducted during 1934.
59. Clerical Medical (1924, 1930, 1940) Investigation into the Society's assets.
60. NPI (1938) Report of Investment Committee, 13 Dec.
61. Clerical Medical (1938) Board minutes, 26 Jan.
62. Source: Mortgage investment records of Clerical Medical and NPI.
63. Clerical Medical (1933) Board minutes, 8 March.
64. Clerical Medical (1933) Board minutes, 31 May.
65. Clerical Medical (1936) Board minutes, 27 April.

66. B. Whitehouse (1964) *Partners in Property*, Birn, Shaw, London, p. 62.
67. B. Whitehouse (1964) *Partners in Property*, Birn, Shaw, London, p. 60.
68. Prudential (1926) Board minutes, 13 May.
69. Eagle Star, Report on General Fund properties, Undated ledger.
70. Including the lease on a property forming part of the development, which Woolworth's already rented from the Prudential.
71. Prudential (1935) Board minutes, 25 April.
72. Prudential (1936) Board minutes, 6 and 13 Feb.
73. Prudential (1936) Board minutes, 27 Aug.
74. Prudential (1936) Board minutes, 11 June.
75. See section 2.6 for a fuller discussion of A.H. Bailey's canons of insurance company investment.
76. J. Humphries (1987) Inter-war house building, cheap money and building societies: the housing boom revisited. *Business History*, **29**.
77. Source: Prudential (1919–1939) Board minutes.
78. Hillier, Parker (1936) Superannuation Fund minutes, 4 May.
79. Hillier, Parker (1938) Superannuation Fund minutes, 30 June.
80. Hillier, Parker (1950) Superannuation Fund minutes, 15 Nov.
81. Hillier, Parker (1951) Superannuation Fund minutes, 12 Jan.
82. Letter to Healey & Baker (19 Sept. 1938) Montague Burton papers, Box 129.
83. Letter to Healey & Baker (27 Sept. 1938) Montague Burton papers, Box 129.
84. Letter to Healey & Baker (16 Nov. 1938) Montague Burton papers, Box 129.
85. Letter to Healey & Baker (21 Nov. 1938) Montague Burton papers, Box 129.

War and recovery 1939–1954 | 5

5.1 THE SECOND WORLD WAR

The onset of war led to a virtual halt in property market activity. Fears of war had resulted in a weakening of market conditions from the early months of 1938 and, following the commencement of hostilities, property values dropped dramatically, particularly in London. The position of property companies and speculators was made worse still by the variety of new costs arising from wartime conditions. The war had an immediate impact on property investment income, due to legislation which raised income tax on rental income, and necessitated the payment of War Damage Contributions and capital expenditure on air-raid shelters. Under the terms of the Civil Defence Act, 1939 landlords were liable to construct shelters for all buildings except those which had less than 50 occupants or were let to only one tenant. The cost of shelters was reduced under the Act by an allowance equal to income tax at the standard rate and could be passed on to tenants in the form of increased rents during the ensuing 10 years (though this usually proved difficult, if not impossible, in practice).

The payment of government War Damage Contributions had a substantial impact on net income from property, while offering some degree of protection against losses resulting from war damage. Contributions paid by Clerical Medical up to July 1941, £14 450, reduced the net yield on its property portfolio by an estimated 0.5%.[1] The War Damages Act provided £200 million in compensation for war damage and a further £200 million was later promised by the government, though payments were usually insufficient to cover the replacement cost of damaged buildings.[2] The Finance Act, 1940, further reduced rental income by making income tax payable on any excess of income received by the owner over and above that of the Schedule A assessment (such income had previously escaped taxation).

While property investment had been unfavourably affected by the war other types of asset also suffered from a reduction in net income. An analysis undertaken by Clerical Medical showed that the gross yield on its property portfolio during the middle of the war-time property slump, in 1942 – 5.0% (4.7% net) –

was significantly higher than the gross yield for its overall portfolio – 4.3%. Property therefore appears to have performed relatively well, compared to other assets, during the Second World War.

Property income also appears to have dropped only moderately (with the exception of war-damaged buildings) in the war years compared to the late 1930s. Table 5.1 shows net yields on book cost for several categories of property held by Clerical Medical for the years 1938–1945, together with figures for the property portfolios of Legal & General and the National Provident Institution. The table shows that yields on book cost fell to a low-point in 1942, recovering somewhat during the latter years of the war. The picture painted by the figures is less severe than might be expected, given the massive destruction of property and economic dislocation during these years. The average yield for Clerical Medical's properties during the war years – 4.9% – was only 0.9% below that for 1938, while even during the worst year the yield was only 1.4% below the 1938 level; a substantial fall, but not a disastrous one.

Table 5.1 Net yields on institutional property holdings: 1938–1945

	1938	1939	1940	1941	1942	1943	1944	1945	1939–1945 Average
Society's offices	4.2	4.4	4.4	4.8	4.5	4.0	4.1	4.6	4.4
City and West End	5.7	4.2	3.4	3.2	3.6	3.7	3.3	3.5	3.5
London suburbs	5.0	5.2	5.1	5.1	5.1	5.0	5.0	5.0	5.1
Provinces	5.1	5.2	5.1	5.1	5.1	5.1	5.1	5.1	5.1
Slough (Residential)	9.1	8.4	8.7	8.5	8.8	8.5	8.5	8.5	8.6
Total freehold	5.9	5.2	4.9	4.8	5.0	4.9	5.2	5.3	5.0
Leaseholds	5.7	5.8	5.3	3.4	4.1	4.3	4.7	4.9	4.6
Total (Clerical Medical)	5.8	5.4	5.0	4.4	4.7	4.7	5.0	5.2	4.9
Total (Legal & General)	4.7	3.6	2.5	2.0	2.3	2.4	5.2	5.5	3.4
Total (NPI)				5.1	4.4	4.5			

Sources: Clerical Medical (1943 and 1945) Asset Investigation Reports; Legal & General (1938–1945) Estates Dept. Annual Reports; NPI (1942–1944) Board minutes.

The figures for Legal & General show a much sharper fall in income, but represent a smaller and less diversified portfolio than that of Clerical Medical, with a greater proportion of properties in London (Clerical Medical's central London properties experienced much more severe income falls than those in the provinces). Figures are only available for the National Provident Institution for the years 1941–1943; these indicate yields of similar magnitude to those for Clerical Medical. It therefore appears that insurance company property investment income was reduced significantly, but not drastically, by the war, with average war-time yields on book cost being above those available from other assets.

Property companies had a less favourable tax position than insurance companies, and thus experienced much more severe falls in income. While the yield on Second Covent Garden's portfolio did not fall appreciably during the war years, earnings amounting to 5% in 1941 and 4.5% in 1942, income tax and war-damage contributions proved sufficient to reduce dividends to zero in both years (though dividends were issued from 1943).[3] Income tax hit property companies particularly severely, as unlike all other trading concerns they received no allowance from the Inland Revenue to be offset against income tax in respect of the depreciation of wasting assets.

Jackie Phillips, one of the most prominent inter-war property developers, was made bankrupt by the sudden downturn in the property market following the declaration of war, and died in poverty on Christmas Day 1939.[4] Other property speculators and developers were spared a similar fate due to a voluntary moratorium on loans on the part of the insurance companies, building societies and other lenders, though in a few cases insurance companies did foreclose on mortgages as a result of non-payment.

Despite the extreme political and economic uncertainty which resulted from war-time conditions, and the physical danger to property from Hitler's bombs, a number of far-sighted entrepreneurs began to purchase property during the war. The rationale behind such investment was simple; if the Allies were victorious property values would appreciate substantially during the post-war recovery, while if the Germans invaded Britain it would be of little importance where they had invested their money.[5] Buying property at rock-bottom war-time prices, to benefit from future capital appreciation, was a strategy which formed the foundation of a number of post-war fortunes. Bombing cleared large cental sites which, if consolidated into single plots, could be used to develop larger, more valuable buildings when construction recommenced.

Insurance companies were unable to invest substantial funds in property during the war years, being obliged to put all new funds in government securities under the war-time 'gentleman's agreement' between the insurance companies and the Bank of England. Even the limited funds over which the insurance companies had discretion with regard to investment were channelled away from property. Clerical Medical's General Manager, Sir Andrew Rowell, suggested a strategy of investing uncontrolled funds in property in October 1942, correctly forecasting the post-war boom in property values in a long memorandum he presented to the Society's directors.[6] However, his ideas were not accepted by Clerical's Board, which imposed a moratorium on property investment until February 1945.[7]

The Second World War resulted in a profound (though temporary) change in the political complexion of Great Britain – the success of war-time planning, together with the spirit of social cohesion and the desire for a better post-war Britain, led to a widespread acceptance of a much more interventionist approach to economic management. The landslide victory of the Labour Party in the June 1945 General Election confirmed and consolidated these changes, and the next six years were to see a radical transformation of many aspects of Britain's economic life, most notably the creation of the welfare state and the nationalization of large sectors of industry.

The war-time devastation of many of Britain's towns and cities focused popular attention on post-war reconstruction and, at an early stage in the war, government turned its attention to the attendant problems of town planning, betterment taxation and the location of industry. Town planning had a long tradition in Britain, dating from the work of late nineteenth century radical liberals such as Ebenezer Howard and developed by the Garden Cities and Town Planning Association and the Town Planning Institute. However, the inter-war years had seen little in the way of positive measures towards comprehensive town planning, the two main planning Acts of the period, the Town and Country Planning Act (1932) and the Ribbon Development Act (1935), proving largely ineffective.[8]

During the war years the influence of the town-planning movement in official thinking grew enormously. As early as October 1940 Lord Reith, the Minister of Works and Buildings, was given responsibility for drawing up preliminary reconstruction plans for presentation to the Cabinet.[9] In January 1941 an expert committee was appointed to study the problems of compensation and betterment, which had proved the major obstacles to town planning in the 1930s. The Committee's interim recommendations – involving fixing the value of land on the basis of which compensation would be paid following public acquisition or control at its March 1939 value, the establishment of a central planning authority to control all development of land and the assignment of 'reconstruction areas' in which rebuilding would not be permitted except under license until proper planning schemes were prepared – were accepted by the Cabinet in July 1941. The central planning authority, the Ministry of Works and Planning, was established in February 1942 by augmenting the Ministry of Works and Buildings and the Ministry of Health's town and country planning functions for England and Wales.[10]

Part of the reconstruction planning agenda set out by Lord Reith in the early months of 1941 included instructions to the local authorities of Britain's blitzed cities to plan boldly and comprehensively.[11] Coventry was one of the first cities to respond to this challenge, adopting an ambitious reconstruction plan, involving the division of the city's centre into zones based on function, surrounded by an inner ring road. Coventry's new shopping centre was to be a traffic-free precinct, possibly arcaded, and made up of six or seven storey buildings.[12] However, this plan, in common with those of other local authorities which included precinct shopping (or other attempts to change the character and/or location of shopping centres), met with vigorous opposition from local traders. Retailers were particularly hostile to the idea of pedestrianization; the inclusion of precinct shopping in Southampton's reconstruction plan led major retailers, including Marks & Spencer, to argue that full vehicular access was essential to maintain a shopping area's importance.[13] The arcade concept was also strongly opposed; though a long-established feature of British urban retailing, arcades had generally been shopping areas of secondary importance, the major multiples preferring High Street pitches. In both respects the planners, rather than the retailers, were to be proved correct, pedestrianized and (later) covered shopping centres eventually gaining the acceptance and support of retailers during the following decades.

The fundamental replanning of city centres had other important implications for retailers. Traders who had premises in blitzed shopping centres faced substantial financial loss as a result of the legislative framework governing compulsory purchase and compensation. The War Damage Act, 1943 made owners of war-damaged premises eligible for a 'cost of works payment', covering the full cost of rebuilding. However, if the premises in question were compulsorily purchased with regard to a local authority planning scheme under the provisions of the Town and Country Planning Act, 1944, the owner would receive only a 'value payment', equivalent to the premises' estimated March 1939 value (with a supplementary addition of up to 30% in the case of owner-occupiers). Many retailers, and local authorities, were concerned that this would leave traders unable to finance the reconstruction of their shops on the council's new sites given current soaring building costs.[14]

Comprehensive replanning of blitzed shopping centres also usually entailed a move to bigger centres with larger shops. These would necessarily command higher rents than their pre-war counterparts, accelerating the (already well-established) trend towards domination of shopping centres by the major multiples. In this respect it is interesting to note that while smaller local traders objected to the re-siting of Bristol's shopping centre, the Multiple Traders Federation supported this aspect of the city's reconstruction plan.[15]

Reconstruction planning was generally based on local authorities acting as ground landlords for the new shopping centres, retailers being able to secure only leaseholds on their shops. This provided further grounds for complaint from traders at the public enquiries held with regard to the reconstruction plans.[16] The investment implications were obvious; retailers would lose their valuable freeholds for compensation which would not reflect their true value and gain premises which, while possibly more spacious, would cost more to rent and would not have the added investment value that a freehold interest conferred.

However, as the result of raw materials and building labour shortages, together with competing priorities and long delays in gaining ministerial approval for reconstruction plans, extremely little city centre rebuilding took place until the early 1950s. Financial stringency led to government reluctance regarding the granting of substantial compulsory purchase powers, retailers' fears regarding the loss of their sites often proving unfounded. Contrary to Correlli Barnett's claims that reconstruction planning entailed the sacrifice of

Britain's real economic needs to eutopian 'New Jerusalem' reconstruction aims,[17] other economic priorities in fact greatly reduced the scope, and pace, of central area reconstruction.

5.2 PROPERTY INVESTMENT AND THE CAPITAL MARKET 1945–1954

Meanwhile, the outcome of the war-time investigations into post-war planning – embodied in the Uthwatt Report on compensation and betterment, the Scott Report on land utilization in rural areas, together with the report of a Royal Commission established in 1937 to investigate problems arising from the changing distribution of Britain's industrial population (the Barlow Report, 1940) – had led to the introduction of one of the most ambitious pieces of post-war planning legislation, the Town and Country Planning Act, 1947.

The 1947 Act brought almost all development under government control by making it subject to planning permission. Planning powers were transferred from District Councils to County Councils, who were given the job of preparing development plans for each area, under the national supervision of the Ministry of Town and Country Planning. Development rights and the development value of land were effectively nationalized by the Act, leaving landowners with their existing (1947) use rights and land values.[18] Compensation for the loss of development rights was to be paid from a national fund, while developers had to pay a levy amounting to 100% of any increase in land values resulting from new development.

The imposition of a 100% tax on property development removed any incentive to develop property. A system of building licences, capital issues controls and other regulations, together with an acute shortage of building materials, placed further obstacles in the way of the developer, with the result that there was little development of new commercial property during the 1945–1950 period outside a few narrow areas, such as buildings for government occupation. By 1949, less than 1% of the City's war-damaged area had been at least temporarily rebuilt.[19]

While these conditions restrained new development they led to an unprecedented boom in investment in existing buildings, which had been made scarcer by war-time bombing and did not have to face competition from newly developed property. Some three and a quarter million properties were damaged or destroyed during the war; most of these were houses, though the total included 75 000 shops, 42 000 commercial properties and 25 000 factories.[20] War damage was concentrated in London, and a number of ports and important industrial centres, such as Bristol, Hull, Portsmouth and Coventry. Within London the worst hit area was the City, with a third of buildings totally destroyed. Furthermore, requisitioning led to many commercial enterprises losing the use of their premises for some time after the war.

Meanwhile, the demand for commercial property had grown substantially. Office work expanded from 6% of total employment before the war to 16% in 1951, reflecting the growth of government, the tertiary sector and administrative jobs within manufacturing.[21] A similar growth in demand had occurred in the retailing sector; by the end of 1954 the value of GDP accounted for by the distributive trades had increased by more than one-third, in real terms, compared to its 1938 level. The substantial demand thus created was met by virtually static supply; inevitably property prices escalated.

These conditions made property investment a highly lucrative sector; most of Britain's largest property companies, such as Land Securities, MEPC and Hammersons were founded during the years 1944–1954. Land Securities, Britain's largest property company, was a tiny company with assets comprising three houses and less than £20 000 in Government securities when it was acquired by Harold Samuel in the spring of 1944. By March 1952 its assets were valued at over £11 million, such was the scope for expansion in the early post-war years.

This period saw an intensification of property investment activity by the insurance companies which had been active in the sector during the 1930s and a substantial expansion in the number of insurance companies that undertook property investment. During these years property continued to constitute a relatively high-yielding fixed-interest security, which was of considerable attraction to institutional investors at a time when cheap money, and legislation which encouraged companies to limit share dividends, led to a shortage of assets which offered attractive yields.

However, the attraction of fixed-interest securities relative to equities was already beginning to diminish. From 1947 to 1954 the annual rate of return on UK ordinary shares averaged 8.7%, while rising interest rates resulted in a negative average return on consols of –1.3% per annum. Property performed less well than equities, though considerably better than consols, during these years, with an average return of 5.6%.[22] Property offered a positive capital return over this period, though this was extremely small in comparison to the income return, averaging only 0.5% per annum.

The volume of insurance company property investment increased substantially, both in value and as a percentage of total net investment, from 1946 to 1951, falling back somewhat during the next three years.[23] The real level of investment was substantially higher than that for the 1930s, but only in 1951 and 1952 did the percentage of total insurance funds invested in property exceed peak inter-war levels. There was also a very limited amount of pension fund property investment during this period, but even by the end of 1953 pension fund property holdings were estimated to amount to only £16.75 million, 1.69% of total pension fund assets and 5.85% of total insurance company property holdings. Some pension fund investment in property had occurred even during the inter-war period, though this was on an extremely small scale and involved only a small number of funds.

Section 10.2 provides estimates of the magnitude of investment in property by institutional, and other, investors. Individual entrepreneurs and the 'traditional' institutions appear to have formed a larger proportion of total investment during this period than in later years. This was probably due to the fact that a great deal of investment activity still concerned relatively small properties, the average value of investment properties growing substantially over the following decades. Furthermore, the attraction of the sector to the individual entrepreneur was greater in this than in subsequent periods, due to the greater opportunities for making substantial profits via relatively short-term speculation.

An indication of the different investment strategies pursued by various categories of property investor is provided by a ledger prepared by the estate agency Jack Rose & Co. in 1949/1950. The ledger was compiled by Jack Rose, who went on, with his brother Philip, to establish the property company Land Investors Ltd. Rose contacted a very large number of property investors, or their agents, and listed the categories of property which they would consider purchasing, according to an alphabetical code. The ledger appears to achieve a fairly comprehensive coverage, including almost all the prominent individuals, companies, institutional investors and market intermediaries which were active in the property market at this time. The contents of the ledger are summarized in Table 5.2.

In Table 5.2 investors have been divided into seven categories;[24] 445 of the 452 investors listed could be categorized in this way, the remaining seven, for which the type of institution is not known, are omitted from the table. The table shows the proportion of each category of investor that expressed an interest in purchasing various classes of property. The average number of property classes in which each category of investor was interested, the standard deviation of the percentage of investors in each category that were prepared to purchase each class of property and a 'risk ratio' statistic for each category of investor are also given.

The risk ratio was calculated by assigning a degree of risk to each class of property,[25] based on contemporary evidence regarding the perceived risk associated with each class.[26] Such a procedure provides only a crude measure of risk, as the magnitude of the difference between each risk weighting is much more difficult to determine than the relative order of risk. The figures might, therefore, be best regarded as an indication of which categories of investor pursued relatively risky strategies, rather than a quantification of the degree of risk. The ratios were derived by multiplying the percentage figures for investment in each class of property by the weighting for that class, the resulting figure being divided by the average number of classes in which an interest was registered. The values obtained were then divided by the figure for all categories and the result multiplied by 100 in order to arrive at values which compare the level of risk associated with the investment policy of each category of investor with the average risk level for all categories.

Table 5.2 A profile of the property investment market in 1949–1950

Institution	Insurance companies	Friendly societies	Property and investment companies	Individuals	Estate agents	Solicitors, architects, accountants	Miscell-aneous	Average
Number of investors	25	3	77	207	68	43	22	63.6
Average number of categories in which an interest is expressed	4.7	4.7	2.6	2.4	3.5	2.5	2.3	3.2
Percentage of investors in each category interested in:								
Ground rents	72.0	100.0	9.1	8.7	45.6	30.2	31.8	43.0
Primary shops	84.0	66.7	22.1	17.4	58.9	58.1	22.7	47.1
Secondary shops	52.0	33.3	27.3	29.0	35.3	37.2	18.2	33.2
Offices	92.0	100.0	41.6	15.5	54.4	30.2	27.3	51.6
Housing estates	44.0	0.0	46.8	30.9	35.3	25.6	31.8	30.6
Flats	60.0	33.3	45.5	24.2	30.9	14.0	13.6	31.6
Vacant and reversionary shops	40.0	66.7	16.9	15.5	36.8	25.6	18.2	31.4
Speculations	0.0	0.0	19.5	35.3	13.2	14.0	27.3	15.6
Small houses	0.0	0.0	7.8	23.7	7.4	2.3	4.6	6.5
Single shops	4.0	0.0	6.5	20.3	11.8	4.7	9.1	8.0
Factories	12.0	0.0	6.5	3.9	2.9	0.0	4.6	4.3
Sites	0.0	0.0	10.4	7.3	7.4	7.0	13.6	6.5
Short leaseholds	4.0	66.7	0.0	2.4	2.9	0.0	4.6	11.5
Standard deviation	34.4	39.6	15.9	10.5	19.8	17.3	10.1	16.7
Risk ratio [a]	92.3	72.2	121.6	130.3	101.5	102.4	111.5	100.0

Source: Ledger compiled by the estate agents Jack Rose & Co. in 1949–1950.

[a] See p.xx for definition.

Table 5.2 shows that insurance companies made up only a small proportion of investors in the property market at this time, though they accounted for the majority of property purchases by value.[27] They pursued relatively low-risk strategies; of the 25 institutions in this class, 72% invested in ground rents, the least risky property asset, and 84% invested in the second least risky class, prime shops. However, none of the insurance companies mentioned invested in the two riskiest classes – sites and speculations.

Property and investment companies formed the largest category of institutional investor in the sample.[28] These were more specialized in their investment requirements, the average number of property classes in which they were interested being 2.60 compared to 4.68 for the insurance companies. Property company investment was much more evenly spread between classes than was the case for insurance offices, however, the standard deviation being less than half that for the insurance companies. They were also prepared to invest in higher-risk property classes.

The largest single category of investor was composed of individual people, who made up over 45% of the sample. The average number of property classes in which members of this group were interested was the lowest of any category of investor, a function of both specialization and the small scale of operations of most individual investors. The influence of limited funds on investment behaviour is illustrated by the office category, in which only 15.46% of individuals were interested compared to 92% of insurance companies. Individuals had the highest risk statistic of any group, but as some of the riskier property categories, such as residential property and small shops, were small scale in nature, there was some element of necessity in their high degree of investment activity in these sectors.

This was not true, however, for the 'speculations' category, in which there was a far higher degree of interest among individuals than for any other category of investor, indicating that the private investor was more concerned with maximizing potential income than the institutions, even at the expense of a greater degree of risk. Greater interest in 'speculations' on the part of individuals might also reflect their ability to reduce *ex ante* risk by entrepreneurial activity, for example, by finding a tenant for a vacant property before purchasing the property.

The 'miscellaneous' category represents a diverse group of investors, including multiple retailers, traditional institutions, a bank and a pension fund. The investment requirements of this group were relatively well dispersed between property classes, due to the diverse nature of the institutions included in this category. The two 'market intermediary' categories – estate agents and solicitors, and architects and accountants – purchased property for clients. Their investment requirements therefore reflected those of a number of institutions with very different investment interests, as was reflected in low standard deviations between classes and risk statistics that were very close to the average. The relatively large number of solicitors in the register (which made up most of the solicitors, architects and accountants category), indicates that they still played an important role as property market intermediaries at the end of the 1940s.

While only 4.3% of investors in the above sample expressed an interest in industrial property, there was more interest in this sector on the part of insurance companies than any other group. An important reason behind the insurance companies' increased interest in industrial property was the extreme scarcity of newly-developed commercial buildings. Much less stringent controls were placed on industrial property development during this period than on commercial development; from 1948 to 1954 commercial property accounted for only 9.6% of all private sector non-residential construction activity, compared to 45.3% during 1955–1964.[29]

The most important source of new investment properties for the financial institutions, however, was the stock of commercial property that had been built prior to the Second World War. The lack of competition from new developments, and the slowness of most institutional investors to realize the threat which inflation posed to fixed-interest stocks, allowed multiple retailers and other companies to sell their property portfolios on a leaseback basis to financial institutions on terms which were to prove extremely advantageous during the following years. A great deal of money could be raised from property portfolios using the leaseback system, at a low, fixed, rate of interest and without the need to repay the principal. Such opportunities were exploited most spectacularly by corporate raiders such as Charles Clore, who found that companies could be acquired for less money than could be subsequently raised from the sale and leaseback of their property assets, as outlined below.

In order to explore the factors influencing the property investment market during these years in more detail it is necessary to examine the experience of individual investors. One of the most active property investors during this period was Legal & General, which was second only to the Prudential among Britain's largest purchasers of investment property during the late 1940s and 1950s. The resumption of Legal & General's property investment programme began in the last months of 1944, with the purchase of five shops in London, Walthamstow and York, for a total of £1.5 million. This marked the start of a period of rapid expansion for Legal & General's property portfolio, which was to make Legal & General one of the largest property owners in the UK by the early 1960s.

All aspects of direct property investment were dealt with by the Society's Estates Department which was managed, from 1946 to the early 1970s, by the Joint Chief Estates Surveyors, John Crickmay and Arthur Green. Rapid expansion of property holdings led to similarly rapid growth of this department; the number of staff it employed, which stood at 11 on the eve of the Second World War, had increased to 55 by the end of 1947. By this time the Society owned 218 properties, with a total book value of over £11 million. The establishment of a managerial and administrative structure to deal with property acquisition and management was a necessary cost of undertaking a substantial volume of property investment, and it is notable that the institutions which were active in the property market in the immediate post-war years generally had property departments already established as a result of their inter-war investment activities. Institutional developments in the inter-war period, such as the setting up of insurance company property investment departments and the emergence of a national commercial property market via the growth of nationwide market intermediaries, were essential prerequisites to the rapid post-war expansion of the commercial property market. Another, related, post-war legacy of the 1930s' property boom was the development of considerable human capital with the necessary expert knowledge of the property market. This is evident from Oliver Marriott's list of 110 people who made over a million pounds from property between 1945 and 1964, a large proportion of whom began their careers as commercial estate agents, or in other property-related professions, during the inter-war years.[30]

A Legal & General Board paper of June 1951 considered investment policy in the light of experience since the Second World War. By this time Legal & General's property portfolio amounted to 20% of total invested funds (compared to an average of about 6% for all insurance companies in 1949), while other property-based securities, ground rents and mortgages, accounted for a further 4 and 11% of total assets, respectively.[31] In discussing further investment in property, the paper split current holdings into property on leases of over 50 years, with a book value of £12 054 000, and property on shorter leases, with a book value of £15 703 000, illustrating an early realization of the importance of property with the prospect of rent revision in the short or medium term to future income growth. The report's conclusion noted that the purchase of property and ordinary shares should be subject to a broad decision regarding what percentage of funds should be invested in assets offering a hedge against inflation, i.e. assets in these two classes. As early as 1951 Legal & General viewed property as an asset capable of income and capital appreciation rather than a fixed-interest security. It took most other institutional investors several more years to reach this conclusion.

Legal & General's Board encouraged the company's active property investment policy. Many insurance company directors had strong reservations regarding the suitability of property as an investment medium for insurance funds, however, favouring more traditional areas of investment. Despite the lifting of the moratorium on property investment in February 1945 some members of Clerical Medical's Board continued to be extremely wary of property as an outlet for the Society's funds. Many of its Directors had commercial experience and understood stock exchange investments better than property, while some favoured gilt-edged securities due to their safety, even though their yield was, at times, lower than the net rate of interest on which the Society's policies were based. Few, if any, understood the property market.

The General Manager, Sir Andrew Rowell, presented a number of papers to the Society's Board, arguing in favour of an active property investment policy.

The longest of these, submitted to the Board in December 1949,[32] attempted to justify why Clerical Medical had a much larger proportion of assets invested in property than the 'giant' insurance offices, such as the Prudential, and why that proportion should be further increased. Rowell argued that Clerical Medical had superior investment opportunities to those available to the largest insurance companies, principally as a result of its smaller size. This made it possible for a number of actions, which could not materially effect the investment pattern of a company of the size of the Prudential, to substantially influence Clerical Medical's asset distribution and the returns on its investment portfolio. These included the establishment of friendly relationships with individual 'dealers' in various sections of the property market who had introduced a flow of investment opportunities in recent years which were more than adequate to absorb available funds. Rowell then stated that other insurance offices did not invest in property to the same extent since they lacked Clerical's market connections. Without such connections investors had to fall back on the services of estate agents which Rowell described as 'a very disappointing and unattractive source of introductions'.[32]

Rowell illustrated his argument by providing a list of the properties which Clerical Medical had sold since the war. In almost every case an insurance company was either the buyer or an unsuccessful bidder. Rowell stated that this was remarkable because these represented not the 'cream' but the least attractive of Clerical's properties and the eagerness of other insurance companies to buy them therefore showed that when opportunities arose other insurance offices were eager to invest in property.

This memorandum reveals much of the nature of the early post-war property market and Clerical Medical's strategy in that market. In a highly imperfect market, such as that for property, a network of individual dealers[33] provided an important mechanism for the transfer of market information. Clerical Medical was able to tap into such a network and use it to take advantage of the imperfect nature of the property market, picking up the bargains which inevitably arose under such conditions.

The Society invested a substantial proportion of new funds in the property sector in the late 1940s. Transactions included the financing of property development via a series of joint development companies, as outlined in section 5.3, a practice which did not become widespread among the financial institutions until the late 1950s. This unorthodox, and highly profitable, policy was pursued despite pressure from some directors to adopt a more cautious investment policy. The degree of caution Rowell encountered from Clerical Medical's board in the early post-war years is demonstrated by a note dated August 1947 in which he stated, when advocating the sale of gilts:

> ... if they once more go substantially better there is a grave danger that the Board will once more decide that they are the finest investment possible and refuse to sell. I personally would rather sell them now, even if I felt much more convinced than I do about a rise, than risk having to keep them.[34]

Such differences in investment outlook were also reflected in policy regarding the Society's ordinary shares; Rowell believed that they were likely to appreciate in value and wanted to retain some of those which the Directors wished to sell. In the same note he stated 'As for equities, Sir Frances Humphrys has several times suggested the sale of any equities still showing a profit on cost, but I have stalled each time.'[34] Rowell's early appreciation of the need to build a substantial equity element into insurance company portfolios was shared by managers of other funds which invested substantially in commercial property in the immediate post-war years, such as Legal & General and the Church Commissioners.

The National Provident Institution re-entered the property market in 1947; as with its earlier ventures into property, in 1895 and 1932, a period of inactivity was followed by substantial purchases. Acquisitions in this year included several industrial properties, the first purchases in this sector by NPI; £332 789 was spent on industrial properties, out of total direct property investment of £530 182.[35] Other insurance companies, including Clerical Medical and Eagle Star, also began to purchase industrial property during the late 1940s, partly as a result of the scarcity of newly-developed commercial property mentioned above.

The limited supply of commercial property also led a number of insurance companies, including NPI, Clerical Medical and Eagle Star, to invest substantially in the residential sector during this period. The wisdom of purchasing residential property, and commercial property on long, fixed-rent leases at low yields, was called into question by a number of these companies only a few years later, however.

By 1951 NPI was beginning to regret some of the purchases it had made during the years 1947–1950. Properties purchased during this period of cheap money depreciated in value as interest rates, and property yields, rose in the early 1950s.[36] Similar problems were experienced by Clerical Medical. The Society's 1952 asset investigation report showed that the market value of its freehold buildings stood at £1 948 000, £130 373 below their book cost. However, the leasehold properties made up for some of this deficit, the resulting shortfall between market value and book cost for the overall property portfolio amounting to only £14 524. In 1953 the market value of Clerical Medical's property portfolio once again exceeded book cost, though its freehold properties continued to show a slight deficit. This depreciation in values for low-yielding properties let on long leases at fixed rents might be an important factor behind both the decline in insurance company property investment from 1952 to 1954 and the pressure to build an equity element into property acquisitions which emerged during the 1950s.

In addition to difficulties caused by low yields, as a result of badly-timed purchases, residential properties presented further problems for the institutions, as they often incurred heavy management and maintenance costs which could not be passed on to tenants due to rent control legislation. As a result most financial institutions had ceased to purchase such property by about 1951, and the 1950s saw a sharp decline in the proportion of institutional property holdings accounted for by residential property.[37]

The insurance companies were not the only institutional investors to take an active interest in commercial property during this period. A number of other institutions, such as the Church Commissioners and the National Coal Board Superannuation Fund, also began to examine opportunities for investment in this sector. This marked the start of a trend towards an increase in the number, and range, of institutional investors in the property market, which was to grow considerably during the 1950s and 1960s.

The Church Commissioners was established in 1948 as the result of a merger of the two main organizations which managed the assets of the Church of England – the Ecclesiastical Commissioners and Queen Anne's Bounty. The Commissioners inherited a large property portfolio from their predecessors, particularly the Ecclesiastical Commissioners. While the Ecclesiastical Commissioners had pursued a relatively inactive property investment policy the Church Commissioners soon set about the task of transforming the character, and profitability, of the Church's property holdings.

During the Commissioners' first year of existence a comprehensive survey of their resources and commitments was undertaken, which revealed that the income from the Commissioners' portfolio was diminishing and that this downward trend was likely to continue.[38] In order to prevent a continuation of this trend the Commissioners moved towards a policy of investing in assets that would maintain their real value over time.

The decisive influence behind the adoption of this policy was Mortimer Warren (later Sir Mortimer Warren), a chartered accountant who had joined Queen Anne's Bounty in 1927 and was appointed as Financial Secretary to the Church Commissioners. Warren eventually pursuaded the Commissioners to adopt two policies which were key to their success in expanding the Church's income during the following decades. The first was the sale of gilt-edged stock and reinvestment of the proceeds in ordinary shares. The second was a similar rationalization of the Commissioners' property portfolio, involving the sale of properties let on long leases and the redeployment of the funds raised in properties let at market rents, with leases that offered the prospect of future rent increases. The policy, which marked a break with the Church's traditionally conservative approach to its investments, gradually won acceptance due to strong support from the Archbishop of Canterbury, Archbishop Fisher,[39] and to a general realization that something had to be done to improve the income from the Commissioners' portfolio.

Warren's original recommendations were fairly modest. In September 1948 he presented a report on the Commissioners' financial position to the Board of Governors. The Commissioners' stock exchange assets were examined, in the light of advice from the government broker that while the proportion of government securities in the stock exchange portfolio, 59%, was not excessive, 5% of funds might be invested in the ordinary shares of the 'very best' commercial and industrial companies[40] so as to increase income. Further investment in the shares of insurance companies and banks, and first class preference shares, was also advocated. It was recommended by Warren that this advice be accepted, since it would bring about 'the increase in income so sorely needed', and ordinary shares had 'the great advantage of increasing in capital value inversely with the progressive fall in the value of money'.[40]

These comments show that as early as 1948 the likelihood of persistent inflation was an important factor in Warren's investment thinking. Warren also extended this philosophy to property, recommending in the same report that obtaining increases in income be taken as the dominant factor in all estate management questions. The danger of inflation continued to dominate the Commissioners' investment thinking during the following years, an Estates and Finance Committee report to the Board of Governors, of April 1951, stating 'The cost of living has broken free of its ceiling ... It will be wrong for the Commissioners to establish any acceptable ultimate policy which does not include stable minimum income for the parochial clergy as a foundation.'[41]

In the late 1940s the bulk of the Commissioners' property holdings still constituted the ancient estates whose administration the Ecclesiastical Commissioners had taken over during the nineteenth century. Of the gross property income for the year ending 31st March 1950, 19% represented agricultural holdings and woodlands, 45% was from urban properties let at ground rents and 36% was composed of rack rented urban properties.[42] The Commissioners' agricultural holdings were reduced from 282 518 acres in 1949 to 217 000 acres by the end of March 1954, with a view to stabilizing holdings at 210 000–220 000 acres. This reduction was much more moderate than had been originally advocated by Warren, in a second major review of investment policy, in 1951. Warren called for a change in the Commissioners' attitude towards their property portfolio establishing, as a fundamental principle, that the Commissioners no longer had a public duty to own real estate.

A property investment policy based purely on monetary criteria would, he argued, involve a reduction in the Commissioners' agricultural estates from 250 000 acres to approximately 100 000 acres, together with the sale of some residential properties.[43] While the Commissioners were never prepared to allow monetary considerations to be the sole determinant of their agricultural investment policy (as is discussed in section 6.2), Warren's recommendations did lead to a substantial rationalization of their urban property assets, with the redeployment of funds from ground rents and residential estates to commercial property.

This period also saw some limited investment in property on the part of the pension funds. One of the earliest major funds to enter the property market was the NCB Superannuation Fund. The NCB established a superannuation fund for all salaried staff in 1947. A further scheme, for the NCB's industrial workers, had been negotiated and approved by the end of 1951, and started operation in 1952.[44]

Property was brought to the attention of the fund's managers due to a series of approaches from agents offering particular properties. Its suitability was debated in August 1949, with the circulation of a memorandum by one of the fund's managers, A.E. Horton, stating that several propositions had been received involving multiple shops offered on a 99 year leaseback basis, yielding about 4% net.[45] Several objections to investment on these terms were raised by the Deputy Chairman of the NCB Staff Superannuation Scheme's Committee of Management, H.W. Naish, in reply to this memorandum. These centred around the danger of shops becoming obsolete before 100 years was past, due to age or future changes in the geographical location of prime shopping pitches; the possibilities of nationalization and adverse legislation; and the absence of any means of adapting income to changes in market rents, interest rates and other factors. He concluded that if current market yields for such property offered a net pre-tax income of 4% on purchase cost, he would rather not invest in this type of asset. He suggested that 4.5%, to provide an extra margin for administrative expenses and sinking fund payments, might be an acceptable minimum yield. Even then he would have grave doubts regarding this type of investment.[46]

Doubts voiced by Naish were echoed by the objections of other members of the committee, leading to a delay in the fund's entry into the property market. By June 1950 the fund's problems in meeting its investment targets had increased, however. While target holdings of gilts were set at 47% of total funds, they currently represented 57.1% of the portfolio, their low yields depressing overall income.[47] In the light of the shortage of investments other than gilts, property appeared increasingly attractive and, despite initial doubts, by March 1952 the NCB superannuation funds were making their first property purchases.

However, Naish continued to voice concerns regarding the fixed-interest nature of sale and leaseback arrangements. In a note dated 29th July 1952 regarding the purchase of factory property he argued that a 'comparatively short term of leases would give opportunities to increase the rents several times in the course of a hundred years to take account of the inflation which, unless history does not repeat itself, will inevitably take place over the period'.[48] By May 1954 stronger doubts regarding the desirability of investing in properties let on very long leases, at fixed rents, were being expressed by the managers of the two funds. In a letter to Naish, J. Davidson stated that:

> I personally dislike tying up our rack-rented properties in very long leases. Perhaps my dislike is, to some extent, instinctive or you may say irrational, but I think that if history repeats itself and inflation continues as it has done, we should retain the right to get possession of our properties much sooner than in 99 years – my view is that generally 21 years would be long enough.[49]

However, the professional advice received by the two funds, from Mr Aubrey Orchard-Lisle of Healey & Baker, who was appointed as property consultant to their Joint Investment Sub-committee in 1953, advocated the investment of the bulk of property funds in shops and offices, which were usually offered on long leases at fixed rents.[50] In July 1954 Naish wrote a note regarding this advice, arguing that tying up funds in properties let on long, fixed-rent, leases was unwise. He advocated greater investment in freehold shops let on short leases as 'our investment would have more of the character of an equity investment ...'.[51]

Thus, the NCB superannuation fund, along with the Church Commissioners and Legal & General, were all looking for ways to build an equity element into property investment by 1954. However, perceptions of the dangers of inflation, and the need to build income growth into investment portfolios, were by no means universal by this time. Many investment managers still thought in terms of initial yields and the security of capital invested rather than future income growth, and did not regard inflation as a long-term phenomenon which had to be taken into account in investment decision-making. This lack of inflationary expectations (despite several years of inflation) on the part of some investors was recalled, with regard to the investment policies of the Second Covent Garden Property Company, by J. Max Keyworth:

> Nobody unfamiliar with conditions in the early 1950s can expect to understand the attitude to money in those times ... Firstly it was established that from the Napoleonic Wars until the outbreak of the First World War there was virtually no inflation ... although there had been inflation again in the Second World War, the possibility of its continuation here in time of peace could be ruled out; until about 1952 many considered deflation more likely, and therefore a guaranteed fixed income was thought to be a good thing. How else can one explain the numerous 99-year leases at fixed rents which were negotiated at this time?[52]

The attractions of property as an investment medium resulted from it providing a relatively secure investment, with a yield which was substantially in excess of that of gilt-edged stock. Capital issues controls reduced the supply of mortgage, debenture, preference and ordinary share issues while increasing the supply of investment property, since companies which wished to raise capital and were constrained by these controls found selling property on a leaseback basis to be an even more attractive means of raising finance than was the case in the 1930s.

The institutions which invested heavily in property during the late 1940s usually found that their early acquisitions, purchased at what were low yields compared to those prevailing during the 1950s and having rents which were fixed for so long as to make them virtually fixed-interest assets, depreciated in value as interest rates rose during the following years. However, these early investments allowed them to develop experience in the property market, and links with property developers and dealers, which were to prove extremely valuable during the 'property boom' when property was transformed from a fixed-interest security to an equity asset. These early institutional property investors were also usually the first to introduce the innovations which built an equity element into investment property, and thus benefited more from the boom conditions of 1954–1964 than the later funds to enter the market.

5.3 THE FINANCIAL INSTITUTIONS AND THE PROPERTY ENTREPRENEURS: 1945–1954

One of the most notable features of the commercial property market in the two decades after 1945 was the large number of individual fortunes that were made from property. In his classic book on the entrepreneurs who made those fortunes, *The Property Boom*, Oliver Marriott identifies no fewer than 110 people who, he estimated, made over a million pounds from real estate between 1945 and 1965, most of whom started their careers without any capital 'beyond the odd hundred pounds'.[53] These included many of the most prominent entrepreneurs of the era, such as Charles Clore, Jack Cotton and Harold Samuel. Conversely, there were very few examples of business failure in the property sector during this period, in contrast to its experience during the 1970s and 1980s. Why were the rewards to entrepreneurship so great in this sector?

The nature and importance of entrepreneurship is a topic which has, until the recent past, been largely neglected in economic theory, with a few notable exceptions.[54] Knight[55] viewed entrepreneurship as the process of bearing risk and uncertainty, the economic reward for which was profit. Schumpeter[56] famously emphasized the importance of innovation as the main entrepreneurial function, perceiving risk-bearing to be the role of the capitalist rather than the entrepreneur. More recently, Casson[57] has emphasized the judgemental role of the entrepreneur, taking decisions regarding the allocation of scarce resources.

While these definitions cast light on the function of entrepreneurship none provide a satisfactory explanation as to why the rewards to entrepreneurship were so high in the commercial property market relative to other sectors of the British economy during the first two post-war decades. The massive profits made by property entrepreneurs during this period do not appear to have been a function of the degree of risk involved; the riskiness of many property ventures was very low, especially to the entrepreneur, who often staked little if any of his own capital. Nor do high profits reflect the difficulty of innovation in the sector; the majority of these innovations were simple financial or legal arrangements that could be implemented at very little cost. Equally, there were few particularly difficult judgemental decisions; working out the viability of a property development project is a judgemental task which, though requiring considerable skill, requires less judgement than entrepreneurship in many areas of manufacturing, sometimes involving the development of new products, processes and markets.

In order to explore the factors which made the rewards to entrepreneurship so high in this sector it is necessary to examine the character of the post-war property entrepreneurs, and of the innovations they introduced. The percentage distribution, by original occupation, of the 110 people identified by Marriott is given in Table 5.3. The table shows that their backgrounds were concentrated in a narrow range of property- and trading-related occupations, over 85% of individuals whose previous employment could be classified falling into these groups. Less than 3% of property millionaires came from the financial sector, despite the close involvement of this sector with the property development industry. A further notable feature of these individuals was their ethnic composition; it has been estimated that at least 70% of the people in Marriott's list were Jewish, the vast majority of whom were second generation immigrants.[58]

Table 5.3 Original occupations of property millionaires, 1945–1965

Profession	Percentage
Estate agent	40.0
Solicitor	10.0
Other property-related professions	15.5
Trading (excluding the above)	15.5
Banking, accountancy and insurance	2.7
Miscellaneous	10.9
Unclassified	5.5
Total	100.0

Source: O. Marriott (1967) *The Property Boom,* Hamish Hamilton, London, pp.267–9.

During the 1930s commercial estate agency was a new and relatively open sector. It had not yet quite attained the status of a profession, but it also lacked the lengthy training for qualifications, and social barriers to entry, that were characteristic of the professions. It was, therefore, of considerable attraction to people wishing to enter the professional classes but lacking the resources to undertake costly training. The Jewish community in London, like many immigrant communities, had a higher degree of perceived social mobility than the bulk of the population, most of whom had very fixed ideas regarding career possibilities, based on the jobs of their older relatives and immediate peer group. Furthermore, this community contained a large number of people from trading-related backgrounds who found that they had the necessary aptitude to make successful commercial estate agents.

In the post-war years it was natural that those who had already amassed considerable knowledge of the property market would have the best chance of success as developers. As for the substantial number of property millionaires

THE FINANCIAL INSTITUTIONS AND THE PROPERTY ENTREPRENEURS

with trading-related backgrounds outside this sector, they were attracted to property as it was the area of trading that offered the highest profit margins during this period. At a time when the property market was still relatively unsophisticated the skills the developer needed most were primarily dealing skills: a knowledge of the importance of prime locations, market conditions, loopholes in the legal framework, and, perhaps most importantly, the ability to arrange transactions in such a way as to secure the highest possible margin of final capital value and revenue over costs. People from dealing-based occupations therefore had skills of considerable value in this sector, and those with property dealing backgrounds were in an even better position to exploit available market opportunities.

The years between the onset of war in 1939 and the removal of building licence restrictions in 1954 constituted the most regulated and restricted period in the history of the commercial property market. Paradoxically, these years offered great opportunities for making a fortune from property which were surpassed only in the following 'property boom' decade.

These opportunities were the result of two main factors:

1. A severe imbalance between supply and demand for commercial property, arising from war-time bombing and stringent post-war controls on development activity.
2. The introduction of a legislative and regulatory framework in the early post-war years which could be exploited by those with the necessary skills to perceive, and take advantage of, loopholes in the system.

The importance of the entrepreneur in such a market is determined by the extent of market imperfection; profits are made by exploiting market imperfection and, by so doing, reducing its extent. During the late 1940s the complex regulatory framework governing the property sector offered considerable scope for such activity. The entrepreneurs made their fortunes by the application of relatively simple, but original, innovations designed to exploit the new conditions.

A thorough understanding of the many government controls introduced during and following the war gave two main advantages. Firstly, there were advantages *vis-à-vis* others in the property market who had a less thorough understanding and did not, therefore, appreciate the economic implications of transactions they entered into. Secondly, there was the possibility of finding loopholes which could be exploited to earn large profits by undertaking activity against the spirit, but not the letter, of legislation.

An example of the first type of advantage involved the early growth of Britain's largest property company, Land Securities. Harold Samuel recognized that a provision of the Town and Country Planning Act, 1947, that blocks of flats which had been requisitioned as offices during the war could continue as offices without the payment of any development charge, offered great scope for profitable investment activity. Samuel found that many owners, including corporate bodies, had no knowledge of this provision and were, therefore, prepared to sell such buildings at prices which reflected their much lower residential value.[59]

An example of the second type of advantage concerns another provision of the Town and Country Planning Act, 1947. While this Act placed stringent controls on development, 'errors of drafting' allowed any war-damaged building to be restored to its exact pre-war shape without any planning permission, even if it had been virtually destroyed.[60] Those who realized the importance of this provision were therefore able to undertake development at a time when development activity was otherwise strictly controlled.

There were also, of course, considerable opportunities to earn large profits by actions against both the spirit and the letter of legislation. For example, building licenses were unnecessary if a war-damaged building was certified as a 'dangerous structure' by a district surveyor. This offered considerable scope for corruption, as the border between safe and unsafe structures was vague. There is a good deal of anecdotal evidence indicating that the extent of bribery of officials during this period was considerable; for example, Marriott tells of one district surveyor who apparently received so many bribes by way of alcohol that he was seldom sober. However, Marriott concluded that there is no reason to suppose that the extent of such practices was any greater in the property sector than in any other area of business which was similarly restricted by government during this period.[61]

I.M. Kirzner has suggested that incentives for entrepreneurial activity are reduced by the imposition of a regulatory framework on a market, due to distortions in the price system which regulation brings about.[62] The above evidence suggests that entrepreneurial incentives may, in fact, be greatly enhanced by those very distortions. However, the welfare effects of entrepreneurship in such a market appear to be largely negative. Some of the innovations implemented during this period, such as the first example given above, imply a zero sum game, profit being earned at the expense of others in the market who are less well informed of the true value of their assets. Others, such as those illustrated in the second and third examples, imply a negative sum game, profits being earned by thwarting government legislation designed to direct resources according to economic priorities and social welfare considerations. The 1945 Labour Government, which had intended to bring greater equity to the property development process, instead made millionaires of the property entrepreneurs.

The entrepreneur usually faces special problems with regard to raising capital. As Casson noted, since the entrepreneur has an unorthodox view of the market it may be difficult to persuade banks and other lenders that this view is correct.[63] In this respect the property entrepreneurs were particularly fortunate, since theirs was an area which at least a few major financial institutions understood and could assess on the basis of a project's merits, rather than on their assessment of the entrepreneur's character and background.

THE FINANCIAL INSTITUTIONS AND THE PROPERTY ENTREPRENEURS

Some of the most important innovations used by property entrepreneurs during this period were connected with the manipulation of government capital issues legislation and would not have been possible without the cooperation of the financial institutions. During this period companies could only raise £10 000 (later raised to £50 000) in a single year via capital issues without permission from a government body known as the Capital Issues Committee. The Capital Issues Committee had been set up at the outbreak of war; it was used during the early post-war years as part of a range of direct controls, most of which had also been evolved during the war years, which sought to concentrate resources into export- and import-saving industries, in line with the government's most pressing short-term economic policy objective of improving the balance of payments.[64]

Capital issues controls applied to mortgage funding. However, companies were free to sell their properties to a number of individuals who could then mortgage them, raising the maximum sum allowed by the Capital Issues Committee on mortgage borrowing each year, and benefiting from the differential between the yield on the property purchase and the mortgage interest rate. The borrowing limits of individual investors were thus effectively purchased by companies in need of funds, selling properties which could only be paid for by the very mortgages which they were constrained from raising on their own account due to the legislation.

Many insurance companies that were still reluctant to buy investment property were happy to grant mortgages on such property. Individual entrepreneurs were, therefore, able to make considerable profits from the purchase of real estate that these companies would not buy as direct investments, using mortgage funds they provided. Harold Samuel, the founder of Land Securities, found that he could buy offices in the best City locations at a yield which was significantly greater than that charged on mortgage by the insurance companies, who loaned him the bulk of the necessary purchase finance. As these properties were let to highly respectable tenants, sometimes the government, they were an extremely safe investment. Leases were long, insuring an uninterrupted long-term income stream, and as rent is a prior charge on income, paid even before a company's debentures, the risk of income loss was slight. For a property let to the government, for example, the security of the investment was, in this respect, equal to that of gilts. However, while the risk of default was the same the yield was much higher.

Mortgage finance was also of tremendous importance in that it offered possibilities of gearing which magnified capital appreciation, as is neatly illustrated by the following example, given by Marriott:

> Assume a man buys a property for £100,000. He borrows £80,000 and puts up £20,000 of his own money. Inflation and demand boost the value of the property to £160,000, 60% more than it cost. The buyer has a profit

on his £20,000 of £60,000, a capital appreciation of 300%. He has only to offset this against relatively small interest charges.[65]

Rising property prices, and the fact that these were usually not reflected in company balance sheets which valued property at book cost, provided another lucrative means of property investment: the application of the sale and leaseback technique to the acquisition of entire companies. The years 1952–1956 saw the first British hostile take-over boom. Heavy corporate taxation, together with capital issue controls, had led to a reduction in the proportion of company profits distributed to shareholders from 52% of gross trading profits in 1938 to 20% in 1952. As share prices are strongly influenced by dividends, the shares of many companies had fallen substantially in relation to profits; by 1952 company profits had risen to 3.2 times their 1938 level, while dividends stood at only 1.2 times their pre-war value.[66] Companies which were particularly vulnerable were conservatively-managed concerns with large liquid, or uncommitted property,[67] assets, which could be used to generate funds for further take-overs.

Property assets were especially attractive, as funds could be raised on them, without recourse to the Capital Issues Committee, via leaseback deals. Thus, property-rich sectors, such as retailing and hotels, were particularly vulnerable to take-over. For example, in the year beginning March 1954 alone Sir Isaac Wolfson, one of the leading take-over preditors, acquired about 350 shops for his Great Universal Stores Group by buying up retailing concerns and selling their property assets to insurance companies on a leaseback basis.[68] An essential ingredient of such expansion was the existence of a market for the property assets acquired, so that their value could be easily realized. The insurance companies provided this market for leaseback property. That they failed to fully exploit their market power when making such deals was one of the most important factors behind the large number of individual fortunes made in the sector, as discussed in the conclusion to this chapter.

Legal & General was one of the most important providers of funds for property investors during this period. One of their largest sale and leaseback transactions resulted from Charles Clore's acquisition of J. Sears & Co. in 1953. Douglas Tovey, of Healey & Baker, brought the proposition to Clore. The attraction of the company was its property portfolio, with a market value of £10 million, compared to a book value of £6 million.[69] Sears was a large and successful shoe manufacturer and retailer with 920 stores, operating under several names. Its share value was little different from its 1930 level, while its property portfolio had appreciated in value significantly since the 1930s. Furthermore, its current profitability was not reflected in its share price, due to dividends being held down to about one-fifth of the company's earnings.[70] Viewed narrowly as the acquisition of a property portfolio the deal made a great deal of sense. The value of the store chain as a going concern added further to its attraction and leaseback finance allowed Clore to release the capital tied up in the property assets without losing the use of the stores.

THE FINANCIAL INSTITUTIONS AND THE PROPERTY ENTREPRENEURS

Having successfully acquired Sears via a hostile take-over bid (something that was still unusual at the time) Clore realized the value of Sears' properties by selling its High Street premises, mainly to Legal & General, on 99 year leases. Clore raised £10 million by doing so, retaining a further £3 million worth of factory, warehouse and retail properties.[71] While the company gained a great deal of capital in this way, in the long-term it lost only very limited property rights to the premises which were sold; inflation and buoyant conditions in the retailing sector soon drove market rents to levels well above the fixed rents payable on the leased-back stores.

Sale and leaseback deals played an important role in the rapid expansion of Legal & General's property portfolio after 1945, allowing the Society to invest very substantial sums in a few large transactions. For example, in 1954 alone a total of 313 properties were purchased on a leaseback basis from the Sears Group in a single deal, at a rent calculated to yield 5.5% on costs, with a single rent review at either the 50th, 60th or 70th year of the lease. In addition to Sears, deals were also done with Montague Burton Ltd, Prices Tailors, Henry's Stores, Express Dairy, New Day Furnishing Stores Ltd, Jacksons Stores and a number of other multiple retailers.

However, while the above deals were conducted on a fixed-rent basis, by the early 1950s Legal & General were already beginning to look for ways in which they could participate in the income and capital growth of property assets and development projects in which they invested. An early example of such an arrangement resulted from Legal & General's search for a new head office building. The property developer Sir Aynsley Bridgeland had begun to piece together a five acre site in the heart of the City, bounded by Queen Street, Cannon Street, Queen Victoria Street and the Mansion House, in the 1940s. This site was developed by means of two joint companies set up between Sir Aynsley and Legal & General, Legenland and Cantling, which were incorporated in 1949 and 1955, respectively. Legal & General provided long-term mortgage finance for the two companies and owned 50% of their shares. Thus, it had both an equity and fixed-interest stake in the companies. Securing such an equity stake in property deals was rarely an important objective for institutional investors during these years, though when equity participation was sought it could be obtained, usually to the great benefit of the financial institution, as is shown by the example of Clerical Medical.

In April 1947 Clerical Medical's Board agreed the provision of substantial finance to Brixton Estate Ltd.[72] A novel funding mechanism was worked out between Clerical's General Manager, Sir Andrew Rowell, and Brixton, which circumvented the Capital Issues Committee and at the same time provided Clerical with a substantial stake in Brixton's new developments. A private subsidiary company, Brixton Development Ltd, was created by Clerical Medical (which held just over 50% of its £50 000 share capital) and Brixton Estate. The Society would finance the company by way of loans up to an initial maximum of £150 000, at an initial interest rate of 3.5%, with provision for later increases. The detailed running of the company would be a matter for Brixton Estate, though the Society would have the right to nominate a majority of its Directors.[73] Ownership of a majority stake in Brixton Development allowed Clerical Medical to provide unlimited loan finance, since loans to subsidiaries were beyond the control of the Capital Issues Committee. A gentlemen's agreement was reached under which Brixton Development Ltd was to be sold to Brixton Estate when the controls ceased, at a price to be agreed at the time.

In explaining the motives behind the founding of this company, Clerical's 1949 asset investigation report stated that they were to enable Brixton Estate to expand its activities, to secure for the Society the benefit of Mr Meighar-Lovett's energy and experience, and to obtain a 'more than 50%' share in the profits resulting from Brixton Estate's new business. Brixton's managing director, Percy Meighar-Lovett, had bought a number of sites for future development at extremely low prices during the war and in the early post-war years, when building licence restrictions made it appear unlikely that it would be possible to develop them for some time. By making this arrangement the Society secured not only the benefit of Meighar-Lovett's skill as a property developer but also a stake in these sites, which were to prove valuable assets during the following years.

In the same month the Society considered a proposal by the developer W. Bernstein, whose business activities the company was already funding via mortgages. Bernstein's forte was the purchase and redevelopment of bombed-out theatres. These made good office sites, being centrally located and of suitable size. Bernstein had originally been a fur trader and claimed to have made and lost a fortune of £250 000 in that business before he was 20. Jim Pegler, who was later to become General Manager of Clerical Medical, described Bernstein's talents as a property developer as follows:

> [He] had a very good eye for a property development proposition, and what was at least equally important in those days he had a remarkable knack of getting licences for development from the appropriate Government department. We do not know how he succeeded where others failed, and rightly or wrongly we decided that provided there was no evidence of impropriety it was not our business to go sleuthing to discover his methods.[74]

Bernstein also had considerable negotiating skill. As Pegler recalled, when the Fortress Property Company later developed office blocks in Manchester, Bernstein 'made history by pursuading businessmen of the City that £1 per sq. ft. was now the going rate for first-class office accommodation there and in Liverpool'.[75] The scope for such negotiation was much greater at this time than in later decades, due to the unsophisticated nature of the property market.

Bernstein proposed that the Society purchase four war-damaged properties in conjunction with him on the basis that, immediately after the transaction, he would enter into a contract for their repurchase at a price equal to the purchase cost plus 2.5%. He also agreed to pay the Society a proportion of his net profit when he sold them and to offer it 'first refusal'. In October 1948 Clerical Medical formed the Fortress Property Co. in conjunction with Bernstein. As with Brixton Development Ltd, Clerical Medical held just over 50% of the shares. The Fortress Property Co. conducted much of its early business for the government, Bernstein's skill in dealing with government departments proving extremely useful in this area. By the end of 1949 Fortress had amassed a property portfolio valued at £788 103.

In December 1949 Clerical Medical bought Bernstein's 49% interest in the company for £162 250, making it a wholly-owned subsidiary. The company's major assets, three sites intended for government buildings, were then transferred to the Society. The profit margins realized on these investments are illustrated by the example of one site, First Avenue House, High Holborn, which Clerical sold to Percy Bilton Ltd. Even though construction work had not begun the sale realized a massive profit of £253 584 on total costs of £416 416. This rate of profit over so short a time may seem very high, but it included not only the site and work prior to construction, but also the all-important building licence.

The Society further increased its interests in property by purchasing a controlling interest in the Oriel Property Trust Ltd, a subsidiary of Shop Investments Ltd. A financing agreement was reached on the same basis as that with Brixton Estate, the Society providing loan finance to a development-orientated subsidiary of an established property company, in which it held just over half the equity. This also proved a very profitable venture, Oriel's accounts for the 15 months ending 31st December 1948 showing earnings on the ordinary capital, after property tax, of 22.4%, equivalent to an annual rate of 17.9% per annum.[76] In 1949 Clerical formed another property subsidiary, the Citadel Property Co. Ltd, in conjunction with Mr R. Graham, a furniture dealer who was able to make use of his extensive business connections in the Newcastle area to acquire properties on attractive terms. Once again the formula of majority share ownership by Clerical Medical and extensive provision of loan finance was used to fund his activities.

Most commercial property development that took place during this period of stringent development controls was permitted by government according to one of three criteria. Properties which might collapse, or otherwise endanger the public, if they did not receive attention were given top priority. 'Dangerous structure notices' were issued, enabling the owner to obtain a building licence.[77] Firms which were involved in economic activity considered essential to Britain's post-war recovery, such as manufacturers producing goods for export, were also given a high priority for any redevelopment work which was necessary to expand capacity. Finally, there was a considerable demand for new property for the government's own use. Developers built office space for government occupation under what became known as 'lessor schemes', permission for development being granted in return for the developer's agreement to let the building to the government at a pre-arranged rent.

Clerical Medical's policy of forming joint development companies with property developers allowed them to obtain lessor scheme properties as direct investments, in addition to revenue from shareholdings in, and loans to, these companies. The Society's links with the developers Percy Bilton and William Bernstein, who undertook several lessor scheme projects, were particularly important in this respect. Clerical's purchases of lessor scheme buildings proved to be extremely profitable, as was shown by its asset investigation report for 1950. Of the properties directly owned by the Society and leased, wholly or mainly, to HM Office of Works, those bought after the building had been completed and the rent had become payable, involving no construction risk, gave an average gross yield on costs of 5.1%. Two properties leased to the government which entailed a full construction risk for the Society proved to be far more profitable. Numbers 56/60 Conduit Street had been built at a cost of £185 897 and produced an estimated total rent of £13 500, offering a gross yield on costs of 7%. If valued at a 4.75% yield basis, which the Society thought appropriate, the market value of this building was about £82 000 in excess of costs. The other property in this category, Fortress House, Saville Row, was built at a total cost of £489 104, while the proceeds of its expected sale to Legal & General amounted to over £800 000.

In 1954 Clerical Medical established contact with another property development company, the Arndale Property Trust Ltd. This company had been formed by Messrs Arnold Hagenbach and Sam Chippindale. Hagenbach was a retired caterer while Chippindale was a qualified surveyor and an experienced property dealer.[78] The company specialized in rebuilding shopping centres in cooperation with local councils. It was agreed to provide Arndale with mortgage finance, up to a maximum of £450 000. Interest was set at 4.75% for the first deal and at 1% above the gross yield on 2.5% consols at the date of approval for each subsequent mortgage.

The innovative feature of this arrangement, however, was the granting of an option for Clerical Medical to purchase 133 336 ordinary shares in Arndale, at a price of seven shillings and six pence each, to be exercised in two years' time.[79] This constituted the earliest use of what was to become a very widespread equity participation technique in the early 1960s, the granting of an option to purchase a certain number of shares in a property company, over a given period, at a pre-determined price, in return for the provision of mortgage finance.

Clerical Medical had thus successfully pioneered two of the most important arrangements by which institutional investors gained an equity stake in property companies in return for providing them with capital, the joint development company[80] and the share option arrangement, several years before these techniques became widely adopted.

The Society's success in investing substantial amounts in small property development companies was in contrast to its relative failure when applying this policy to small manufacturing enterprises. Clerical provided considerable capital to the Electrical Radiological and Instrument Company (ERIC), which had made equipment for government contracts during the war and intended to break into the radio and gramophone markets in the late 1940s. This and the Society's other ventures into funding small non-property companies in the early post-war period, Cecil Tress & Co. Ltd and Tress Engineering, ended in failure and financial loss. Clerical's one successful non-property subsidiary, the P.E.G. Investment Co. Ltd, was engaged in an area that the Society understood and had experience in: financial investment. While the Society stuck to funding companies in areas in which it had expert knowledge, property and investment, it was successful. When it diversified into other areas of venture capital, involving markets and products of which it had little experience, its efforts were not profitable.

The period from 1945 to 1954 was, in general, one of lost opportunities for the insurance companies with regard to their dealings with property entrepreneurs. While they were more than willing to supply funds, on mortgage or via sale and leaseback arrangements, very few of them considered a more active role in the property boom, involving the establishment of joint companies of the type discussed above. Credit controls gave the insurance companies a great deal of market power and had they sought to take advantage of this by insisting on equity participation in the property companies they funded, and the inclusion of rent reviews when purchasing property in leaseback deals, they would have obtained a significant share of the profits of the ventures they financed. Sale and leaseback transactions involving long leases and fixed rents provided the property entrepreneurs with ample funds, but left the financial institutions with only a series of virtually fixed-interest securities, which were to depreciate steadily in value as a result of inflation and rising yields.

This apparant lack of interest in profit maximization on the part of the insurance companies occurred despite a trend towards life assurance contracts which offered clients benefits which were directly linked to returns from invested funds. A survey of the 30 largest life assurance companies, conducted during 1956, showed that by the end of 1953 two-thirds of the companies examined had at least 40% of policies, by value, in 'with-profits' contracts.[81]

The explanation for the lack of a more 'entrepreneurial' approach to property development funding on the part of most insurance companies appears to be linked to their perceptions of future economic conditions. While a few institutional investors appreciated the dangers posed by inflation many still operated in a static world, where inflation was either ignored or assumed to be only a temporary phenomenon. There was also a considerable degree of inertia regarding the adaption of established mechanisms for conducting property transactions to take account of inflation, and rising rents and property values. Institutional investment managers generally accepted the basic terms on which property transactions were offered, terms which had been designed by the property companies and their agents to maximize their financial gain. Negotiations generally centred around initial yields rather than the possibility of securing an equity stake for the financial institution, as most investment managers were not yet thinking along these lines. Thus, some part of the high returns accruing to individual entrepreneurs from real estate investment between 1945 and 1954 resulted from a lack of entrepreneurship among the financial institutions which financed the post-war property investment boom.

Finally, it is necessary to consider the social costs and benefits of the relationship between the financial institutions and the property sector during this period. At a time when capital issues were strictly controlled by government, the use of the sale and leaseback arrangement, and the other techniques outlined above, acted to reduce the effectiveness of the government's investment control policy. For the property development sector this was of limited importance, since building licenses and other controls proved effective restraints on development activity which did not meet with official approval. However, for retailing chains, and other 'property-rich' companies, leaseback finance offered a means of raising money that was beyond the reach of the Capital Issues Committee.

In addition to channelling capital, and therefore resources, to sectors which the government gave a low priority in its recovery policy, these techniques earned fortunes for entrepreneurs whose only economic function was to exploit loopholes in the regulatory framework, mortgaging properties that they had purchased from capital-starved occupiers who were not themselves able to raise substantial finance in this way. As such, some of the profits of the early property tycoons could be best described as 'rent-seeking behaviour'[82] rather than entrepreneurship. However, responsibility for this situation must lie with government and the officials who designed the capital issues controls. Omitting leaseback finance from these controls left such a gaping loophole in the regulatory framework that a substantial flow of funds to otherwise restricted sectors via this mechanism was inevitable.

REFERENCES AND NOTES

1. Clerical Medical (1941) Investigation into the Society's Assets. These payments were treated as capital expenses in the accounts. Had they been taken against income, rather than capital, they would have had a greater effect on net yields.
2. B.P. Whitehouse (1964) *Partners in Property*, Birn, Shaw, London, p. 13.
3. J.M. Keyworth (1990) *Cabbages and Things*. Privately published, p.157.
4. O. Marriott (1967) *The Property Boom*, Hamish Hamilton, London, p. 43.
5. O. Marriott (1967) *The Property Boom*, Hamish Hamilton, London, p. 45.
6. A.H. Rowell for Clerical Medical (1942) Note regarding the Society's investment policy.
7. Clerical Medical (1945) Board minutes, 7 Feb.
8. J.B. Cullingworth (1988) *Town and Country Planning in Britain*, 10th edn, Unwin Hyman, London, p. 5.

REFERENCES AND NOTES

9. F.J. Osborn and A. Whittick (1977) *New Towns: Their Origins, Achievements, and Progress*, L. Hill, London, p. 44.
10. In Scotland these powers were retained by the Secretary of State for Scotland.
11. J. Hasegawa (1992) *Replanning the Blitzed City Centre*, Open University Press, Buckingham, p. 11.
12. J. Hasegawa (1992) *Replanning the Blitzed City Centre*, Open University Press, Buckingham, p. 32.
13. J. Hasegawa (1992) *Replanning the Blitzed City Centre*, Open University Press, Buckingham, p. 103.
14. J. Hasegawa (1992) *Replanning the Blitzed City Centre*, Open University Press, Buckingham, p. 41.
15. J. Hasegawa (1992) *Replanning the Blitzed City Centre*, Open University Press, Buckingham, pp. 119–20.
16. J. Hasegawa (1992) *Replanning the Blitzed City Centre*, Open University Press, Buckingham, pp. 93 and 104.
17. C. Barnett (1986) *The Audit of War*, Macmillan, London.
18. J.B. Cullingworth (1988) *Town and Country Planning in Britain*, 10th edn, Unwin Hyman, London, p. 16.
19. P. Cowen *et al.* (1969) *The Office: A Facet of Urban Growth*, Heinemann, London, p. 162.
20. B.P. Whitehouse (1964) *Partners in Property*, Birn, Shaw, London, p. 13.
21. D. Cadman and A. Catalano (1983) *Property Development in the UK – Evolution and Change*, E & FN Spon, London, p. 4.
22. See Table 10.19.
23. See Table 10.2.
24. The division of investors into separate categories, and the assignment of 'risk ratios' to their investment policies, were undertaken as part of the analysis of the data for the present study and were not included in the original ledger.
25. The weightings assigned are as follows (a larger number indicating a greater degree of risk): ground rents, 1; prime shops, 2; offices, 3; secondary shops, housing estates, flats, vacant and reversionary shops, factories, 4; small houses, single shops, 5; speculations, sites, 6. No weighting was assigned to short leaseholds, as this would vary according to the type of property on which the lease was secured. The weightings were determined in order to indicate differences in the order of risk, rather than the magnitude of risk.
26. This evidence included text books on property investment, such as R.B. Sunnocks (1935) *Investment in Property*, Banbury Publishing Co., London and R.M. Lester (1937) *Property Investment*, Pitman, London, plus evidence from the records of the institutional investors examined during the course of this study.
27. See Tables 10.1 and 10.5.

28. Measured according to the number of investors rather than the amount invested.
29. Sources: 1948–54, M.C. Fleming (1980) Construction and the related professions, in *Reviews of United Kingdom Statistical Sources, Vol. 12* (ed. W.F. Maunder), Pergamon Press, Oxford; 1955–64, CSO (various issues) Annual Abstract of Statistics.
30. See section 5.3.
31. Legal & General (1951) *Investment Policy*, Board paper, June.
32. A.H. Rowell for Clerical Medical (1949) *Notes on Investment Policy*, a memorandum, 14th Dec.
33. These would be mainly composed of property developers and speculators.
34. Clerical Medical (1947) *An Appreciation Of The Society's Investment Position*, Aug.
35. NPI (1948) Assets Report.
36. NPI (1951) Assets Report.
37. See Table 10.7. This decline was mainly due to the absence of new purchases, rather than substantial sales.
38. Church Commissioners (1949) Annual Report, 31 March, p. 4.
39. O. Marriott (1967) *The Property Boom*, Hamish Hamilton, London, p. 81.
40. Mortimer Warren for the Church Commissioners (Sept. 1948) A report on the financial position and related matters, in *Board of Governors Minute Book, Vol. 1*, 20 Oct. 1948.
41. Church Commissioners (1951) Board of Governors minutes, 11 April.
42. Church Commissioners (1950) Annual Report, 31 March, pp. 8–9.
43. Church Commissioners (c. 1965) *The Church Commissioners Index* – an historical review of investment policy. Sales were, in the event, confined to the Commissioners' lower-quality estates, or those that did not form part of large consolidated holdings.
44. W. Ashworth (1986) *The History of the British Coal Industry, Vol. 5*, Clarendon, Oxford, p. 536.
45. PRO (1949) Coal 23.347, note dated 30 Aug.
46. PRO (1949) Coal 23.347, note dated 1 Sept.
47. PRO (1950) Coal 23.347, note dated 17 Aug.
48. PRO (1952) Coal 23:347, note from H.W. Naish to J. Davidson, 29 July.
49. PRO (1954) Coal 23:347, letter from J. Davidson to H.W. Naish, 27 May.
50. PRO (1954) Coal 23:347, memorandum by A. Orchard-Lisle, July.
51. PRO (1954) Coal 23:347, note from H.W. Naish to J. Davidson, 7 July.
52. J.M. Keyworth (1990) *Cabbages and Things*. Privately published, p. 163.
53. O.Marriott (1967) *The Property Boom*, Hamish Hamilton, London, p. 2.
54. See T.A.B. Corley (1993) The entrepreneur: the central issue in Business history?, in *Entrepreneurship, Networks and Modern Business* (eds J. Brown and M.B. Rose), Manchester University Press, Manchester, pp. 11–29.
55. F.H. Knight (1921) *Risk, Uncertainty and Profit*, Houghton Mifflin, Boston.

REFERENCES AND NOTES

56. J.A. Schumpeter (1934) *The Theory of Economic Development*, Harvard University Press, Cambridge, MA.
57. M. Casson (1982) *The Entrepreneur*, Robertson, Oxford.
58. S. Aris (1970) *The Jews in Business*, Cape, London, pp. 186–7.
59. O. Marriott (1967) *The Property Boom*, Hamish Hamilton, London, p. 49.
60. O. Marriott (1967) *The Property Boom*, Hamish Hamilton, London, p. 4.
61. O. Marriott (1967) *The Property Boom*, Hamish Hamilton, London, p. 48.
62. I.M. Kirzner (1980) The primacy of entrepreneurial discovery, in *The Prime Mover of Progress: The Entrepreneur in Capitalism and Socialism. Papers on the Role of the Entrepreneur* (eds I.M. Kirzner *et al.*), IEA, London, pp. 23–6.
63. M. Casson (1982) *The Entrepreneur*, Robertson, Oxford, p. 332.
64. H. Mercer, N. Rollings and J.D. Tomlinson (eds) (1991) *Labour Governments and Private Industry: The Experience of 1945–51*, Edinburgh University Press, Edinburgh, pp. 37 and 76.
65. O. Marriott (1967) *The Property Boom*, Hamish Hamilton, London, p. 3.
66. G. Bull and A. Vice (1961) *Bid for Power*, 3rd edn, Elek, London, p. 30.
67. Property assets which had not already been used as security for loans.
68. G. Bull and A. Vice (1961) *Bid for Power*, 3rd edn, Elek, London, p. 97.
69. D. Clutterbuck and M. Devine (1987) *Clore: The Man and His Millions*, Weidenfeld & Nicolson, London, p. 64.
70. W. Mennell (1962) *Takeover: The Growth of Monopoly Capital in Britain, 1951–61*, London, p. 31.
71. D. Clutterbuck and M. Devine (1987) *Clore: The Man and His Millions*, Weidenfeld & Nicolson, London, p. 70.
72. See section 4.3 for details of the earlier financial links between these two companies.
73. Clerical Medical (1947) Board minutes, 2 April.
74. J. Pegler (1981) Unpublished memoirs, p. 27.
75. J. Pegler (1981) Unpublished memoirs, p. 28.
76. Clerical Medical (1948) Asset Investigation Report.
77. J. Rose (1985) *The Dynamics of Urban Property Development*, p. 150.
78. J. Pegler (1981) Unpublished memoirs, p. 43.
79. Clerical Medical (1954) Board minutes, 5 March.
80. While the Prudential had used this technique to fund a hotel development prior to the Second World War, Clerical Medical is the first known insurance company to form a joint development subsidiary with a property company as a vehicle for conducting a number of development projects, rather than as a means of funding a single development to assist a potential occupier of the property in question.
81. J. Johnson and G.W. Murphy (1956–57) The growth of life assurance in the U.K. since 1880. *Transactions of the Manchester Statistical Society*, **1956–7**, 45.
82. An economic term for behaviour which aims to earn abnormal profit by the creation of a monopoly, or other economic activity which increases profits by restricting, or exploiting restrictions on, competitive conditions.

6 | The property boom 1955–1964

6.1 INTRODUCTION

The Conservative government which came to power in 1951 reacted against its predecessor's emphasis on comprehensive town and country planning, reversing, or toning-down, much of the planning legislation of the 1940s.[1] This was part of a wider move away from economic planning; the complex system of direct controls developed during the war years, which had already been substantially reduced by the time Labour left office, was dismantled and economic management was left largely in the hands of fiscal and monetary policy. In December 1952 Harold Macmillan, when introducing the bill which became the Town and Country Planning Act, 1953, set the new tone, stating 'The people whom the Government must help are those who do things: the developers, the people who create wealth ...'.[2]

The Town and Country Planning Act, 1953 removed the 100% development levy introduced by the 1947 Act and in November 1954 the other major impediment to property development – building licenses – were abolished, following several years of their gradual relaxation. The new political climate also tilted the balance of town centre reconstruction policy in favour of the private sector. For example, when, in 1954, Coventry Council decided to develop a section of its shopping precinct itself, rather than award the contract to the shop developer Ravenseft, it was overruled by the Minister.[3]

The abolition of building licenses marked the end of the era of government development controls after the war and almost a decade of post-war restrictions. Meanwhile, building materials shortages, which would have restricted development during the previous decade even in the absence of controls, had substantially eased. A boom in property development resulted which was to last for 10 years, until legislation once again restricted office development, in and around London, in 1964.

The boom was based on institutional finance and individual talent. Many fortunes were made in property during this decade, by entrepreneurs who often had little initial capital and relied on the financial institutions to fund their oper-

PROPERTY INVESTMENT AND THE CAPITAL MARKET 1955–1964

ations. This chapter examines the evolution of the property investment and development markets during the boom years, and attempts to discover why the profits made from commercial property during this period were so high, and persisted for so long, in a competitive market with few visible barriers to entry.

6.2 PROPERTY INVESTMENT AND THE CAPITAL MARKET 1955–1964

The property boom years saw a substantial rise in the level of both direct institutional investment in property and the provision of development finance. Direct insurance company investment in land, property and ground rents averaged £576 million in real (1990) prices during 1955–1964, an increase of 81.4% compared with the average level for 1946–1954.[4] Pension fund direct property investment rose even more dramatically, average net investment during this period amounting to £138 million, a 496% increase over the average for the previous nine years. In addition, a very substantial volume of institutional funds was channelled to the property development market via share acquisition, and mortgage and debenture finance.

This expansion in institutional property investment occurred alongside rapid growth in the inflow of funds to the institutions. While the real level of insurance company property investment was far higher during this period than for 1946–1954, the average proportion of new insurance company funds invested in this sector had increased much more modestly, from 9.7 to 11%. Other major factors behind the expansion of institutional property investment during this decade were an increase in demand for investment funds, as a result of the lifting of property development restrictions, and the transformation of property from a fixed-interest security to an equity investment, at a time when the institutions were becoming increasingly aware of the effects of persistent inflation on the real value of fixed-interest assets.

By the mid-1950s the dangers of persistent inflation were already being noted by a few astute investors. With inflation reaching levels of 4–5% per year the real value of a property let on a 999 year lease at a fixed rental would be far less, in 20 years' time, than one let at prevailing market rents. Such considerations prompted the most important innovation in the property investment market during the post-war years, the rent review. Rent review clauses provided for upward-only revisions in the rent paid on buildings at regular intervals specified within leases. A few isolated instances of transactions which had rent review-type clauses can be found before the mid-1950s; for example, in November 1951 a syndicate of insurance companies, comprised of Eagle Star, Friends Provident, Law Union & Rock and the Atlas Insurance Co., considered a proposal to purchase a Lewisham department store on a leaseback basis. The property was to be let on a full repairing lease for 100 years with a rental that was fixed for the first 42 years 'and thereafter by agreement or arbitration'.[05] However, such clauses were rare and were not part of a concerted attempt to build an equity element into property transactions.

The introduction of the rent review proper can almost certainly be dated to 1955. By May of that year it was still virtually unheard of; the minutes of a partners meeting of Hillier Parker on the 23rd May 1955 recorded that:

> Mr Edgson enquired what instances the partners may have met of leases at rack rents with rental tied to the value of gold or other monetary standard and with powers for review of the rent on changes taking place in such value standards. A client wished such an arrangement. The partners' experience all pointed to the idea having been found impracticable.[6]

By the end of the year, however, at least one of Britain's leading financial institutions, the Church Commissioners, was seeking rent review clauses for its new property acquisitions. Two of the first voices to be raised in favour of rent reviews as a general principle of institutional property investment were those of George Bridge of Legal & General and Eric Young of the NCB Pension Fund. The two combined forces to persuade property developers and retailers to allow rent review clauses to be built into leases.[7] Their main difficulty was that most of the other funds which were active in the property market, though few in number, were happy to continue doing business on the current prevailing terms of 99, or even 999, year leases, with no rent review. Progress was eventually made, however, first with rent reviews at the 33rd and 66th years of leases, and later at progressively shorter intervals as the practice became more widespread and the effects of inflation more widely appreciated. George Bridge coined a phrase that was frequently quoted in the press and elsewhere, summing up the advantages of property, with rent reviews, as an investment, 'as a hedge against inflation, bricks and mortar reign supreme'.

A Board discussion of Legal & General's property investment policy, conducted in March 1955, illustrates the main reason why the Society had difficulty in introducing the rent review in the mid-1950s; many of their competitors were not yet concerned about purchasing property on the customary long, fixed-rental, leases:

> Mr Raynes [the Society's Actuary] said that we should be more choosy in the type of property we buy, and he was not at all enamoured by the long leases showing returns of 5¼% or 5½%. The reply to this was that we had to be guided by what our competitors would pay ... The case of Rego Clothiers was mentioned where Great Universal Stores had offered the business to us on a 99 year lease and would not consider a shorter term. We eventually agreed to purchase this to show 5½% return on these terms. Mr Raynes still expressed his disapproval. The General Manager replied that it might not be wise to let a deal like this go by. Mr Wykes was in favour of such transactions going through because it also brought a considerable amount of fire insurance. Even at 5¼% it showed up better than investment in Government stock and it was decided that each transaction should be treated on its own merits.[8]

However, during the following years, Legal & General was successful in getting rent review clauses built into the leases of its new property acquisitions. In 1959 an analysis of Legal & General's property portfolio was undertaken, dividing all properties into 'equity' (properties where rents could be increased within 35 years by the renewal of leases or rent review), 'non-equity' (properties where the rent could not be increased until 35–50 years' time) and 'debentures' (properties offering no prospect of rent increase for more than 50 years). The book value of properties in each of the three categories was £39 453 000, £12 939 000 and £27 948 000, respectively – almost half the portfolio offering the prospect of increased rents within 35 years.[9] Over the next few years the proportion of funds invested in non-equity properties diminished and the yield differential between equity and non-equity property widened; during 1963 purchases of equity property amounted to almost £11.5 million, with an initial yield of 6.82%, while just over £1 million was invested in non-equity property at an initial yield of 8.41%.

Legal & General also began to invest in overseas property during the 1950s. This formed part of a wider policy of investing insurance funds in those countries in which they had originated, a policy also followed by other insurance companies with extensive overseas interests. This overcame problems regarding currency restrictions and earned favourable publicity and goodwill in the countries concerned. Legal & General was particularly active in property investment in Australia and several African countries; techniques which had proved successful in the British property market, such as the sale and leaseback, were introduced to Southern Africa by the Society.

By the end of the 1950s insurance company property holdings were still concentrated among a few companies with well-established property departments; George Bridge estimated that in 1960 more than half of total insurance company property assets were held by four companies.[10] The two largest of these, Legal & General and the Prudential, had amassed vast property empires during the 15 years following the Second World War; in 1960 Arthur Green stated that they were probably the two largest property holders in Britain, having more valuable property portfolios than even those of the Crown and the Church.[11] Managing their massive property holdings necessitated the establishment of large and highly specialized property departments; Legal & General's Estates Department, the administrative structure of which is shown in Figure 6.1, employed 97 staff by 1957.

Until the early 1960s direct property investment continued to offer a yield which was in excess of that on any other major category of insurance company asset, other than preference shares, as indicated by an analysis undertaken by Legal & General, reproduced in Table 6.1. However, as the equity element in investment property grew the yield differential between property and fixed-interest stocks narrowed, becoming negative during the early 1960s.[12] The income growth potential of investment property had become sufficiently great to more than compensate for its greater risk, lower marketability, indivisibility, higher management costs and other disadvantages compared to gilts.

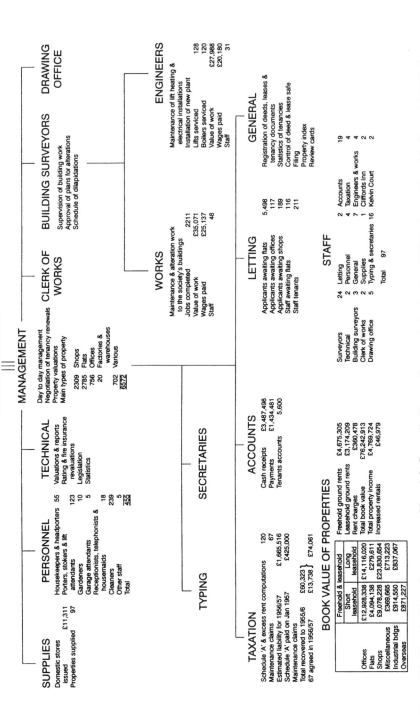

Figure 6.1 The administrative structure of Legal & General's Estates Dept. at the beginning of 1957. Source: Legal & General (1957) Estates Dept Annual Report.

Table 6.1 New investment by Legal & General during 1961

	Investment (new) (£)	Sales and redemptions (£)	Net increase £	–Average Yield–	
				1961 %	1960 %
British Govt. Securities	–	290 000	– 290 000	–	6.075
Municipal and County Sec.'s	175 000	995 000	– 820 000	6.175	5.863
Commonwealth Govt. etc.	770 000	165 000	605 000	6.188	6.175
Debentures	5 960 000	280 000	5 680 000	6.563	6.313
Preference shares	43 000	103 000	– 60 000	7.063	6.775
Ordinary shares	8 020 000	1 010 000	7 010 000	4.250	3.613
Total Stock Exchange	14 968 000	2 843 000	12 125 000	5.300	4.367
Mortgages	15 508 000	1 543 000	13 965 000	6.388	5.988
Property	17 178 000	160 000	17 018 000	6.725	6.325
Total	47 654 000	4 546 000	43 108 000	6.163	5.542

Source: Legal & General (1962) Estate Dept. Report.

Legal & General's investment policy was constrained by the rapid inflow of new funds. Having a sufficiently difficult task in placing new money in good investments the Society, together with most other insurance companies, had little scope for the rationalization of its existing portfolio by the sale of assets and reinvestment in more lucrative areas. The Church Commissioners, in contrast, had a virtually static fund, new investment being contingent on the sale of existing holdings. During this period it was at the forefront of attempts to introduce an equity element into property investment in order to safeguard the real value of its limited funds.

The Commissioners' property expertise had been augmented in 1954 when Sir Malcolm Trustram Eve (later Lord Silsoe), a lawyer who had been chairman of the War Damage Commission and the Central Land Board, became First Church Estates Commissioner, bringing with him 'a reputation as one of the shrewdest business brains in the country'.[13] Trustram Eve was a keen advocate of the policy initiated by Warren in the late 1940s, involving selling off the Commissioners' fixed-interest securities and low-yielding property assets, and reinvesting the proceeds in ordinary shares and commercial property. His appointment led to an acceleration of this process.

In 1955 the Commissioners were one of the first investors to introduce the rent review. An Estates and Finance Committee Paper, dated 30th September 1955, suggested that:

... attempts should be made to obtain an interest in the equity of some of the new buildings now being erected, and by taking advantage of the expected increase in rental values insure against the expected depreciation in the value of money in the next 100 years ...

The proposal is that the Commissioners should purchase a new building and grant back a long lease at a rent at least twice covered by the full net income of the property, and that, at intervals of say 25 years, the rent should be increased by a proportion of the excess of the then full net income over the full net income at the time of purchase. In this way the Commissioners will obtain a share in the increased rents received by their lessee over the period of his long lease.[14]

Within a few months the rent review, which was hitherto virtually unknown in the property world, had become an important feature of the Commissioners' investment policy. In the Commissioners' Annual Report for the year to 31st March 1956 it was stated that:

During the year the Commissioners widened the scope of this type of [property] investment ... they endeavoured to obtain a greater share in the anticipated increase in rental values in the future by granting shorter leases, or by granting rent revision clauses in long leases. They also purchased some properties let on short leases where a reversionary increase in income may be expected.[15]

In January 1958 it was decided to dispose of those buildings bought earlier for investment reasons and let on 999 year leases at fixed rents, the proceeds being reinvested in property with rent review clauses.[16]

In their 1978 study of landownership in Great Britain, *Capital and Land*, Massey and Catalano identified three forms of large-scale commercial landownership in Britain: former landed property, industrial landownership and financial landownership. Former landed property, under which they included the Church Commissioners, was distinguished from the other two categories as:

Land for this group is not one sector for investment, like any other, chosen simply on the basis of its potential economic return, and with no commitment beyond that, ... but is an integral part of a wider role, in which considerations other than 'return on capital invested' are of real importance ... 'Pulling out of land' for this group would not be a question of simply transferring their money capital. It would also involve a change in social role.[17]

While the above discussion of the Commissioners' investment activities would appear to contradict this view, the Commissioners' policy towards their agricultural holdings suggests that Massey and Catalano's thesis is valid. The Commissioners' thinking on this subject is illustrated in a memorandum by Sir

Three leading inter-war shop developers, Edward Lotery (left), Simon Marks (centre) and Montague Burton (right).

Top, Bush House, London's first 'skyscraper'; bottom, a typical Montague Burton shop development, including two shops let to other tenants and a first-floor billiard hall.

Left, The Pantheon Oxford Street, Marks & Spencer's largest inter-war store development. Copyright Marks and Spencer PLC; right, the Odeon Southsea; the expansion of the Odeon cinemas was partly funded via property deals with insurance companies. Copyright Portsmouth City Council.

Left, Edward Erdman, one of the leading inter-war commercial estate agents; right, Sir Andrew Rowell, General Manager of Clerical Medical, who developed some of the key property company equity finance innovations during the early post-war years.

Two of the leaders of the early '50s takeover boom, Charles Clore (left) and Isaac Wolfson (right).

Britain's first precinct shopping centre, Coventry.

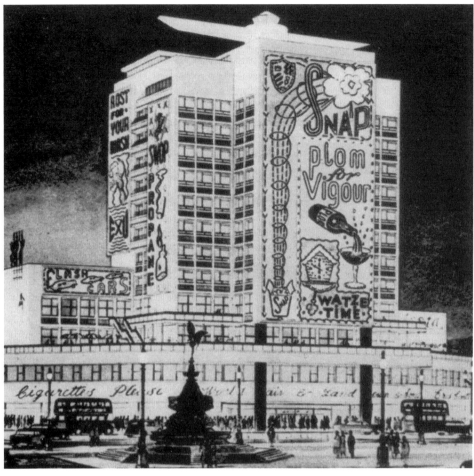

Jack Cotton's ill-fated proposal for the Monico site, Piccadilly Circus.

Left, Jack Rose, a major post-war office developer and author of several books on property valuation and the property industry. Centre, Max Rayne, with whom the Church Commissioners set up their first property development company. Right, Harold Samuel, the most successful property tycoon of the post-war era, who built up Land Securities as Britain's largest property company.

Canary Wharf, the most ambitious property development project launched in Britain this century.

The merchant developers. Top left, Godfrey Bradman (Rosehaugh); top right, Stuart Lipton (Stanhope and Greycoat); bottom left, Tony Clegg (Mountleigh); bottom right, Trevor Osborne (Speyhawk).

Richard Rogers' Lloyds of London Building

Terry Farrell's Embankment Place.

Left, Sydney Mason, Chairman of Hammersons; right, Hammersons' Brent Cross development, Britain's first regional shopping centre.

Aubrey Orchard-Lisle (above left) and Douglas Tovey (above right) of Healey & Baker, who set up some of the most important property deals of the post-war boom on behalf of their clients.

Opposite, Centre Point, Britain's most notorious 1960s office block, together with its designer, Richard Siefert (left), the leading office architect of the property boom. Centre Point photo by Peter Altman.

Two pathbreaking developments. Left, Fountain House, Fenchurch Street, which marked a new style in post-war office design (reproduced with permission of the Guildhall Library); right, a model of the Elephant and Castle Shopping Centre, together with its architect, Paul Boissevain.

PROPERTY INVESTMENT AND THE CAPITAL MARKET 1955-1964

Mortimer Warren, dated 17th November 1959. At the end of March 1959 their agricultural landholdings amounted to over 216 000 acres. It was estimated that, while they offered a yield on book value of 3.9%, the yield on estimated market value would be less than this, possibly below 3%. Even when calculated on the most favourable yield basis possible, agricultural land compared unfavourably with the Commissioners' other assets, as is shown in Table 6.2.

Table 6.2 The Church Commissioners' assets as at 31st March 1959

	Book value (£M)	Yield (%)	Book value (%)	Income (%)
Stock exchange securities	122.94	5.47	56.4	55.4
Mortgages	28.02	5.00	12.8	11.5
Agricultural land	11.67	3.90	5.3	3.7
Farms in hand and woods	0.59	4.15	0.3	0.2
Urban property	54.93	6.59	25.2	29.2

Source: Church Commissioners (1959) Investments in agricultural estates, Memorandum by Mortimer Warren, Estate and Finance Committee papers, 17 Nov., C.C. 95055, part 18/19.

Despite rising rents in recent years, which were expected to be accelerated by the recent Agriculture Act, it was stated that:

> On financial considerations alone, these investments in agricultural land have for many years been poor value, and the income of the general fund could be much increased if they were sold, with reinvestment of the proceeds in urban properties having an equity element, and in Stock Exchange shares.[18]

It was estimated that income from these holdings might be as much as doubled by their sale and the reinvestment of the proceeds. Nor was retention justified on capital appreciation grounds:

> It has often been remarked that agricultural property is a sound hedge against inflation. The history in England of average capital values and average rents per acre over the past eighty years does not lend much support to this view. It is probable that urban properties with an equity element are a better hedge against inflation, both for capital and income.[18]

Financial considerations merited the sale of the Commissioners' agricultural holdings. However, these were not the only considerations. It was stated in the same paper that:

Although the financial arguments would indicate the desirability of selling all land, there are important considerations to the contrary. There is the whole history of inherited Church lands, and the accumulated wealth which has stemmed from that inheritance; there is the prestige enjoyed by the Church from holding these lands; there is the national duty to support agriculture.[18]

In 1962 a further paper on the Commissioners' agricultural landholdings, by D.A. Collenette,[19] echoed Warren's views.[20] He stated that while the investment performance of agricultural land was markedly inferior to that for the Commissioners' other assets, a policy of comprehensive sales was not advocated, since:

It would not be practicable for all the agricultural property to be sold – however advantageous from the standpoint of income. The implications are obvious; the ancient link between the Church and the land; flooding the market; 'letting down' British agriculture and the landlord–tenant system; and so on.[20]

Despite allowing non-monetary factors to influence their policy towards agricultural land the Commissioners did achieve substantial success in transferring the bulk of their other assets from low-yielding or fixed-interest investments to equities and commercial property with an equity element. An Estates and Finance Committee paper of November 1961 divided the Commissioners' commercial property holdings into three categories:

(a): Property let on 999 year leases, which produced a net income of £55,000.

(b): Property let on leases with more than 50 years unexpired, excluding those in category (a), producing a net income of £1,090,000.

(c): Property let on short leases, or long leases with rent reviews, producing a net income of £1,850,000.[21]

A programme of selling the remaining properties in categories (a) and (b) was commenced in January 1962, and was virtually completed by the end of March 1963.[22] The money raised was used to finance property development via CEDIC.[23] By 1964, when the Brown Ban on office development in and around London led to a reduction in property development activity, the Church Commissioners had achieved substantial success in increasing income from their property portfolio. During the 10 years to 1963/64 net capital invested in the Commissioners' estates had risen by less than 5% of the book value of the portfolio, while net income had risen by 78%. Meanwhile, the number of urban properties owned had fallen from 50 740 in 1954 to 6800 10 years later.[24]

INSTITUTIONAL INVESTORS AND THE PROPERTY DEVELOPERS 1955-1964

The Church Commissioners and Legal & General were among the first institutions to move towards building an equity element into property acquisitions, something which did not become common among the financial institutions until about 1960. For example, as late as 1959 Eagle Star bought properties on leaseback at fixed rents, and even gave the vendors an option to repurchase over a certain time, usually before the 15th year of the lease, at a price only slightly above the purchase cost. Such options were usually exercised as property values escalated during the 1960s.

The investment performance of property during the years 1955-1964 was much better than that for gilts, but poorer than that for ordinary shares, the average return for the three major asset classes being 7.0, 0.7 and 10.2%, respectively.[25] Despite rapid rises in market rents the long periods over which rental income was static prevented institutional investors from benefiting from these increases; Jack Rose estimated that during the years 1954-1964 City office rents increased at an average compound rate of approximately 8% per annum,[26] while the capital return on investment property averaged only 1.4% per annum.[25] However, by the end of this period the average time to rent review for institutional property portfolios had fallen to about one-third of its 1954 level – from about 75 to 25 years.[27] Property's attractions as an equity asset were at last beginning to rival those of shares.

6.3 INSTITUTIONAL INVESTORS AND THE PROPERTY DEVELOPERS 1955-1964

In contrast to the immediate post-war years the decade from 1955 saw an unprecedented boom in construction activity, as the imbalance between the supply and demand for commercial property was finally corrected. In addition to pent-up demand for property, arising from the lack of new development during the previous 15 years, the continued growth of the service sector, which expanded by 24.2% in terms of real GDP from 1955 to 1964, ensured a high level of demand for newly-developed property. During 1955, the first year following the removal of building license controls, the value of commercial property construction output amounted to £1032 million in real (1980) prices, compared to £109 million in the previous year. For the 1955-1964 decade as a whole the value of commercial construction output averaged £1847 million, compared to an average figure of only £90 million for the previous nine years.[28]

Growth was particularly rapid in the City office market; many of the country's largest companies sought prestige headquarters at the heart of Britain's capital at a time when clerical and managerial employment were undergoing rapid expansion. Office building in the London region accounted for an estimated 80% of all office floorspace built in England and Wales from 1945 to 1962, as is shown in Figure 6.2.[29] The 1955-1964 office building boom saw a mushrooming of City office space; it is estimated that from 1948-63 the net floorspace of Central London offices had risen by 50% compared to its 1939 level,[30] in addition to building which replaced offices destroyed during the war.

Figure 6.2 The distribution of post-war office space in England and Wales (over 200 000 sq. ft. built or under construction) 1945–1962. Source: S. Taylor (1966) A study of post-war office developments. *Journal of the Town Planning Institute,* **52**, 56.

Those entrepreneurs who made fortunes from meeting this demand for office space were generally those who had founded their property empires during the 1945–1954 'property investment boom'. The skills required by the property entrepreneur during the development boom were, like those needed during the investment boom, primarily dealing skills, together with an understanding of the locational factors which determine the value of a site. The technical aspects of development could be dealt with by hiring architects and other specialist personnel, in a sellers' market which generally provided basic office buildings designed to maximize floorspace within planning restraints, and minimize costs. Those entrepreneurs who had proved successful property speculators during the 1945–1954 investment boom therefore generally found little difficulty in adapting to the new conditions.

The office boom made fortunes for a large number of individuals;[31] it also led to the property developer being cast in the popular mind as the arch villain of British capitalism. There are a number of factors which contributed to the property developers' low public esteem. Firstly, there were a number of flaws and loopholes in planning controls, which the profit-maximizing developer naturally exploited. The most serious example of legislation which did not match up to its task concerned the system of limiting building densities via 'plot ratios'. Plot ratios, which determined the legally permissable maximum floor area of buildings that could be put on a site of given size, were set by the London County Council under a scheme designed by the planner Professor, later Lord, Holford, in order to control the maximum density of workers in each area of the capital.[32] They were intended as a set of theoretical limits on the density of workers in given areas, none of their designers apparently considering that developers would ever want to build as much as the plot ratio allowed.[33]

This was, of course, just what the developer was most interested in achieving; he often succeeded in this task via the use of architects that specialized in constructing buildings to meet such criteria. The most successful of these architects, Richard Seifert, built up a practice which had a turnover of £20–30 million by the mid-1960s[34] and designed London's most enduring, and notorious, monument to the property boom, Centre Point.

A further legislative loophole which exacerbated the problems caused by defects in the plot ratio system was the notorious 'Third Schedule'. Under Schedule Three of the Town and Country Planning Act, 1947 it was possible to enlarge a building by up to 10% of its cubic content. The intention behind this was to allow owners to make minor improvements to existing buildings without having to pay any development charge. However, it was used for a very different purpose by developers, who demolished existing buildings and built new ones with 10% greater gross volume. Since old buildings generally had higher ceilings and a larger proportion of space taken up by internal walls, passages and staircases than their steel-framed 1950s counterparts, their replacement with up-to-date buildings, with 10% greater cubic capacity, added considerably more than 10% to the floor area and number of workers accommodated.[35] It also allowed developers to exceed plot ratio limits by immediately adding 10% extra to new buildings. It was not until 1963 that the Third Schedule was revoked; together with the plot ratio controls it allowed developers to treat as minimums what London's planners had regarded as theoretical maximum densities. To criticize the developer for acting in this way was, as Marriott noted, 'like criticising a giraffe for having a long neck'.[36]

Another source of bad publicity for the development industry concerned the demolition of historic buildings to make way for new office blocks. The 1950s saw a substantial increase in public concern regarding the loss of Britain's building heritage. This concern had a long pedigree, the Society for the Protection of Ancient Buildings (SPAB) having been founded by William Morris as early as 1877, its protection being extended to Georgian London with the establishment of the Georgian Group as an offshoot of SPAB in 1935.

The demolition of much of historic London which had survived Hitler's bombs during the mid-1950s led to further moves to fight indiscriminate redevelopment. For example, in 1957 the developer Felix Fenston was drawn into the public eye when the actress Vivian Leigh protested against his plans to redevelop the St James Theatre as an office block by interrupting a debate on the proposed development in the House of Lords (leading to her removal by Black Rod) and by holding a protest in Trafalgar Square.[37] Her campaign failed to save the theatre, but served to focus public attention on the negative effects of London's growing office boom.

The Civic Trust was also established in 1957 – an urban equivalent of the Council for the Protection of Rural England. This organization successfully led the battle to prevent Jack Cotton developing his controversial Monico Piccadilly Circus project, one of the conservation lobby's first major victories against the developers.[38] The following year saw the establishment of the Victorian Society, marking a substantial reduction in the vintage of buildings deemed worthy of protection.

However, conservation considerations still carried little weight in official circles. Indeed, the two most notorious demolitions of the era, the Euston Arch (1961) and the Coal Exchange (1962) were both initiated by public authorities undertaking transport developments and were approved by Conservative ministers. While the Coal Exchange represented a real example of the conflicting needs of conservation and necessary development, the cost of rebuilding the Euston Arch, estimated by government at a mere £190 000 (which could easily have been recouped by an office development above the new Euston Station) reveals the extent of official apathy; even the demolition contractor was said to be shocked at the decision.[39]

In addition to the above instances of conflicts between developers' legitimate private interests and the social costs of those activities, the boom also brought to light some instances of straightforward corruption. Rachmanism, while being almost exclusively confined to the residential property sector, served to cast developers in general in a bad light. The case of the developer John Gaul, who ran Sun Real Estates, a public property company with assets of £3 million, who was fined £25 000 in 1962 for living on the earnings of a Soho prostitute, had a similar effect.[40] The most celebrated scandal of the commercial property market, however, the Jasper affair, involved malpractice in an area which was at the very heart of the property boom, institutional finance for property development.

The scandal involved a disastrous conflict of interest on the part of H.H. Murray, the Secretary and Managing Director of the State Building Society, who was also a substantial property developer. Together with F. Grunwald, one of the State Building Society's solicitors, he devised a scheme whereby Murray's property empire was expanded via a series of property company takeovers, using funds borrowed from the State Building Society and channelled via a series of 'Dummy companies'.[41]

In this way a number of property companies were acquired from 1956 through to the demise of Murray's property empire in 1959. These were often purchased in the face of fierce opposition, which entailed paying high prices for their property assets. Meanwhile, mortgage payments were high, since attracting sufficient new funds to the State necessitated offering very favourable terms to depositors.

Eventually they did one deal too many: the take over of Lintang Investments, a large property company owned by the developer Maxwell Joseph. This bid overstretched their resources, representing a substantial overvaluation of Lintang's property assets and costing Murray and Grunwald more money than they could raise.[42] About £3.25 million was illegally made available from the State Building Society to fund the deal, though even this did not allow them to meet the full cost. In September 1959 payments for the Lintang shares were not met, dealings in them were suspended, and the Board of Trade and Fraud Squad launched investigations. The State Building Society was forced to close its doors, depriving its depositors of access to their capital, and both Murray and Grunwald were eventually sentenced to five years in prison.

Despite the above instances the number of financial scandals arising from the 1950s property boom, involving major figures in the industry, were very few, certainly far fewer than were produced in the financial services sector during the 1980s boom. The notoriety of the property developer during this period was as much a factor of the traditional antipathy towards the speculator and the landlord, combined with envy at the huge profits that were obviously being earned by developers and the disdain of many in the British establishment towards any group of outsiders who were making large fortunes by their own enterprise, as any real or alleged abuses which occurred during this period.

There was, however, one fundamental criticism of the 1950s property developer that was fully justified; many, if not most, of the offices constructed by speculative developers during the 1950s and 1960s are widely regarded as among the most ugly commercial buildings ever built in Britain. One important factor behind this was the system of plot ratios mentioned above and the consequent demand for architects who based their designs around maximizing lettable space within these restrictions. Marriott estimated that between half and three-quarters of speculative office buildings erected in the Greater London area during the property boom were designed by 10 firms of architects.[43] These specialized in functional buildings within a given price limit, usually the lowest possible price. As a result speculative office blocks were produced at a cost which might be as much as 60% lower per sq. ft. than would be the case for a building commissioned by an owner-occupier.[44]

This system ensured the rapid and economic provision of new offices. It did not, however, result in particularly attractive, durable or well-constructed buildings. As Jack Rose, one of the most active office developers during this period, recalled, designing buildings so as to maximize profits usually entailed:

> sacrificing elaborate facades in favour of curtain wall construction and limiting floor to ceiling heights to the minimum permitted by building byelaws. Hardly a single building incorporated air-conditioning. Central heating boilers were oil-fired rather than gas fuelled and there were generally insufficient electric power outlets. In essence, the standard of building conformed to the sellers' market.[45]

A few of the more prestigious office developments of the era did incorporate innovative design features. The first major example of an office development with a design which distinguished it from its 1930s predecessors was Fountain House, Fenchurch Street, designed by W.H. Rogers and built by the City of London Real Property Company in 1954–1957. This introduced the New York-style of setting a tower at one end of a low-built podium, a design made economically viable by London's plot ratio restrictions (which had previously led to a 'jelly-mould' style of office building). Fountain House was also one of the first British buildings to be built with 'curtain walls', a construction technique involving hanging the external walls (usually made predominantly from glass) from the concrete floors like curtains. During the 1960s this was to become the typical method of office construction, being adopted in many lower-quality developments as a result of its low cost.[46]

The decade of the 'property boom' saw a transformation of the funding links between financial institutions and property developers, paralleling the transformation in the nature of property as an investment medium. Prior to 1955 the two usual avenues for obtaining funds from an insurance company for property investment or development were borrowing on mortgage or selling the property via a sale and leaseback arrangement. Both of these operated to the benefit of the entrepreneur, rather than the funding institution, under conditions of inflation and rising property values.

Funds could be borrowed on mortgage to a maximum of two-thirds of the value of a development, but this was assessed on the basis of the value of the building on completion rather than its construction cost. As the final value might be typically 50% above costs in the mid-1950s, the developer might be able to fund the entire project using money borrowed on mortgage without committing any of his own funds. Thus, each deal provided the developer with an equity stake equal to half the value of the project. More importantly, this ability to undertake development without tying up any of the developers' own capital meant that there were effectively no financial constraints on the rate of growth of his activities, providing the financial institutions were prepared to supply funds on this basis. In this respect Britain differed markedly from continental Europe, where most financial institutions would only advance money on the basis of two-thirds of the site and building cost, rather than estimated market value on completion, of a development.[47]

During the late 1940s and early 1950s alternative investment mechanisms, which allowed the institutions to participate in the equity of developments they funded, had been successfully applied to property transactions by a few pioneering institutions, as discussed in section 5.3. The 1955–1957 credit squeeze greatly accelerated trends that were already under way, rather than bringing them about. Rising interest rates during the squeeze led to the danger of capital depreciation for fixed-interest stocks being more widely appreciated among the financial institutions. The credit squeeze also altered the balance of power between the institutions and the developers. The insurance companies, which faced less stringent controls over lending than the banks, found they faced less competition in the market for the provision of development finance and were able to dictate their own terms to developers. Such changes also affected the balance of power in the market for completed properties; it is interesting to note that those institutions, such as the Church Commissioners and Legal & General, which pioneered the building of an equity element into direct property investment were also generally those that were first to insist on equity participation in development funding arrangements.

While raw materials shortages and building licenses had restricted property development prior to 1954, the government now sought to control 'undesirable' investment in sectors not directly contributing to exports, such as property, via credit restrictions. In his April 1955 budget the Chancellor, R.A. Butler, made the mistake of giving out £135 million in tax cuts at a time of economic boom, with nearly twice as many vancancies as men out of work.[48] The resulting inflation and balance of payments difficulties led the Treasury to attempt to reduce the pressure of aggregate demand via a more restrictive monetary policy, together with some fiscal measures.

Interest rates rose from 3% at the beginning of 1955 to 5.5% in 1956.[49] In July 1955 the Chancellor instructed the clearing banks, via the Bank of England, to bring about a 'positive and significant reduction' in bank advances. During the following years further restrictions on bank lending were imposed, together with an increase in the proportion of Capital Issues Committee application rejections, from 2.9% in 1955 to 12.0% in 1956 and 15.0% in 1957.[50] Bank rates eventually rose to 7%, following a run on Britain's gold reserves during the summer and autumn of 1957. However, it was the reduced availability, rather than the increased cost, of credit which provided the major threat to the prosperous property development sector.

The reduction in Capital Issues Committee approvals and bank advances had a double effect on institutional finance for property development. Firstly, it increased the demand for such finance by restricting the availability of alternative sources of funds. Secondly, it increased the attraction of supplying finance by reducing the supply of an alternative outlet for insurance company funds, i.e. ordinary shares. The lack of new issues made it increasingly difficult for insurance companies to invest as much of their funds in the equity market as they would have wished, at a time when equity investment was becoming increasingly popular among the financial institutions as a result of a growing emphasis on the need to build capital and income appreciation into investment portfolios. This may have accelerated the trend towards property investment and development funding being transformed into equity-type investments, property acting as a substitute for ordinary shares. Thus, while developers were forced to rely more heavily on the insurance companies as a result of the credit squeeze, the same conditions made the financial institutions more receptive to such propositions.

Thus, the number of joint development companies, share option agreements and other equity participation arrangements between the financial institutions and the property entrepreneurs began to multiply. These arrangements brought the two sectors into a close relationship, a number of institutions gaining representation on property company boards. Their influence was often beneficial, providing useful advice at a time when many property companies were experiencing rapid growth, with its attendant problems of increasingly complex administrative, financial and managerial coordination.

Financial innovation provided one of the main avenues of entrepreneurship in the property sector during this period. The most appropriate definition of entrepreneurship, with regard to both phases of the post-war property boom, is the exploitation of market imperfection. During the 'investment boom' of the immediate post-war years, entrepreneurship largely involved the circumvention or exploitation of restrictive government legislation covering development and capital issues. During the following 'development boom' phase the government's reliance on credit and capital issues controls as its main means of controlling investment made financial innovation of even greater importance. An innovation can be seen as a new technique which creates market imperfection by its appearance, as a new market equilibrium can only be reached when it is widely diffused. An innovator is any person or institution that adopts the technique before it has become fully diffused. The persistence of abnormal profits in a market, and thus the rewards to entrepreneurship, are determined by the strength of barriers to its diffusion. This is the essence of the entrepreneur's economic role, the exploitation of market imperfection, which, by doing so, reduces the extent of that imperfection.

The conservatism of most institutional investors acted as the main barrier to the diffusion of these financial innovations, as is discussed below, maintaining abnormal profits both for the developers and for the more enterprising financial institutions. Eventually, however, favourable press coverage led to these innovations becoming 'fashionable' and their growth, which had previously been a trickle, became a flood.

Even before the advent of the credit squeeze, financial institutions were beginning to realize the threat that inflation posed to the real value of invested funds, and the high rewards offered by the property sector, for which they were already the main source of long-term funding. These factors, and not the later influence of the credit squeeze, stimulated the initial development of equity partnership funding mechanisms. The pioneering activities of Clerical Medical in developing these techniques were discussed in the previous chapter. Two other notable pioneers in this area were Legal & General and the Church Commissioners.

Legal & General had established a joint development company, Legenland, with the developer Sir Aynsley Bridgeland as early as 1949, as discussed in section 5.4. Other prominent developers with whom they forged close links included Jack Cotton (with whom they formed a joint development company in 1955 to develop the Monico site in Piccadilly) and Harold Samuel, the founder of what was to become Britain's largest property company, Land Securities.[51]

Table 6.3 provides a summary of Legal & General's largest property funding arrangements, involving mortgage finance or the purchase of property on a leaseback basis, undertaken prior to November 1960. By this time, Legal & General's major funding transactions, and commitments, had reached over £120 million. Of these the majority were property acquisitions rather than mortgages. Legal & General did not enter into partnership arrangements with developers unless mortgage or leaseback business had already been undertaken with them, the conventional leaseback and mortgage business conducted by the Society in the late 1940s and early 1950s acting as a base for launching more complex partnership arrangements in the decade after 1954.

In addition to partnership deals with developers, a large number of developments were undertaken directly by Legal & General from the late 1950s. During 1960 a total of 16 development projects were at the construction or planning stage;[52] the number of developments under construction had increased to 31 by 1965. In March 1960 George Bridge stated that the Society's direct developments constituted a greater proportion of its total development programme than partnership arrangements with property companies.[53]

Three other insurance companies were also active as direct developers at this time: the Prudential, Norwich Union and the Pearl, together with one other institutional investor, the Church Commissioners. Direct development by institutional investors was rarely of a 'speculative' nature during this period, a tenant being arranged before development commenced. However, it does demonstrate that the degree of involvement in the property boom by the financial institutions formed a continuous spectrum, from the provision of fixed-interest mortgage finance with no equity stake at one extreme to undertaking development in their own right at the other.

During the mid-1950s Clerical Medical continued its policy of equity participation in property development companies, along the lines established during the late 1940s. The Society's established links with property companies brought

Table 6.3 Major leaseback transactions and funding arrangements undertaken by Legal & General to November 1960

	Property purchases (1)	Property commitments [a] (2)	Mortgages outstanding and commitments [b] (3)	Total granted and committed (1) + (2) + (3)
Developers				
Sir Cyril Black	2 014 383	–	509 297	2 523 680
Sir Aynsley Bridgeland	300 000	–	4 540 155	4 840 155
Charles Clore	11 184 269	62 500	9 866 200	21 112 969
Jack Cotton: Monico	2 977 671	14 750	–	2 992 421
Others	8 443 292	1 160 000	6 560 250	16 163 542
F.D. Fenston	50 000	–	567 000	617 000
Hugh Fraser	8 188 000	10 237 000	4 935 000	23 360 000
R.W. King	–	–	724 575	724 575
Edward Lotery	218 932	–	404 280	623 212
Harold Samuel	232 850	–	19 457 350	19 690 200
Archie Sherman	551 014	423 150	2 365 321	3 339 485
Harry Sherman	263 850	–	1 486 531	1 750 381
Bernard Sunley	1 250 000	265 000	1 043 500	2 558 500
D.A.R. Tovey	221 400	181 000	454 830	857 230
Major Webb	–	–	2 238 500	2 238 500
Retailers				
British Home Stores	618 590	355 000	–	973 590
EMI Ltd	340 000	–	–	340 000
Great Universal Stores	143 491	18 000	–	1 270 491
Henry's Stores	338 000	–	–	338 000
Hide & Co. Ltd	498 250	–	–	498 250
International Tea Stores	1 494 500	508 000	–	2 002 500
Joseph & Louis Littman	4 609 681	–	–	4 609 681
MacFisheries Ltd	235 900	–	–	235 900
Macowards Ltd	875 000	–	–	875 000
Montague Burton Ltd	5 105 692	–	–	5 105 692
New Day Furnishing Ltd	378 000	–	–	378 000
Phillips Furnishing Co.	282 000	–	–	282 000
Sixty Minute Cleaners	168 475	–	–	168 475
Swears & Wells	793 475	–	–	793 475
United Drapery Stores / Prices Tailors / Others	1 494 542	–	–	1 494 542
Total	54 380 217	13 224 400	55 152 789	122 757 446

Source: Legal & General (1960) Table, dated 18 Nov., appended to Board minutes, 14 Dec. No date is given for the earliest transaction, but it is almost certain that the table refers to transactions undertaken from 1945 to 1960. Values are not adjusted to allow for inflation.

[a] 'Commitments' represents investments which have been agreed to, but not yet undertaken.
[b] Includes mortgage debentures.

it further business; in October 1955 another mortgage funding and share option deal was agreed with Arndale. Clerical Medical's relationship with Arndale illustrates the problems that could arise with property company management at this time, property entrepreneurs that were highly talented in their central function (property development) often proving less skilled in areas such as financial management and administrative coordination.

Arndale's problems were very similar in nature to the more famous case of City Centre Properties,[54] fortunately with much less disastrous results. J. Pegler, who represented the Society on the Arndale Board, began to have doubts about the running of Arndale due to two main problems. While the managing director, Sam Chippindale, had a shrewd sense for property deals Pegler considered that he had no good 'financial sense'.[55] The company was consequently entering into development commitments which required a great deal of capital without ensuring that the necessary finance was available at an acceptable rate of interest. This was a problem of increasing magnitude, due to rising interest rates and the tightening of credit restrictions in the mid-1950s. The other problem concerned the relationship between the company's two founders, Arnold Hagenbach and Sam Chippindale, who were unable to get on with each other, resulting in a severe breakdown in communications within the company.[56] Eventually these problems were resolved by finding a company to take over Arndale. It was arranged that Town & City Properties would acquire the company, since its driving force, Barry East, appeared to be capable of working with Chippindale and ensuring that the company's commitments were kept within the limits of available finance.[57]

Clerical Medical forged two further major equity links in the late 1950s, with Town & City Properties and Real Estate & Commercial Trust Ltd. By the end of 1957 the Society's property interests were so wide-ranging that its direct property holdings, with an estimated market value of £4 889 000, made up only a fraction of its total property-based assets[58] of £16 549 000.[59]

The Church Commissioners were one of the most active and innovative investors in the property development sector in the decade after 1954, forming a series of joint development companies. The first of these was established in December 1955 with the developer Max Rayne, to develop two sites in Eastbourne Terrace, which had been part of the Commissioners' Paddington Estate until it was divided into smaller estates in the 1950s. A 150 year building lease was granted to a company in which the Commissioners, via their holding company, and Rayne held equal shares, the Commissioners financing development via mortgage loans.[60] The Commissioners held shares in their joint development ventures via a holding company, the Church Estates Development and Improvement Company Ltd (CEDIC). The formation of this company had been under consideration since the early months of 1955, though it was not finally approved until the joint venture with Rayne had been agreed.

The initiative for this first joint development project came from Max Rayne, the Commissioners having previously intended to sell the Eastbourne Terrace sites,[61] though the fact that CEDIC had been under discussion prior to this agreement suggests that the Commissioners had already begun to think about conducting deals along these lines before Rayne approached them. The resulting development received acclaim in the architectural press and was also extremely profitable. It cost £1.75 million, while the profit to the joint company was £5.8 million by 1966. As the deal entailed the Commissioners providing all development finance it was only necessary for Rayne to commit £1000 to the joint company while his property company, London Merchant Securities, made a profit of £2.9 million on the deal. In his discussion of this development in *The Property Boom*, Marriott concluded 'What had Max Rayne done to achieve this profit? In essence he had produced an idea ... Such was the value of a developer.'[61]

In the relatively unsophisticated property market of the 1950s this was only one of many instances in which relatively simple ideas, applied before they had become widely accepted, resulted in substantial profits for those who pioneered them. However, the greatest long-term beneficiaries from this episode were the Commissioners, who applied the joint development company formula many times over the following years, usually with highly profitable results. By the end of March 1965 CEDIC held shares in 27 subsidiary companies. Developments undertaken by these companies had cost a total of £38.3 million, £31.3 million of which had been provided by the Commissioners. Partners included George Wimpey & Co., Wates Ltd, Berwin Estates (Bahamas Ltd) and City Centre Properties, the Commissioners providing the finance and sometimes, though not always, the land, and the developers contributing their expertise.

The terms on which equity participation agreements were made involved a trade-off between complete fixed-interest funding, which would be preferred by the developer, and the largest possible equity stake for the financial institution which was consistent with an acceptable degree of risk, which was the aim of the institutions. For example, in January 1961 a proposal to form a joint development company with Jack Cotton's City Centre Properties was discussed by the Commissioners. The scheme involved the redevelopment of 101/104 Piccadilly as shops, offices and flats. The proposal originally put to the Commissioners was that they should advance the finance for the development at 5.75% and take 20% of the equity. City Centre were told that the Commissioners would consider the proposal if the interest rate on the finance was raised to 6% and the equity stake to 25%. City Centre would not agree, stating that their policy was to grant only 20% of the equity to institutions and exceeding this limit for the Commissioners might lead to other institutions asking for similar terms. The Church Commissioners replied that a 25% participation was their normal minimum. A deal was finally reached whereby the Commissioners received only 20% of the equity but City Centre paid 6.2% interest on the finance, the estimated overall yield to the Commissioners, 6.9%, being the same as that for 6% interest and a 25% equity stake.[62]

In addition to development projects conducted under the auspices of CEDIC the Commissioners also conducted other development schemes during this period, including residential development of some of its agricultural land, industrial property development on bombed-out sites owned by the Commissioners in Finsbury and a shopping development in the New Town of Basildon. These projects offered high returns, with estimated yields of up to 14–15% in some cases.

By the late 1950s the financial world was beginning to appreciate the value of a funding link with a financial institution to the growth prospects of a property company, the announcement of such a move often resulting in rising share values. Such a deal could, therefore, be of immediate benefit to both the financial institution and the property company, as by the very act of making the arrangement the institution made capital appreciation of the shares it had obtained extremely likely. For example, in 1957 an agreement was made between Pearl Assurance, one of the pioneers of equity links with developers, and Baranquilla Investments, a property company run by two prominent developers, Harry Hyams and Felix Fenston. This gave the Pearl a 10% equity stake in the company and an option to purchase a further 10% in return for £1 750 000 debenture finance.

At the beginning of 1957 the company's shares were worth 10*s* each. By March of that year, following the announcement of several development schemes and the all-important finance agreement, they had risen to 32*s* 6*d*, an increase of over 220%. The Pearl had been both a cause and a beneficiary of this rise.[63] This ability to alter risk and return on an investment by the very act of undertaking it is a central feature of much development activity, as was discussed in section 3.3.

Many insurance companies did not transform their property funding arrangements to an equity basis for some years after such deals were adopted by the institutions discussed above. However, by the early 1960s the bulk of British insurance companies, and other financial institutions which participated in property development funding, were switching from conventional fixed-interest finance to equity participation agreements. There were two main reasons behind this. Firstly, the property share boom of the late 1950s and early 1960s made share option deals appear very attractive. Secondly, such arrangements were becoming 'fashionable'.

'Fashionability', sometimes referred to as the 'herd instinct' by financial commentators, is here defined as the attraction of an investment due to considerations other than its perceived risk and expected return. In the mid-1950s equity participation arrangements were regarded as a new and untried technique, involving the financial institutions in a much more direct relationship with the property company sector than almost all of them had hitherto taken. While the buoyancy of the sector might indicate the advantages of such arrangements, insurance company investment managers had also to take other factors into consideration. Operating in a bureaucratic business environment they faced a certain asymmetry in decision-making; the rewards of success for an investment manager pursuing a novel and profitable strategy might be smaller in magnitude than the penalties for failure should such a strategy prove unsuccessful. The incentives were therefore balanced against new policies that had not yet gained general acceptance among the financial community.

The training and background of investment managers also contributed to the slow diffusion of innovation. Most insurance company investment managers were actuaries, trained in stock exchange investment but having little formal knowledge of the property market, which many of them regarded with some distrust. Furthermore, many insurance companies still retained power over all major investment decisions at board level. Insurance company boards, which were usually dominated by people without specialist investment skills, were generally conservative in outlook and did not view novel schemes for funding the activities of 'shady' property entrepreneurs with enthusiasm. Novel property funding mechanisms therefore went against the grain of the conservative corporate culture of the British insurance industry.[64]

By the early 1960s, however, equity partnerships had become fashionable, receiving favourable comment in the financial press and earning a reputation for those institutions that undertook them as dynamic and far-sighted investors. The balance of factors not strictly connected with expected profitability and risk therefore began to tilt in favour of participation in such schemes so that institutions did not appear to have been 'left behind'. Thus, a large number of institutional investors, many of whom had much less knowledge of property development than those who had pioneered property partnership agreements, climbed on the property 'bandwagon'.

In 1960 the government once again felt obliged to apply the brake to economic expansion, in the light of an approaching balance of payments deficit,[65] credit restrictions being applied from April of that year. This second credit squeeze was anticipated by the property world and led property companies to establish lines of long-term credit with the financial institutions, which would constitute prior commitments on the part of the institutions, and would therefore be unaffected by credit controls. Credit arranged by three of the largest property companies, Land Securities, City Centre and Capital & Counties, alone amounted to £52 million.[66] Such arrangements proved extremely useful, as in July 1961 the Chancellor instructed the banks and insurance companies not to lend funds for certain classes of commercial activity, particularly targeting property development as an area for which funds should not be loaned.[66] However, the 1960–1961 credit squeeze proved to be of shorter duration, and less importance to the property market, than that of 1955–1957, the flow of funds to the property sector soon resuming its upward trend.

It is difficult to measure accurately the total insurance company commitment to the property sector via mortgage lending and share ownership, as these were usually not disaggregated from total mortgage and share holdings in published statistics. However, evidence provided to the Radcliffe Committee by the

INSTITUTIONAL INVESTORS AND THE PROPERTY DEVELOPERS 1955–1964

British Insurance Association gives figures for 1957, reproduced in Table 6.4. The table shows that mortgages to property companies, and other commercial borrowers, amounted to just over 25% of all mortgage business in 1957.

Table 6.4 Mortgages held by members of the British Insurance Association in 1957

	£M
House purchase	410
Agricultural	14
Commercial	185
Overseas	80
Total	689

Source: *Committee on the Working of the Monetary System, Report* (Aug. 1959) Cmnd 827 of 1959.

An analysis undertaken by Brian Whitehouse, concerning the financial arrangements entered into by insurance and property companies towards the end of 1963, indicated that the 21 insurance companies he could obtain details for had advanced, or pledged, finance totalling £351 million in agreements with property companies and held, or had an option on, over £60 million of property company shares as a result of such agreements.[67] This compares with total insurance company mortgage and ordinary share holdings of £1297 million and £1732 million, respectively. The Whitehouse figures provide only a very rough guide to insurance company holdings, as full details of some arrangements had not been released and some had not been publicly announced at all; for example, Clerical Medical's property company interests amounting, at the end of 1964, to £5 926 000 of mortgage finance (advanced or pledged) and shares with a market value of £3 942 000 are not included. The figures therefore indicate that property company mortgages amounted to at least 27.5% of all insurance company mortgages, and property company ordinary shares amounted to at least 3.7% of all insurance company ordinary share holdings at the end of 1963, both figures underestimating the true amounts by a considerable margin.[68] Furthermore, as equity participation could be obtained by means other than share holdings, such as obtaining a share in the equity of particular properties, the figures underestimate the total volume of insurance company funds directed to property companies by an even greater margin.

Tables 6.5 and 6.6 show the growth of equity participation arrangements between financial institutions and property companies from 1946 to 1963. Only arrangements with public property companies have been included, as those entered into with individual developers, such as Cantling,[69] sometimes involved the development of property for the insurance company's own use, and, as data on the formation of such companies is less comprehensive than that for deals with public companies, their inclusion would increase the margin of error of the figures. Even some arrangements with public property companies were not publicly announced, though Tables 6.5 and 6.6 should be sufficiently accurate to at least indicate broad orders of magnitude.

Table 6.5 Equity partnership arrangements between insurance companies and property companies, 1939–1963

Year	Number of arrangements	Net increase	Number of insurance companies
1939	3	0	4
1946	3	0	4
1947	5[a]	2	4
1948	5	0	4
1949	5	0	4
1950	5	0	4
1951	5	0	4
1952	5	0	4
1953	5	0	4
1954	6[b]	1	4
1955	7	1	5
1956	9	2	7
1957	11	2	8
1958	13	2	8
1959	20	7	10
September 1961	58	51	22
1963	106	55	29

Sources: B.P. Whitehouse (1964) *Partners in Property*, Birn, Shaw, London; Finance for property companies (1961) *Investors Chronicle*, **213**, 15 Sept, 916–19; information from Clerical Medical archives. *Note*: The table does not include joint development companies formed between institutions and particular developers, such as the Fortress Property Co. or Cantling, which did not involve a public property company. Nor does it include arrangements between financial institutions and companies developing property for their own occupation.

[a] First joint development company (formed by Clerical Medical and Brixton Estate Ltd).
[b] First share option agreement (between Clerical Medical and Arndale Ltd).

Those arrangements shown for the years 1946 and 1947 were all initiated during the inter-war period, an indication of the degree of continuity between the inter-war and post-war property investment markets. From 1948 to 1954 the only insurance company to enter into such arrangements with public property companies was Clerical Medical. During the following five years the number of insurance companies undertaking equity participation deals increased gradually, much more rapid growth occurring during 1960–1963. Until the end of 1959 no institutions other than insurance companies had entered into agreements of this type, with the single exception of the Church Commissioners. By 1961 competition in this area had considerably widened. The first major financial link between a property company and a pension fund was announced in July 1960; by the end of 1963 there were 14 such agreements, and a number of banks and other investors had also entered this market.

Table 6.6 Equity partnerships between various organizations and property companies, 1955–1963

	Insurance companies	Pension funds	Church Commissioners [a]	Banks [b]	Other	Total
1955	7	0	0	0	0	7
1956	9	0	2	0	0	11
1957	11	0	2	0	0	13
1958	13	0	5	0	0	18
1959	20	0	6	0	0	26
September 1961	58	5	17	3	1	82
1963	106	14	38	9	15	176

Sources: B.P. Whitehouse (1964) *Partners in Property*, Birn, Shaw, London; Finance for property companies (1961) *Investors Chronicle*, **213**, 15 Sept., 916–19; information from Clerical Medical archives; Church Commissioners (1956–1964) Annual Report. *Note*: The figures in the 'Total' column for 1961 and 1963 are less than the sum of those in the columns for the various classes of institution. This is due to some equity arrangements being conducted jointly between two categories of institution, such as a pension fund and an insurance company, which together provided finance to a property company.

[a] Figures refer to 31st March of the following year, except for 1961 where they refer to 31st March of the year given.
[b] Includes merchant banks.

The deterioration in the terms obtained by insurance companies and other financial institutions as a result of this new competition led them to favour equity options (giving the financial institution the option to buy shares at some future date, at a pre-determined price) rather than joint development companies. At a time of falling purchase yields on property company shares such deals offered the opportunity to obtain shares at yields which were expected to be higher than those currently prevailing in the market. Furthermore, options minimized downside risk, as there was no obligation to exercise them should conditions have become unfavourable in the intervening period. Options were also generally favoured by developers, as they allowed time for the developments being funded to become income producing before the company's equity was diluted.[70] This became increasingly important during the 1960s, as a result of declining profit margins on development schemes.

The spread of property partnership arrangements during the early 1960s also resulted in a rapid increase in the number of public property companies, from 111 at the end of March 1958 to 169 four years later. This increase was partly at the behest of the financial institutions, who were reluctant to gain an equity stake in private property companies due to the difficulty of marketing the shares obtained, as described in the following account by Jack Rose:

> In 1962 after a long and eminently satisfactory association with the Legal and General ..., we met with a refusal to confirm their hitherto usual promise to grant a mortgage on a project that the bank had agreed to finance temporarily. We sought an interview with the General Manager who told us of a major change in the company's policy which he assured had been adopted by all the other insurance companies and institutions acting as mortgagees.
>
> Having witnessed the extent of the profits made by property developers ... the lenders wanted not only interest on the loans but also a stake in the property. The going rate for this stake was 10% of the shares of the company set up to do the deal. We had not a minute's hesitation in agreeing to this but our offer was not enough. It was explained that since the remaining 90% of the shares in the company would be owned by the developer, the chances of selling a 10% stake to any one else would be slim and probably the only purchaser would be the developer who could set his own price.
>
> The insurance company whilst quite prepared to do the business, now insisted that the developer was to be a public company whose shares were quoted on the Stock Exchange thus offering at all times the opportunity for the insurance company to find a wider market should it want to sell its stake.[71]

Jack Rose and his brother Philip decided that in the light of these circumstances their business might be severely affected if it remained a private company and launched Land Investors Ltd as a public company in 1963.

While the initial equity participation arrangements, conducted during the mid to late 1950s, appear to have been very profitable, those institutions which forged their equity links in the early 1960s fared less well. Even the *Investors Chronicle*, one of the most bullish advocates of property shares during these years, noted in a review of such arrangements, in September 1961, that:

> By the beginning of 1960 many of the insurance companies whose knowledge of property and property companies can gently be described as limited, felt the need to ride the new property bandwagon and certainly some of the links are to say the least curious.[72]

The explosive growth of equity participation arrangements between 1960 and 1963 had fuelled the price of property shares, contributing to a massive, but temporary, boom in property share values. Between March 1958 and March 1962 the total market value of property shares had increased almost eightfold. Marriott estimated that only about £120 million of this increase was attributable to the extra 58 companies that obtained a quotation during these years, the market value of the companies quoted throughout the period having increased by 560% in only four years.[73] Meanwhile, the property market was rapidly becoming overheated. The City's office floorspace had increased by 11% between 1957 and 1960,[74] and it was by no means certain that this further wave of developments would find tenants at prevailing market rents.

Property shares peaked in May 1962 and during the following five years the market value of quoted property companies fell steadily, as is illustrated in Figure 6.3, which shows the market capitalization of all property companies, and the average capitalization per company, during the 10 years from March 1958. At the peak of the share boom property shares had become so highly priced that dividend yields for many companies stood at well under 2%. Insurance companies had rushed into the property company sector at the very time when property shares were trading on what proved to be their least attractive basis.

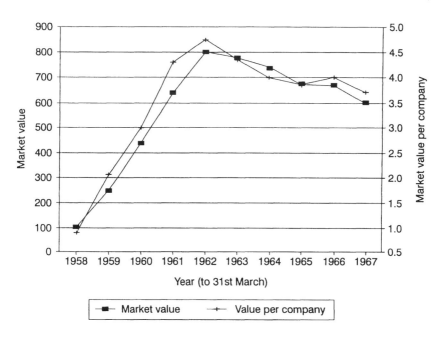

Figure 6.3 The market capitalization of quoted property companies and the average market capitalization per company (£M), 1958–1967. Source: O. Marriott (1967) *The Property Boom,* Hamish Hamilton, London, p. 274.

While those few institutions which acquired their stake in the property boom in the mid to late 1950s generally profited from their agreements, often to a considerable extent, those which jumped on the bandwagon in the early 1960s were less fortunate. The profitability of a development project depends on four factors:

1. The relationship between the construction cost of the building in question and its value on completion.
2. The rate of interest on development finance.
3. The yield basis on which the property is sold by the developer.
4. The funding mechanism used to raise the development finance.

A useful mathematical framework for analysing the relationship between factors 1–3, given funding via conventional mortgage finance, was developed in the early 1960s by Charles Gordon and Garry Arnott.[75] The relationship between development cost for a property and its value on completion was represented by Gordon and Arnott by dividing final value by development cost, the resulting figure being known as the Capital Conversion Factor. In the mid to late 1950s a Capital Conversion Factor of 1.5 could often be obtained on developments, the developed property having a value 50% in excess of costs.

Achieving a factor of 1.5 was of crucial importance, as at this level the development could be financed entirely from mortgage funds, which were generally loaned to a maximum of only two-thirds of the estimated value of the development on completion. This meant that none of the developers' own capital had to be tied up in the project, the Capital Security Multiplier (outlined in Chapter 3) being infinite. However, by 1962 declining margins on development projects had reduced the Capital Conversion Factor to a figure nearer 1.2.[76] A lower Capital Conversion Factor would necessitate a greater proportion of income from sources not directly connected with the property portfolio, such as share issue, for a company to expand at a given rate, according to the model of corporate growth outlined in Chapter 3. This may have increased the willingness of property companies to sell shares to the institutions in return for finance, a greater volume of share issue becoming necessary to sustain growth rates as the Capital Conversion Factor fell.

The relationship between the Capital Conversion Factor and the other variables affecting the developers' profits is illustrated by the following hypothetical example. A £100 000 development produces an income of £10 000 on completion and is valued at 15 years purchase, £150 000. The rate of interest on the mortgage, calculated at two-thirds of the development's final value, £100 000, is 6%. The developers' income, net of mortgage interest, is therefore £4000, a 6.67% return on the one-third equity slice of the development and a 0.67% return on the other two-thirds of the development which is mortgaged. The ratio of the developer's net income from the development to the construction cost, 4% in this case (named the Income Conversion Factor by Gordon and Arnott), provides a measure of the initial income received from developments in relation to the Capital Conversion Factor, the initial yield on the development and the rate of interest on the mortgage. Table 6.7 illustrates Income Conversion Factors for properties valued at 10 years purchase and 15 years purchase, at varying interest rates and Capital Conversion Factors.

Table 6.7 The Income Conversion Factor for properties valued at 10 and 15 years purchase

Capital Conversion Factor	Valued at 10 years purchase Mortgage interest rate (%)					Valued at 15 years purchase Mortgage interest rate (%)				
	5.5	6.0	6.5	7.0	7.5	5.5	6.0	6.5	7.0	7.5
1.1	6.97	6.60	6.23	5.87	5.50	3.30	2.93	2.57	2.20	1.83
1.2	7.60	7.20	6.80	6.40	6.00	3.60	3.20	2.80	2.40	2.00
1.3	8.23	7.80	7.37	6.93	6.50	3.90	3.47	3.03	2.60	2.17
1.4	8.87	8.40	7.93	7.47	7.00	4.20	3.73	3.27	2.80	2.33
1.5	9.50	9.00	8.50	8.00	7.50	4.50	4.00	3.50	3.00	2.50

Source: C. Gordon and G. Arnott (1962) Some observations on property financing. *Investors Chronicle*, **215**, Property Supplement, 23 Feb., xvii.

During the early 1960s the Income Conversion Factor fell appreciably due to two factors. Firstly, intensified competition in the development market, and a reduction in the imbalance between the demand and supply of commercial property, had lowered the Capital Conversion Factor from 1.5 to about 1.2, as discussed above. Secondly, the rate of interest on mortgage finance rose. Interest rates, as reflected in the yield on consols, averaged 5% during 1957–1959; in 1960 this rose to 5.25% and, in 1961–1962, to 6.25%. The final variable, the years purchase figure on which developments were valued, remained fairly stable during these years.[77] Given a Capital Conversion Factor of 1.5 in 1957 and 1.2 in 1962, the increase in mortgage interest rates (assumed to be 1% above the yield on consols) would imply, for a development valued at 15 years purchase, a fall in the Income Conversion Factor from 4 to 2.2%. Furthermore, if the developer's own capital invested in the project is given a notional interest rate equivalent to that charged on the mortgage the fall becomes even steeper, from 4 to 0.75%. The initial income return on property development was being squeezed to such an extent that it had almost disappeared.

In 1964 Brian Whitehouse referred to a Capital Conversion Factor of 1.2 as the minimum at which the developer would be prepared to conduct business, as at that rate the capital value created by the developer would be only equal to his initial capital input, given mortgage funding.[78] It therefore appears that by 1962 both the initial income, and capital, returns on property had been reduced to such a point that development was only marginally worthwhile in terms of initial gains.

The sudden popularity of equity partnership deals during the early 1960s is all the more remarkable given that insurance companies had information at their fingertips which showed that profit margins on development projects were falling. The margin between the cost of a development and its estimated value on completion was routinely used by insurance companies in their assessment of the maximum amount of mortgage funding they would provide for particular development projects. The steady decline in this margin, from 50% in the mid-

1950s, to 20% by 1964, should, of itself, have alerted them to the fact that profits in the sector were undergoing a substantial decline.

This fall in development margins was reflected in the profitability of the equity participation deals undertaken towards the end of the boom. One hundred equity participation agreements entered into over the previous eight years were analysed in *The Investors Chronicle* in March 1967, to assess their profitability for the financial institutions concerned. In 52 cases the ordinary shares acquired were worth less than their purchase price, or the share options had been abandoned. In 15 cases the options were still available, but would be unprofitable if taken up. In four cases options had been arranged so that the price paid would yield 5.5%; they would therefore probably be profitable if exercised. The remaining 29 cases had resulted in a profit for the institution, though some shares in this category had not appreciated in value. As Marriott stated 'A ratio of success of 29% is a poor rating and shows how badly timed was the insurance companies' awakening to the property boom.'[79]

The massive inflow of institutional funds to this sector from 1960, via the equity participation arrangements discussed above, had led to a correction of the imbalance between demand and supply in the development market, removing the super-normal profits which had resulted from that imbalance. Those institutions which had gained from the property boom were the small group of insurance companies such as Legal & General, Clerical Medical and the Pearl, and one other institution, the Church Commissioners, which had the courage and foresight to enter into equity partnerships with developers before such arrangements became fashionable, when profit margins were high and competition was low. Such conditions were never to arise again in the property development market.

REFERENCES AND NOTES

1. E. Reade (1987) *British Town and Country Planning*, Open University Press, Milton Keynes, p. 53.
2. P. Cowen *et al.* (1969) *The Office: A facet of urban growth*, Heinemann, London, p. 165.
3. N. Tiratsoo (1990) *Reconstruction, Affluence and Labour Politics: Coventry 1945–60*, Routledge, London, p. 77.
4. See section 10.2.
5. Eagle Star (1951) General Purposes and Finance Committee minutes, 28 Nov.
6. Hillier, Parker (1955) Minutes of partners meeting, 23 May.
7. C. Gordon (1985) *The Two Tycoons*, Hamish Hamilton, London, pp. 58–9.
8. Legal & General (1955) Annual Report on Freehold and Leasehold Properties, to Board of Directors, 2 March.
9. Legal & General (1958) Estates Dept. Report.

REFERENCES AND NOTES

10. George Bridge (1960) Investment of insurance funds. *The Chartered Surveyor*, **92**, March.
11. Arthur Green (Jan. 1960) Notes on a paper by George Lingwood.
12. See section 10.4.
13. D. Sutherland (1968) *The Landowners*, Blond, London, p. 138.
14. K. Ryle, Accountant, for the Church Commissioners. Memorandum for Estates and Finance Committee, *Property Investment Purchases: A share in the equity of a property*, C.C. 95055, part 8/9.
15. Church Commissioners (1956) Annual Report, 31 March, p. 4.
16. Church Commissioners (1959) Estates and Finance Committee papers, 20 Oct., paper E, C.C. 95055, part 18/19.
17. D. Massey and A. Catalano (1978) *Capital and Land: Landownership by Capital in Great Britain*, Edward Arnold, London, p. 66.
18. Mortimer Warren for the Church Commissioners (1959) Investments in agricultural estates, a memorandum, Estate and Finance Committee papers, 17 Nov., C.C. 95055, part 18/19.
19. Collenette was a member of the Commissioners' Estates and Finance Committee.
20. Church Commissioners (1962) Sales of agricultural land, Estates and Finance Committee paper, 17 July, C.C. 95055, part 28/29.
21. Church Commissioners (1961) Estates and Finance Committee paper, 21 Nov., C.C. 95055, part 25/26.
22. Church Commissioners (1964) Annual Report, 31 March, p. 7.
23. See section 6.3.
24. Church Commissioners (*c.* 1965) *The Church Commissioners Index* – an historical review of investment policy.
25. See section 10.6.
26. J. Rose (1985) *The Dynamics of Urban Property Development*, E & FN Spon, London, p. 154.
27. See Table 11.1.
28. Sources: 1948–54, M.C. Fleming (1980) Construction and the related professions, in *Reviews of United Kingdom Statistical Sources, Vol. 12* (ed. W.F. Maunder), Pergamon Press, Oxford; 1955–64, CSO (various issues) *Annual Abstract of Statistics*. As the figures for 1954 and 1955 are based on different (government) sources there may be some degree of distortion, though this is not likely to be great.
29. S. Taylor (1966) A study of post-war office developments. *Journal of the Town Planning Institute*, **52**, 55.
30. I. Alexander (1979) *Office Location and Public Policy*, Longman, London, p. 63.
31. See section 5.3.
32. O. Marriott (1967) *The Property Boom*, Hamish Hamilton, London, p. 31.
33. P. Cowen *et al.* (1969) *The Office: A facet of urban growth*, Heinemann, London, p. 166.

34. O. Marriott (1967) *The Property Boom*, Hamish Hamilton, London, p. 32.
35. O. Marriott (1967) *The Property Boom*, Hamish Hamilton, London, p. 171.
36. O. Marriott (1967) *The Property Boom*, Hamish Hamilton, London, p. 30.
37. O. Marriott (1967) *The Property Boom*, Hamish Hamilton, London, p. 6.
38. L. Esher (1981) *A Broken Wave: The Rebuilding of England 1940–1980*, Allen Lane, London, pp. 72–3.
39. L. Esher (1981) *A Broken Wave: The Rebuilding of England 1940–1980*, Allen Lane, London, pp. 139–40.
40. O. Marriott (1967) *The Property Boom*, Hamish Hamilton, London, pp. 9–10.
41. G. Bull and A. Vice (1961) *Bid For Power*, 3rd edn, Elek, London, p. 270.
42. G. Bull and A. Vice (1961) *Bid For Power*, 3rd edn, Elek, London, p. 274.
43. O. Marriott (1967) *The Property Boom*, Hamish Hamilton, London, p. 27.
44. O. Marriott (1967) *The Property Boom*, Hamish Hamilton, London, p. 29.
45. J. Rose (1985) *The Dynamics of Urban Property Development*, E & FN Spon, London, p. 153.
46. A.R. Goobey (1992) *Bricks and Mortals*, Century, London, p. 24.
47. B.P. Whitehouse (1964) *Partners in Property*, Birn, Shaw, London, p. 139. This marked a change from the situation which prevailed prior to the Second World War, when British insurance companies usually based their mortgage lending on construction costs rather than estimated final value, as outlined in Chapter 3.
48. S. Brittan (1964) *The Treasury Under The Tories 1951–1964*, Penguin, Harmondsworth, p. 177.
49. S. Brittan (1964) *The Treasury Under The Tories 1951–1964*, Penguin, Harmondsworth, p. 49.
50. S. Brittan (1964) *The Treasury Under The Tories 1951–1964*, Penguin, Harmondsworth, p. 50.
51. B.P. Whitehouse (1964) *Partners in Property*, Birn, Shaw, London, pp. 44 and 55.
52. Legal & General (1960) Estates Dept. Annual Report.
53. George Bridge (1960) Investment of insurance funds. *The Chartered Surveyor*, **93**, March.
54. Detailed accounts of the troubled history of City Centre Properties Ltd are given in Charles Gordon (1985) *The Two Tycoons*, Hamish Hamilton, London; and O. Marriott (1967) *The Property Boom*, Hamish Hamilton, London, pp. 132–53.
55. J. Pegler (1981) Unpublished memoirs, p. 43.
56. J. Pegler (1981) Unpublished memoirs, pp. 44–5.
57. J. Pegler (1981) Unpublished memoirs, p. 44.
58. Includes mortgages, property company shares and direct property holdings.
59. Clerical Medical (1957) Asset Investigation Report.
60. Church Commissioners (1955) Lancaster Gate Estate Eastbourne Terrace sites proposed redevelopment, Estates and Finance Committee Papers, 1 Dec., C.C. 95055, part 9/10.

REFERENCES AND NOTES

61. O. Marriott (1967) *The Property Boom*, Hamish Hamilton, London, p. 85.
62. Church Commissioners (1961) Estates and Finance Committee minutes, 17 Jan., paper C, C.C. 95055, part 22/23.
63. B.P. Whitehouse (1964) *Partners in Property*, Birn, Shaw, London, p. 67.
64. The influence of such conservatism on the part of the British insurance industry in inhibiting innovation has been discussed, for an earlier period, in O. Westall (1993) Entrepreneurship and product innovation in British general insurance, 1840–1914, in *Entrepreneurship, Networks and Modern Business* (eds J. Brown and M.B. Rose), Manchester University Press, Manchester, pp. 200–4.
65. S. Brittan (1964) *The Treasury Under The Tories 1951–1964*, Penguin, Harmondsworth, pp. 205–6.
66. B.P. Whitehouse (1964) *Partners in Property*, Birn, Shaw, London, p.105.
67. B.P. Whitehouse (1964) *Partners in Property*, Birn, Shaw, London, p. 180.
68. These figures were calculated on the basis of Clerical Medical's figures and Whitehouse's figures.
69. See section 5.3.
70. B.P. Whitehouse (1964) *Partners in Property*, Birn, Shaw, London, pp. 119–20.
71. Jack Rose (1993) *Square Feet*, RICS Books, London, pp. 114–15.
72. Finance for property companies (1961) *Investors Chronicle*, **213**, 15 Sept., p. 915.
73. O. Marriott (1967) *The Property Boom*, Hamish Hamilton, London, p. 273.
74. D. Cadman and A. Catalano (1983) *Property Development in the UK – Evolution and Change*, E & FN Spon, London, p. 5.
75. C. Gordon and G. Arnott (1962) Some observations on property financing. *Investors Chronicle*, **215**, Property Supplement, 23 Feb., xv–xviii.
76. C. Gordon and G. Arnott (1962) Some observations on property financing. *Investors Chronicle*, **215**, Property Supplement, 23 Feb., xv.
77. See section 10.4. (Years purchase is the inverse of yield.)
78. B.P. Whitehouse (1964) *Partners in Property*, Birn, Shaw, London, p. 115.
79. O. Marriott (1967) *The Property Boom*, Hamish Hamilton, London, p. 40.

7 From Brown Ban to Barber Boom 1965–1973

7.1 INTRODUCTION

On November 4th 1964 Labour's new Minister of Economic Affairs, George Brown, announced the 'Brown Ban', an almost complete ban on office development in and around London, thus marking the end of the property development boom.[1] During 1965 and 1966 office development controls were extended to Birmingham, and then to the whole of the South East and the East and West Midlands. As a result of these restrictions the property investment market began a new phase, in which rising institutional demand for office property was met by a limited supply of new developments.

Rather than damaging the position of property investors, the 'Brown Ban' marked the start of an unprecedented property investment boom, which was eventually to precipitate the most severe crisis experienced by the British banking system during this century. A number of previous accounts of this crisis have explained the 1973/4 property crash and the boom which preceded it in terms of Marxist theory of crises, focusing on declining profit rates in the British economy.[2] In the following two chapters these events are reassessed in the light of the theory of commercial crises set out by the economist Thorstein Veblen, in *The Theory of Business Enterprise*.[3] Veblen viewed commercial crises as speculative phenomena, triggered by disturbances which initiate upward price movements. His theory provides a comprehensive model of the factors bringing about the boom, the reaction of investors and developers, the way in which the boom eventually precipitated a crisis in the sector, and the reaction of the banks, and of government, to the crisis.

7.2 THE PROPERTY INVESTMENT BOOM

Veblen saw crises as being primarily 'phenomena of price disturbance'.[4] The period of prosperity which typically preceded crises was described as arising from 'some traceable disturbance of the course of business'[5] which initiated the price rise. A number of legislative measures introduced by the new Labour government during the mid-1960s had precisely this effect. In addition to the Brown Ban, discussed above, these included a number of changes in the tax system. The Finance Act 1965 introduced Capital Gains Tax on property development, development gains being taxed at a rate of 30%. A more important element of the Act, from the institutional investors' point of view, was the introduction of Corporation Tax, which made direct investment in property more attractive than indirect investment via property company shares, as direct ownership avoided the payment of this tax. This was particularly important for institutions which paid no tax, such as pension funds, or paid tax at a preferential rate, such as life assurance companies. The introduction of Corporation Tax also led to a reduction in equity participation deals between property companies and financial institutions, involving share options or joint development companies, as this would necessitate the payment of tax on dividend income. Attention turned to sale and leaseback as a means of providing institutional funding for property development, reinforcing the switch from equity and loan investment to direct investment in the property sector.

The Land Commission Act, 1967 further weakened the position of property developers. The Act created the Land Commission, which was to buy land required for essential purposes and collect a 'betterment levy' when land was sold or leased, or upon the realization of development value, by carrying out a project of material development. Increases in the current-use value of land were taxed by Capital Gains Tax, as before. The betterment levy was initially set at 40%, with the aim of an eventual increase to 50%. This Act proved cumbersome in practice and was abolished, following a change of government, in 1971.[6] However, by reducing the volume of new development during the late 1960s the Act added to the relative scarcity of investment property, increasing its attraction for institutional investors.

The financial institutions responded to the reduction in development activity, and the new tax incentives, by substantially increasing their direct investment in the property sector. During 1948–1954 the real value of net direct property investment by the financial institutions had grown at an average rate of 2.77%. Growth had accelerated to 8.07% per annum from 1955 to 1964, but became much more rapid in the nine years after the imposition of the Brown Ban, averaging 17.78%.[7] This increase was less impressive in terms of the proportion of new institutional funds invested in property. For example, annual net insurance company property investment increased by 115.6%, in real terms, from 1965 to 1973, while the proportion of new insurance company funds invested in the sector in these two years rose from 13.54% to only 18.46%.[7]

The increase in institutional property investment occurred alongside a fall in initial yields, both in absolute terms and relative to the yield on gilts. This was part of a longer-term trend; during the years 1946–1954 the average yield differential between shop property and gilts had been 1.86%. The average for

1955–1964 was only 0.31%, the differential becoming negative from 1961. This downward trend continued during the years 1965–1973, with an average differential of –1.74%. The 'reverse yield gap' was partly the result of the transformation of property from a fixed-interest stock, lacking the 'risk-free' security of gilts and offering poorer investment characteristics in terms of liquidity, marketability and divisibility, to an 'equity' investment, which had a potential for capital appreciation that made it much more attractive than fixed-interest assets in a world of persistent and accelerating inflation. It was also partly the result of the rapid growth of institutional funds earmarked for the property sector, which outstripped the increase in the supply of new investment properties, thus forcing down yields.

The sellers market for investment property, together with the growing equity element that more frequent rent reviews built into property assets, led to a substantial increase in property values. During this period property performed extremely well, both historically and compared to other investment media. The rate of return on property averaged 15.1% from 1965 to 1973, which compared favourably both with returns on equities and gilts (9.0 and 3.2%, respectively), and with property's performance during any other period for which data are available. Propery achieved this impressive return due to the increase in investment property values, which averaged 9.5% per annum during this period.

The supply restrictions led to rising market rents, though there were signs that the growth of demand for commercial property was beginning to decelerate, threatening a slump in the property sector should supply ever be allowed to grow freely again. The rate of growth of office employment had fallen from an average of 3% per annum in the 1950s to 2.5% in the early 1960s and 1.8% in the late 1960s. This was offset, to some extent, by an increase in the rate of growth of floorspace per worker, which averaged 1% per annum in the late 1960s and early 1970s, though the overall impact of these trends was to reduce the rate of growth of demand for new office floorspace.[8]

However, such fears did not loom large in the thinking of the financial institutions. Veblen stated that the reaction of investors to an initial price disturbance in the value of a commodity might lead to a continuation in the upward price movement. Investors interpret the rise in prices as a trend rather than a once and for all increase, and incorporate this assumption into their business decisions. Once initiated the price rise therefore, to some extent, becomes self-sustaining, investors bid up the price of the commodity in question and perceptions of its increased prospective earnings are reflected in an appreciating market capitalization.[9]

The reaction of the financial institutions to the rise in property values resulting from Labour's new legislation supports this hypothesis. From the mid-1960s the institutions significantly increased their property investment targets. Such was the magnitude of the inflow of new funds that established investors soon began to experience difficulties in placing as much money as they had earmarked for the sector. This led to growing pressure for yet higher targets, which came to reflect not only the proportion of current new funds that the institutions thought should be invested in property, but also investment to make up for the failure to meet targets set in previous years.

Clerical Medical's asset investigation report for 1965 already noted disappointment at the amount of money that the Society had been able to invest in property during the year. This was due to Clerical having been out-bid for a number of buildings. The report noted falling initial yields for investment property, at a time of high gilt yields, which was explained as being due to investors purchasing property at prices which made allowance for continued inflation.

Accelerating inflation was to intensify the property investment boom during the late 1960s and early 1970s. Retail prices, which grew at an average rate of 3.6% per annum during 1965–1967, rose annually by 5.48% from 1968 to 1970 and 8.62% from 1970 to 1973. This increased the attractiveness of property to institutional investors, since property had come to be perceived as a 'hedge against inflation', i.e. an asset which could be expected to experience price inflation equal to, or greater than, the general rate of inflation. As a consequence, institutional property investment rose significantly, pushing up prices in a sellers' market and thereby accelerating the upward trend in property values.

From 1965 to 1967 the rate of return on UK investment property only just managed to out-pace inflation, averaging 1.55% per annum in real terms. But from 1968 to 1970 property returns averaged 10.65% in excess of inflation, and from 1971 to 1973 they rose by 14.20% per annum in real terms. Property's inflationary edge was sharpened during these years by an increase in the frequency of rent reviews. During the early 1960s an interval of 21 years between reviews was typical, this was reduced to 14 years in the mid-1960s, seven years from about 1968 and five years by 1972.[10] Institutional investors exerted downward pressure on the interval between reviews during rent negotiations, trading lower initial rents for more frequent reviews.[11]

The growing frequency of rent reviews led to a significant divergence between initial yields and effective yields[12] on investment property, as is shown in Table 7.1, taken from a paper prepared for the Board of the National Provident Institution in June 1968.[13] The example applied to institutions with the tax position of a gross fund, and was based on an investment with an initial yield of 6%, assuming an annual rise in rental levels of 3% per annum.

A further intensification of demand in the property investment market was seen in 1968, pension funds and property unit trusts competing with the insurance companies for available investments.[14] Of 26 property investment propositions considered by Clerical Medical during this year, which had been approved by the Society's Board, only 50% resulted in transactions; two years earlier 70% had been successful. The National Provident Institution was also having difficulty in finding an adequate volume of new investment properties during the late 1960s; during the 18 months to June 1968 11 properties had reached the stage of consideration by NPI's Board, but only four deals had been undertaken.[13] In the light of the new market conditions both Clerical Medical and NPI decided to lower property purchase yields to allow for the adoption of inflationary expectations in property values.

Table 7.1 The impact of different rent review periods on effective property yields

Rent reviews	Effective yield (%)	Initial yield required for 9% effective yield
Every year	9.0	6.0
Every 7 years	8.5	6.5
Every 14 years	8.1	7.0
Every 21 years	7.8	7.5
Every 35 years	7.3	8.2

Source: NPI (1968) Board paper 1968/28, 27 June.

Clerical's 1969 asset report revealed further evidence that investors were becoming more concerned with capital appreciation than immediate rent. It noted that:

> The philosophy of property investment appears to be changing. Up to a year or so ago property investment was generally regarded as primarily a long term fixed-interest investment with equity participation by way of rent reviews. The emphasis on the equity element in property as a hedge against inflation is at present assuming a more important role. The fixed interest element, although still important, now fills the secondary role.[15]

Inflationary assumptions were even built into the yield basis on which properties were offered for sale; at a meeting of Clerical Medical's Investment Forum in August 1970 it was noted that 'Thus we might be offered an investment of 5% rising to 6.5% which until very recently would have been considered simply as a 5% yield'.[16] These developments were part of a wider trend towards the adoption of inflationary expectations in calculating yields and interest rates. At an F.T./Investors Chronicle Commercial Property Seminar held in the Spring of 1970, G.C. Morley of Shell suggested that the long-term interest rate was 4.5% plus the current rate of inflation. This meant that the yield required on ordinary shares (for which no inflationary allowance was necessary) was 4.5%, while that on property with seven yearly rent reviews was 5%, 14 year reviews 5.5% and 21 year reviews 6%.[17]

The growth of Legal & General's property portfolio during the late 1960s and early 1970s was also restricted by a shortage of properties available on terms that it considered acceptable. During the late 1960s Legal & General's Board had begun to set broad targets for the distribution of new funds between asset categories. A memorandum on future investment policy, dated 31st December 1970, stated that out of total estimated funds available for investment during 1971 (£72 million), 29% should be invested in property, 21% in ordinary

THE PROPERTY INVESTMENT BOOM

shares and the remaining 50% in fixed-interest securities.[18] The 50% split between fixed-interest and 'equity' investments had also been the target for 1970, 25% of total funds being allocated to property and 25% to ordinary shares. Property investment during that year had totalled £8.5 million, less than half the target allocation, however, and the target for 1971 had been increased to make up for this shortfall.

Heightened competition from 'new' property investors was noted in the memorandum:

> Owing to the demand by property bond unit trusts, pension funds and smaller insurance companies the market in properties up to about £250,000 value has become a very difficult one. It is thought therefore that investment should be concentrated although not exclusively on the very large units, say two or three million pounds upwards where competition is usually less intense and where the departmental effort is no greater than in purchasing a property only a fraction of that size. The competitive market in property could also be by-passed through undertaking major development schemes either on our own or in association with well established property companies.[18]

Conditions during the following year confirmed the need to undertake developments in order to obtain good quality properties. A Board paper of July 1971, reviewing investment policy for the second half of the year, noted the increased difficulty experienced in obtaining good properties, even at very low initial yields, concluding 'Future investment in volume depends in part on development projects, either alone or in association with others'.[19]

During 1972 property prices continued to escalate; a property in Wimbledon, which Clerical Medical tried to buy for £302 000 went for £451 000, while another for which the Society's limit was £25 000 went for £51 000.[20] In November the Conservative government introduced a temporary freeze on commercial rents, which was not finally removed by the subsequent Labour government until March 1975. This was a far more serious measure for investors than dividend control on shares, as it prevented the review of rents which had been in force for long periods, allowing some rents to continue at levels set over 20 years previously. Furthermore, a landlord with a restricted rent forgoes the lost income for ever, while undistributed company profits are retained by the company, benefiting the shareholders in the longer term.

The Counter Inflation (Temporary Provisions) Bill of November 1972 had little immediate effect on the property market, though its implications led Healey and Baker to express the view that the market might turn more in the buyers' favour, and the Prudential and the Co-operative Insurance Society refrained from buying in the hope of lower prices.[21] The prospect of more permanent commercial rent control caused greater anxiety however, as the rising trend in prices during the previous years was based on expectations of high future rental growth. Despite these fears the market ended the year in an extremely buoyant condition. As Clerical Medical's asset report for 1972 noted:

> By the end of the year there was some evidence of properties being purchased for the sake of owning property regardless of price or yield. This comment seems particularly true in the case of agricultural land which is now changing hands on terms which indicates investments are being made in the hope of very long term capital appreciation rather than in expectation of any substantial return on the capital invested in the foreseeable future.[22]

Established investors faced ever greater difficulties in placing as much money as they wished in property, partly as a result of the arrival of a variety of new investors to this market. A number of insurance companies which had not previously shown any interest in property began to invest in the sector during these years, often guided by one of the prestigious firms of chartered surveyors such as Richard Ellis or Jones Lang Wootten. Furthermore, the pension funds were rapidly expanding their property interests. In 1965 pension fund net direct property investment amounted to only 43.8% of the value of net insurance company property investment. By 1973 this figure had increased to 97.7%.

In addition, rising property values and the new tax incentives for institutions with preferential tax status to hold property directly, rather than via property company shares, stimulated financial innovation. A variety of new vehicles for property investment were developed from the mid-1960s, opening up the market to investors that had previously been too small, or lacked sufficient expertise, to participate in the sector. These were usually committed to investing in only property or cash, thus lacking the flexibility of the insurance companies and pension funds. The most important of these new investment devices were the property unit trusts.

The first property unit trust, the Pension Fund Property Unit Trust (PFPUT), was launched in March 1966. The idea of setting up a property unit trust, as a means of establishing a pooled property portfolio to enable small- and medium-sized pension funds to invest in property, had been conceived by one of PFPUT's founders, Norman Bowie, as early as 1960. A number of practical difficulties involving the legal status of a property unit trust, managerial procedure and the adaption of the unit trust concept to property, which is less amenable to unitization than stock exchange securities due to problems of indivisibility, illiquidity and performance evaluation, led to long delays in its development however.

The main legislative obstacle concerned the Prevention of Fraud (Investments) Act, 1958. This Act made it an offence to distribute circulars seeking investment in unit trusts, unless the trust was 'authorized' under the terms of the Act. 'Authorized' trusts were not allowed to invest in property, and while no such restriction applied to unauthorized trusts their legal prohibition regarding advertising for investors raised serious practical difficulties.

Bowie publicly launched the idea of a property unit trust in October 1961, in a speech at a conference on pension fund investment held by the Industrial Welfare Society, but despite initial interest on the part of a number of pension funds the next few years saw little substantial progress.[23] Legal difficulties, and a range of problems regarding the detailed working of the fund, presented serious obstacles to its realization. However, the introduction of Corporation Tax in 1965 gave a considerable boost to the project, by offering substantial tax advantages for direct or trust investment in property by financial institutions, compared to indirect investment via property company shares.

In the light of these taxation changes, the PFPUT idea was revived in 1965 and after further negotiation with the Inland Revenue and the Board of Trade, concerning PFPUT's legal and tax position, the Trust was launched in March 1966. Two hundred funds initially expressed interest in joining PFPUT, and an application was made to the Board of Trade for permission to circularize a total of 135 funds. In the first year of operation the Trust received £6.7 million in subscriptions;[24] in the light of this success the founders of PFPUT launched two complementary trusts in 1967, one for charities and one for partly-approved pension funds. The merchant banks also began to launch property unit trusts. The Superannuation Fund Property Unit Trust was launched by Lazards and the estate agents Peter Angliss and Yarwood, and Weatherall, Green and Smith in June 1967. A month later a third trust was launched, the Hanover Property Unit Trust, promoted by Samuel Montagu and Knight, Frank & Rutley.[25] September saw the launch of yet another trust, Mutual Property Unit Trust, by Hill Samuel and Edward Erdman.

By September 1969, when the *Bank of England Quarterly Bulletin* reviewed the growth of property unit trusts in a special article, there were 13 property unit trusts, with assets valued at over £100 million.[26] The vast majority of trust units were taken by the pension funds; it was estimated in 1976 that around 95% of all funds received by property unit trusts from their inception to that date came from pension funds, the remaining 5% being invested by charities.[27]

The launch of another property investment vehicle also occurred in 1966 – the property bond. Property bonds are unit-linked life assurance policies, the units being invested in property. They were aimed at individual investors rather than the financial institutions, but were similar to property unit trusts in as much as they allowed the pooling of funds for collective investment in property. It has been estimated that by 1970 about £30 million per annum was being invested in property bonds.[28] By the beginning of 1974 the number of property bonds had risen to about 30,[29] and funds invested in them amounted to £570 million.[30] Most funds were managed by insurance companies or merchant banks, as was the case with the property unit trusts.

A number of other, less important, vehicles were also developed for pooled investment in property. For example, in 1966 the Church Commissioners' property expertise was extended to the investments of the individual chapters of the Church of England, by the setting up of the Chapter Property Pool in which both the chapters, and the Commissioners themselves, held shares, the Commissioners undertaking administration. By 1974 34 chapters were participating in the scheme and a total of £5.87 million had been invested.[31]

Thus, a number of new investment vehicles, which were locked into property as their only investment medium other than cash, were appearing at a time when established institutional investors were attempting to place ever larger funds in the same market. At the same time user-demand was slowing down as a result of deteriorating economic conditions and other factors, such as changes in office technology. The user-market was kept buoyant only as a result of government development restrictions and when these were relaxed a downturn in market conditions was inevitable. In the event, the government not only relaxed development restrictions but also deregulated the banking sector, facilitating a bank-financed development boom which turned that downturn into a financial crisis.

7.3 THE PROPERTY DEVELOPMENT MARKET 1965–1973

The 1950s property boom, like all property development booms, contained the seeds of its own destruction. By the early 1960s City office floorspace was expanding at such a rate that significant oversupply was inevitable within a few years, reducing rents and leading to a downturn in development activity. In the event this was not allowed to take place, due to the introduction of the Brown Ban. In introducing the ban the government acted not out of any wish to prevent a slump in the property market, but out of concern at the rapid growth of London's working population and the transport problems involved in moving an ever growing workforce to and from London's centre on a daily basis.

As early as 1957 the London County Council had published a *Plan to Combat Congestion in Central London*. It was intended to restrict office development in the City, while encouraging development in outer areas such as Middlesex, Kent, Croydon and Essex.[32] Despite the Council's efforts to restrict office development by reducing the areas zoned for offices and lowering plot ratios, little reduction was achieved, largely due to the 'Third Schedule' loophole discussed in Chapter 6. Furthermore, while local authorities in the areas around London had initially welcomed office development, many soon began to see disadvantages, especially the traffic congestion problems which resulted from the inrush of employment, and introduced their own measures to limit office development during the early 1960s.[32]

Central government began actively encouraging decentralization of office employment in February 1963, with the issue of the White Paper *London – Employment: Housing: Land*. Licensing office development was rejected but the Town and Country Planning Act, 1962 was adjusted to close the Third Schedule loophole, encourage the dispersal of London's civil servants and establish the Location of Offices Bureau.[33] The Location of Offices Bureau provided advice for companies which wished to move out of the capital; by 1969 it had relocated 84 000 people on a voluntary basis, according to its own estimate, though the true figure has been estimated to be as high as 104 000–150 000. The government also made a direct contribution towards decentralization, moving about 25 000 government employees from Central London during 1962–1967.[34]

However, prior to this legislation decentralization had already begun, becoming apparent by the end of the 1950s. The proportion of detached office space constructed in suburban and outer London rose from 22% of all office construction in Metropolitan London in 1955 to 66% by 1960.[35] This appears to have been due to the higher costs of a central location, rapid price and rental inflation for City offices leading occupiers to look at areas such as Middlesex and Croydon. Institutional investors had also begun to think in terms of increasing the geographical spread of their property portfolios by the end of the 1950s; as early as March 1955 Legal & General's Board had discussed the need to widen the geographical distribution of their property portfolio, it being recorded that:

> The General Manager then drew attention to the geographical distribution of the properties, and the chairman mentioned that it appears that there was quite a high proportion of the Society's holdings in Oxford Street and Regent Street. The General Manager replied that we do not own the whole of Oxford Street and Regent Street, this was only a small proportion ... This brought a remark from the Chairman that if an atom bomb dropped what effect it would have on our properties, and the General Manager replied that he thought it was sufficiently spread, to which Mr Norton replied that in his view the atom bomb could do quite a lot of destruction to all our properties in those two streets. The General Manager thought that it would be advisable to get as many properties in the provinces as possible.[36]

Despite the beginnings of a trend towards office decentralization, government concern at the continuing growth of the City increased. Following the 1964 General Election the new Labour Government introduced the Brown Ban[37] on London office development, requiring the issue of Office Development Permits (ODP) for new office building in and around the capital. In August 1965 ODPs were extended to cover the Birmingham conurbation, and in July 1966 the legislation was further extended to cover major parts of southern England, and the East and West Midlands.[38]

I. Alexander estimated that from the introduction of office development controls to 1979, when they were abolished, between 170 000 and 250 000 jobs had been dispersed from Central London, a number equivalent to one-quarter to one-third of Central London office employment at the start of the 1970s.[34] Despite this, Central London office development appears to have been more rapid in the decade following the imposition of the Brown Ban than during the previous 15 years; approvals for new office building led to an estimated average yearly net increase of 0.4 million square metres of office floorspace between

August 1965 and April 1977, compared with an average of only 0.3 million square metres from 1948 to 1963.[39] However, these figures conceal differences within the 1965–1977 period, tight controls on office development during the mid to late 1960s being replaced by a much more relaxed attitude in the early 1970s following the return of a Conservative government.

The ODP system had a number of defects. It resulted in the rapid growth of Central London office rents, market rents in the City rising from typical levels of £2 per sq. ft. in 1963 to £5 in 1969 and £14–18 during the early 1970s.[40] If the Brown Ban drove businesses from the City it did so by lining the pockets of its property owners. Restricting London office development also failed to reduce regional disparities. In the period up to April 1967 three-quarters of jobs relocated by the Location of Offices Bureau had moved to another part of the South East, while only 9% had gone to the least prosperous 'assisted areas'; for three regions, Wales, Scotland and Northern Ireland, the percentage of jobs received amounted to less than 1%.[41]

Furthermore, the Brown Ban, and earlier attempts to restrict the growth of the City, were based on inaccurate data. The decision to restrict office development in Central London was based on a belief that between 1951 and 1961 an annual average of 15 000 people had been added to Central London's employment. This estimate was derived from Ministry of Labour figures, calculated from the number of workers holding National Insurance cards exchangeable in Central London offices. This data was subject to an important bias, since many cards were held at London company head offices but represented employees working outside the Central area.[42] It was not until 1966 that the Industry Tables of the 1961 Census of England and Wales became available, providing a more accurate picture of the growth in Central London employment during the 1950s. This indicated that between 1951 and 1961 annual employment growth averaged only 7000–9000 people, a figure not much more than half that used by the government in its decision to restrict Central London office development, and no greater than the rate of growth of the working population throughout Britain.[43]

As a result of the Brown Ban, and other legislation introduced by the new Labour government, particularly Corporation Tax, property developers faced a much more difficult economic environment than had been the case during the 1950s. Financial institutions now preferred to own property directly, to avoid the payment of Corporation Tax, and had developed a more professional attitude to their funding links with property developers, demanding a significant equity stake in development projects in return for the provision of finance. Property shares continued to fall in value from their 1962 peak during the mid-1960s; as *The Economist* noted 'the glamour has unmistakably departed from the commercial development business'.[44]

The Brown Ban led many developers to turn to the shops market, which appeared to offer high profits and was free of government restriction. The years since 1945 had witnessed important changes in the nature of retail provision, which had been largely suppressed until the mid-1950s by building licenses and other impediments to shopping centre development. The most important of these trends was a move towards self-service retailing. This was pioneered by the Co-operative movement, which opened Britain's first self-service grocery store in Portsmouth as early as 1947.[45] However, it was not until the mid-1950s that the removal of building restrictions allowed the number of self-service stores to become substantial. By the middle of 1957 there were at least 80 supermarkets in Britain; five and a half years later their numbers had increased more than tenfold to about 1000.[46] The larger size of outlet required for self-service retailing generally made the conversion of existing shops impracticable, thus generating a demand for new, much larger, stores. Meanwhile, many non-food multiples also wished to develop significantly larger premises. The demand for retail development was therefore considerable; by 1963 70 large-scale central area shopping development schemes were before the Minister for approval.[47]

From the earliest post-war schemes to redevelop blitzed city centres, shopping development was carried out on the basis of close cooperation between the developer and the municipal authority. This type of development was pioneered by the property company Ravenseft in the late 1940s, their practice of winning the confidence of the local authority as a first step to undertaking development[48] forming the basis of shopping centre redevelopment during the following decades. This arrangement was initially necessitated by local authorities' influence over the issue of building licenses, though following the abolition of building license controls their wide planning powers provided ample justification for its continuation.

During the early post-war years relatively few major developers undertook shop development projects, though by the end of the 1950s growing competition in the London office market led to a switch in activity to this sector, while the Brown Ban produced a further transfer of development activity to shopping schemes. However, shop development never provided as lucrative a vehicle for making the fortunes of a large number of small developers as the 1950s office market. Developing shopping centres proved a much more complex process than office development. In addition to the tricky business of gaining the support of the local authority, success was dependent upon an understanding of factors such as pedestrian flow, proximity to car-parking and public transport and, most importantly, position. Pinpointing a good location is much more important for shops than for offices, a distance of a few hundred yards often having a major influence on market rents. As a result shop development remained something of a specialist field despite the increased competition; during the 1960s the sector was dominated by the 'Big Six' developers: Ravenseft, Arndale, Hammersons, Laing, Murrayfield and Town & City.[49]

Shopping centre developments during the early 1960s continued the 1950s pattern of fully or partly pedestrianized precincts, which had been pioneered in

American suburban centres prior to the Second World War. However, from the mid-1960s larger, covered, shopping centre schemes began to dominate new development. The British covered shopping centre has a similar pedigree to the pedestrianized precinct, being inspired by earlier developments of the same type in the United States. Two large, notorious, projects marked the start of covered shopping centre development in the UK, the Elephant and Castle Centre in London, and Birmingham's Bull Ring Centre.[50]

Both these developments began life, by coincidence, in September 1959. Both embodied a revolutionary concept, the development of a very large number of shops[51] on more than one level in a single shopping centre which was closed to the sky.[52] These were ambitious developments, but unfortunately both failed to avoid many of the pitfalls which face the pioneer. Their problems of pedestrian flow and other design faults, discussed in detail by Marriott in *The Property Boom*,[53] provided expensive lessons in shopping centre design for their developers, Willet's and Laing's. However, these lessons were to prove valuable for developers in general during the following decade, paving the way for a boom in, very largely, successful developments of this type.

The 1965–1973 period saw a boom in retail development, the floorspace of new shopping schemes[54] opened during each year rising from 2176 million sq. ft. in 1965 to 5126 million sq. ft. in 1971 and 5646 million sq. ft. in 1973.[55] In contrast to the 1955–1964 office boom, the 1965–1973 shopping centre boom generated very little public controversy. This appears to have been the result of strict local and central government control on development in this sector, together with the absence of individual white elephants or general oversupply.[56] Only about a dozen centres developed during this period were never fully let, failures being mainly due to errors in location rather than size or design.[57] The covered shopping centre had proved a success and become firmly established as the main form of new shopping centre development.

The development of closed shopping centres typically involved a substantially larger scale of development than the shopping precincts of the previous decade, entailing a greater role for the local authority. The authority's planning powers, which could be used to compulsorily purchase the necessary land and prevent rival centres being developed in the same area, were put to the service of the developer via partnership arrangements in which developers worked closely with municipal authorities. Such arrangements were encouraged by central government; the Ministry of Housing and Local Government's *Town Centre Manual*, published in 1962 and 1967, provided details of how such deals could be organized.[57]

There was a considerable community of interest between the developer and the local authority during this period. Both generally looked for pedestrianized schemes near to a bus station or car-park. Developers were often prepared to throw in non-commercial buildings, such as a library or sports centre, as part of the scheme in order to keep the local authority happy.[57] However, the municipal authorities soon began to realize that they were not making full use of their bargaining position *vis-à-vis* the developers, and in a belief 'they might not be getting their share of the cake',[50] countered the greater experience of the developers by turning for advice to the commercial estate agents which specialized in shop property. These established consultancy services acted as 'poachers turned game keepers'[50], by putting the expertise they had built up in the service of developers to the benefit of the local authorities. As a result local authorities were obtaining significantly better terms from developers by the early 1970s; many owned substantial stakes in shopping developments and some even became shop developers in their own right.

The active role of local authorities in shopping centre development contributed to the lack of oversupply of shopping space. Unlike factory and office development, shopping centres could be refused planning permission on shopping policy grounds in addition to considerations regarding the appropriateness of the site to its use. Questions such as the adequacy of local spending power to support the centre, and its impact on existing shops, are legitimate areas of concern under British planning law.[58] Unlike the USA, Australia and Western Europe, where a significant amount of suburbanization of retailing activity has taken place during the post-war period, British local authorities have (until the 1980s) successfully used these powers to restrict shopping developments to locations of, or near, traditional shopping activity, preventing the decay of such areas.[59] Furthermore, their influence has, in some cases, ensured the inclusion of public amenities in shopping centre projects.

R. Schiller argued that, while local authority planning powers might have modified the character of shopping centres to some extent by bringing non-commercial criteria into the development equation, such modifications have been relatively slight. However, he suggested that delays in development arising from the planning process may have increased the cost of shopping centres substantially, leading to a significant increase in rents. This increase in costs may have been an important factor in keeping less profitable retailers, in areas such as books and other slow moving lines, out of shopping centres, while in the United States the turnover necessary to secure a shopping centre location is substantially less.[60] Delays also increase the obsolescence of shopping centres, completed centres being based on designs which might have been first proposed 10 years or more ago. Data collected by D.J. Bennison and R.L. Davies, based on information from 140 District Planning Departments in late 1978 to early 1979, showed that for a sample of 172 shopping centre schemes built since 1963 the time between drawing up the first plans and completing the centre averaged nine years.[61]

Furthermore, by the very act of restricting shop development, local authorities have maintained shop rents at levels higher than they would otherwise have been. While the waste of resources entailed by oversupply has been avoided, and the decay of traditional shopping centres has been prevented, this has occurred at the expense of higher shop rents which have been passed on to the consumer, to some extent, in the form of higher prices.

Another notable trend of the 1965–1973 period was a growth in development activity by the financial institutions. Although a small number of pioneering institutions, such as Legal & General, Norwich Union, the Pearl and the Church Commissioners, had undertaken direct development activity prior to 1965 it was during these years, and particularly the early 1970s, that a major expansion of development activity by the financial institutions took place. A shortage of sufficient investment opportunities in the market for completed properties, and consequent low initial yields on such investments, led a number of institutional investors either to undertake direct development activity, or to form closer links with property developers, in order to ensure an adequate supply of new investment properties.

Direct developments comprised an important part of Legal & General's property acquisitions during this period, especially in the case of offices; for example, all but one of the office properties added to the Society's portfolio during 1967 were acquired via development. The Society also invested heavily in the development sector via property company shares, holding a higher proportion of shares in this sector than would be expected from its weighting in the FT Actuaries All Share Index. In December 1969 property shares formed 8.4% of Legal & General's ordinary share holdings, by market value.[62]

NPI also turned to the property development market during the 1970s in search of higher yields than could be obtained on conventional property investments. Activity was concentrated on the redevelopment, in conjunction with London Life, of a number of properties owned by NPI. By 1972 the Institution had development projects involving a total commitment over the next six years of just over £1.75 million per annum, a considerable proportion of annual net new money received by NPI, which stood at £5 million per annum.[63] Development projects under consideration by the Institution during the early 1970s offered yields as high as 9–10% – far above those available on conventional property investments.

Clerical Medical also considered undertaking developments during the early 1970s, but found few propositions which it considered attractive. During 1973 it turned to a strategy of increasing its interests in property by offering finance to small property companies in cases where the development under consideration was thought to be good but the financial history of the company was not strong enough to command competitive rates from other institutions. This marked a return to the policy the Society had pursued in the early post-war years; however, the expansion of bank lending to property companies during the early 1970s had led to greater competition in this market and few suitable projects were found. The lack of sufficient market opportunities for direct property investment also caused Clerical Medical to expand its property share portfolio, which rose from 14.5 to 16.6% of its total ordinary shares between the end of 1971 and 1972.

Standard Life continued the policy it had pursued since the early 1960s of investing in the property sector largely through partnership arrangements with various property development companies. By 1968 it owned substantial blocks of shares in eight UK property companies. Many of the properties which entered Standard's portfolio during these years were acquired via leaseback arrangements with property companies with which Standard had close funding links, the equity links it had forged in the early 1960s proving useful during these years of relative scarcity of investment properties. Standard's 1972 *Asset Investigation Report* listed property holdings according to the institution from which they were purchased. The company's UK property assets acquired by conventional market purchases were valued in that year at just over £20 million, while those acquired by sale and leaseback deals had a market value of £56 million.[64]

Eagle Star also turned to property development as a means of investing in property on an acceptable yield basis. In October 1973 the company discussed a proposed scheme to redevelop office properties already in their portfolio which had development potential. It was estimated that £2.25 million could be invested in this way, at an initial yield of 6–8%, compared to prevailing initial yields on completed offices of 5%.[65] Eagle Star also considered another possible means of augmenting its property holdings – property company acquisitions. During 1973 the company attempted to acquire Bernard Sunley Investment Trust Ltd and Grovewood Securities Ltd, with which it had funding links going back to 1961 and 1958, respectively. The take-overs were referred to the Monopolies and Mergers Commission, and by the time they had been cleared the 1974 property crash led Eagle Star to postpone the bids.[66]

Property company take-overs appeared attractive investments during the late 1960s and early 1970s, partly as a result of the effects of Corporation Tax. Corporation Tax hit property companies especially severely, as it was levied on distributed profits and property companies traditionally distributed almost all their income. To maintain dividends following the introduction of this tax, which was initially levied at a rate of 40% on distributed profits, required an average rise in overall profits of 25%, something which property companies had difficulty achieving under conditions of falling development margins.[67]

Corporation Tax encouraged property companies to reinvest profits rather than distributing them as dividends. Thus, assets grew while low dividends depressed share prices. This led to a growing disparity between the market capitalization of a company's shares and the net value of its assets, shares sometimes standing at as much as a 40% discount to the assets they represented. These conditions gave both the financial institutions and other property companies a considerable incentive to acquire undervalued property assets via corporate acquisitions, resulting in a growing concentration of the property company sector. In 1960 the 10 largest property companies represented 20% of the assets of the sector. By 1970 this figure had increased to nearly 50%,[68] while between 1964 and 1970 60 property companies had been absorbed into others.[69]

The shortage of investment properties during the early 1970s provided a further strong incentive for property company acquisitions. Several institutional investors which had been frustrated in their attempts to meet property investment targets for several years considered this an attractive method of acquiring an entire property portfolio in a single deal, thereby making up for their accumulated deficit of property purchases. In 1969 the BP pension fund purchased Western Ground Rents, and in 1973 the Post Office pension fund acquired English and Continental Property, which had gross assets valued at around £95 million. The insurance companies undertook more extensive property company purchases; during 1971–1973 the Prudential, Royal Insurance, Commercial Union and Legal & General all acquired property companies.[70]

While pension funds undertook fewer property company take-overs than the insurance companies many did become heavily involved in property development funding, often entering into deals on terms which were somewhat less prudent than those demanded by the insurance companies. In *That's The Way The Money Goes* financial journalist John Plender drew attention to the heavy commitment to speculative property ventures on the part of a number of pension funds during this period, including those of ICI, Unilever, the Post Office and the Electricity Supply Industry, often involving overseas property development.[71]

One deal in particular illustrates the difference of approach to development funding between the pension funds and the more experienced insurance companies. The Post Office Pension Fund, which was eager to increase its stake in the property sector during the early 1970s, entered into a funding arrangement with Argyle Securities, a property development group which had recently been acquired by the financier James (later Sir James) Goldsmith.[72]

Argyle was involved in a massive scheme to redevelop the Grands Magasins du Louvre, in the centre of Paris, at an estimated cost of just under £30 million. The Post Office entered into a deal with Argyle, acquiring a 50% share of the profits of the scheme in return for jointly guaranteeing the borrowings with Argyle. A few months later, toward the end of 1973, the fund purchased just over 15% of Argyle's ordinary shares, adding to its overall stake in the project.[72] Argyle had launched a take-over bid for Cornwall Properties shortly before this purchase; by the time the take over was completed the property crash had begun, leaving the company with heavy borrowings at a time of falling property prices.

This element of the Post Office's problem with Argyle was largely a matter of bad timing. However, there was another aspect to the deal. The Grand Magasins development had funding guaranteed jointly by both Argyle and the Post Office Pension Fund. Long-term funding, upon completion, was to be by a mortgage, 50% of which would be guaranteed by the fund and 50% by Argyle. However, Argyle had arranged the deal so that it could withdraw from guaranteeing the long-term mortgage if it wished. Such a provision was completely at odds with traditional insurance company property development funding arrangements, which sought to secure an equity stake in development ventures while minimizing any loss should the deal go sour. Instead, the option to withdraw from the scheme, while retaining the benefits should it succeed, was held not by the pension fund but by the property company. Thus, the Post Office Pension Fund had taken on 50% of the potential profits of the venture and 100% of the potential losses.[73]

In 1975 Argyle did, indeed, refuse to guarantee the long-term mortgage and stated that it wished to withdraw from the project. The Post Office pension fund took on the whole development, in addition to paying Argyle £1.5 million under their arrangement regarding termination of the partnership. Thus, Argyle, rather than being faced with what appeared very likely to be a substantial loss on the development, made a profit of £1.1 million.[74] This arrangement was an extreme case, even among the catalogue of investment woes described by Plender, though it is interesting to note that the pension funds came in for much stronger criticism in his account of institutional property investment during the 1970s than the more experienced insurance companies.

The early 1970s witnessed a boom in property development. The intensification of investment activity by the property-investing financial institutions was one cause of this boom, though a far more important factor was the massive increase in bank lending to property developers. The banks' eagerness to loan funds to developers was largely a result of the boom in property values which had taken place during the previous five years. Veblen noted that rising collateral values enabled businessmen in the sector to increase their borrowing since:

> This recapitalization of industrial property, on the basis of heightened expectation, increases the value of this property as collateral ... But during the free swing of that buoyant enterprise that characterizes an era of prosperity contracts are entered into with a somewhat easy scrutiny of the property values available to secure a contract ... not only is the capitalization of the industrial property inflated on the basis of expectation, but in the making of contracts the margin of security is less closely looked after than it is in the making of loans on collateral.[75]

During the early 1970s bank advances to the property sector rose substantially, from £362 million in February 1971 to £2584 million in February 1974, an increase of 614%. This compares with an increase in total bank advances over these three years of only 162%.[76] The rapid inflow of funds to the property company sector fuelled a boom in property share prices; in the year to May 1972 alone property company share values virtually doubled.[77]

The development boom was facilitated by a change in government policy following the return to power of a Conservative government in 1970. During 1965–1969 Office Development Permits had severely restricted the volume of development activity, thereby maintaining the buoyancy of the property market despite demand conditions which were growing progressively less favourable. Restriction resulted in rising market rents for offices, which grew at an average rate of 15.65% from 1966 to 1973. This compares with an average rate of growth of only 7.29% for market rents in the shop sector, which was not subject to these restrictions.[78] Development controls, and consequent higher rents, for offices increased the relative attraction of this class of property for institutional investors; from 1965 to 1973 the proportion of institutional property holdings made up of offices rose from about 45 to 62%.[79]

In 1970 the new Secretary of the Environment, Peter Walker, relaxed controls on office development in order to reduce the upward pressure on rents and allow office space to expand in anticipation of increased demand following Britain's entry to the EEC.[80] An office development boom followed the lifting of these restrictions, to meet a demand for office space that proved to be notably absent during the mid-1970s. From 1966 to 1969 the volume of office construction permitted under office development permit legislation in the South East planning region had averaged 9 273 000 sq. ft. The relaxation of restrictions increased this figure to 26 353 000 sq. ft. during 1970–1973.[81] Provincial centres were also caught up in the boom; office floorspace in the central areas of Leeds and Bristol increased by 25 and 33%, respectively, between 1970 and 1975.[82] The repeal of the Land Commission legislation in 1971 provided a further stimulus to development, by restoring to owners the whole of the development value of their property.[83]

The volume of overall construction activity was closely linked to these changes in government policy; from 1961 to 1964 the average real value of new orders received by contractors for private sector commercial property was £2583 million. Following the introduction of the Brown Ban this fell substantially, to an average of £1987 million during 1965–1970, but the relaxation of controls during the following years led to a dramatic increase in orders, to an average of £3283 million during 1971–1973.[84]

The development boom occurred alongside a continuation of the trend towards declining initial development yields that had been evident since the early 1960s. By the early 1970s the initial rent for completed properties seldom exceeded the annual interest on development costs. Projects still offered a capital gain, but the developer had to find some way to retain the property, and thus avoid paying tax on that gain, while meeting the interest bill.[85] In order to get around this problem property developers increasingly turned to the banks, which had traditionally been an important source of short-term development finance, for medium-term funding. As property values rose sharply during the early 1970s the gap between initial income and interest payments that developers, and the banks, considered acceptable grew, while the period over which they were prepared to accept an income deficit increased.[86] Some property companies even allowed interest payments to exceed their overall current income.[87]

The emergence of a considerable reverse-yield gap between the initial income from a property and the interest rate on funds borrowed to finance its development was sometimes dealt with by 'rolling up' interest payments. This involved interest being added to the outstanding capital each year until the loan was paid off. Paying off a loan on these terms depended on steadily rising rents, something which was threatened by the introduction of commercial rent control, and stable interest rates. The growing practice of borrowing money at variable, rather than fixed, interest rates further increased the property company sector's vulnerability, should future high interest rates coincide with slower rental growth. Another growing practice, borrowing from overseas in foreign currencies, incurred similar risks if Sterling fell significantly in value.[86]

The legal framework governing property company taxation also encouraged greater borrowing. One consequence of the introduction of Corporation Tax was to increase property company gearing ratios, i.e. the ratio of borrowings to assets, as loan stock interest was allowable as a deduction against Corporation Tax, and income used to service interest payments on loans and debentures was taxed at a lower rate.[88] This led to a spate of property company equity reconstructions, unsecured loan stock being substituted for preference shares, in addition to encouraging property companies to raise a greater proportion of new funds via borrowings.[88]

Changes in the way property company assets were valued also served to destabilize the sector during the 1960s and early 1970s. During the early 1960s it became fashionable for stockbrokers to assess the value of a property company on the basis of its net asset value, rather than the relationship between share price and earnings. This represented a significant move from concentrating on present income to the expected future income stream represented in property values, adding to the instability of the sector, as values were far more susceptable to decline under adverse market conditions than income. The imprecision of property valuations accentuated such difficulties since, as Jack Rose later put it 'The gross rental income is factual, the capital valuation of the property is in the mind of the valuer'.[89]

Furthermore, chartered surveyors were becoming increasingly liberal in their valuation practices when calculating the market value of company property holdings for accounting purposes. As Plender stated:

In many cases, chartered surveyors and other valuers with qualifications of varying suitability accepted instructions to value speculative, uncompleted and unlet developments – often just holes in the ground – on the assumption that they would be completed on time and let in full to tenants at current market rents. Property companies would then incorporate these hypothetical figures in their accounts; the only allowance many of them made for the fact that the development was months, or in some cases even years, from completion, was to deduct the estimated cost of putting up the building from the supposed value on completion.[90]

However, fears of the effects of a downturn in market conditions, or an increase in interest rates, or both, on the precarious financial position of the property industry were rarely expressed, expectations of steady or accelerating property value growth being built into investors' calculations. In an environment of rising inflation, property was viewed as one of the few sectors that could be counted on to produce real capital growth. As Margaret Reid observed, property investment and development had become 'the new alchemy'.[91]

The 1971–1973 development boom was largely financed via bank lending to the property company sector. The overcommitment of the banks to property development funding was made possible due to changes in the system of bank regulation. During the 1960s the Bank of England had become increasingly dissatisfied with its system of credit regulation, which relied on the restriction of bank lending and therefore led to the burden of controls falling unduly heavily on the London clearing banks and the Scottish banks. This had encouraged the development of the secondary or 'fringe' banking sector, over which the Bank of England exercised less firm control.[92] The establishment of fringe banks had been facilitated by Section 123 of the Companies Act, 1967, which allowed the Board of Trade to certify that particular companies were carrying on the business of banking, giving them a 'junior' form of banking status.[93] By 1970 87 such certificates had been issued and by 1973 the number had increased to 133.

The development of new financial markets during the 1960s, such as those for local authority temporary money, inter-bank deposits and certificates of deposit, also made the existing system of regulation less effective by weakening the relationship between bank rate and interest rates. This relationship had been administratively determined via a cartel of the London clearing banks, which set deposit and loan rates, but proved less successful in controlling rates in these new markets.[92]

In May 1971 the Bank of England set out suggested changes to the regulatory system in a consultative document, *Competition and Credit Control*, which suggested that the previous system of bank credit regulation be replaced by control via a system of required balance sheet ratios, and requiring banks to make special deposits with the Bank of England.[94] These proposals, together with further measures affecting the banking system, were introduced in September 1971. The new system marked a move away from the quantitative restrictions on lending which had forced developers to turn to alternative sources of finance in the mid to late 1950s. It aimed to control bank lending by putting pressure on bank liquidity using changes in the Reserve Assets Ratio.[95] Interest rate adjustments were also to be used as a major instrument of monetary policy. This was to prove unfortunate, since the precarious financial margins on which developments were currently undertaken meant that a major increase in interest rates (as occurred during the second half of 1973) would turn expected profits on many development projects into heavy losses.

This system led to a rapid acceleration in bank lending, particularly to the property sector. Much of this lending took place via the secondary banks, which borrowed money from the more regulated clearing banks and channelled it to the property sector and other areas. These 'fringe' banks were not closely monitored by the Bank of England, which had established a system whereby the more information the banks gave the Bank of England concerning their finances, the more freedom they were given regarding the categories of business they could conduct. This was described by Plender as 'An Alice in Wonderland system ... whereby the bigger and more reputable the bank was, the more closely it was monitored'.[96]

An expansion of bank lending was a conscious policy of the Heath Government, which hoped to increase output by liberalizing credit. In his March 1972 budget speech the Chancellor, Anthony Barber, announced that he aimed to raise output growth to 5% per annum, an ambitious target given Britain's poor post-war growth record.[97] However, industry was not inclined to expand its borrowing to the required extent, and the funds went, instead, to property development, a sector which was particularly vulnerable should the Heath–Barber 'dash for growth' fail. Furthermore the failure of the rest of the economy to expand as quickly as the property development sector during these years meant that the demand for property was not increasing as fast as the supply, a fact which, in itself, cast doubt on the profitability of the large number of development schemes currently underway.

On August 8th 1972 the Bank of England requested the banks to reduce their lending to property and financial companies, following concern at the rate at which money was entering this sector, and concentrate instead on the provision of finance for industry.[98] However, it no longer had the necessary mechanisms to enforce such requests, as a result of the 1971 changes. Those banks which were prepared to ignore the Bank of England were able to channel the funds of their more responsible, and more regulated, counterparts to this sector by borrowing from them.

Thus, massively expanded and virtually uncontrolled bank lending, assisted to some extent by insurance company and pension fund development finance, were funding a development boom in a property market which was facing stagnant demand, which threatened to turn into falling demand should the expansionist policies of the Conservative Government fail. The banks, insurance companies and pension funds, were all heavily committed to this volatile market and a severe property crash had the potential to bring down the entire financial system. It almost did.

REFERENCES AND NOTES

1. Full stop for London offices (1964) *The Economist*, **ccxiii**, 7 Nov., 616–17.
2. For example, D. Massey and A. Catalano (1978) *Capital and Land: Landownership by Capital in Great Britain*, Edward Arnold, London; H. Smyth (1985) *Property Companies and the Construction Industry in Britain*, Cambridge Univerisity Press, Cambridge.
3. T. Veblen (1965) *The Theory of Business Enterprise*, Kelley, New York (originally published 1904).
4. T. Veblen (1965) *The Theory of Business Enterprise*, Kelley, New York, p. 180 (originally published 1904).

5. T. Veblen (1965) *The Theory of Business Enterprise*, Kelley, New York, p. 194 (originally published 1904).
6. D. Cadman and L. Austin-Growe (1983) *Property Development*, E & FN Spon, London, p. 242.
7. See section 10.2.
8. R. Barras (1979) *The Returns from Office Development and Investment: Centre For Environmental Studies Research Series No. 35*, Centre for Environmental Studies, London, p. 2.
9. T. Veblen (1965) *The Theory of Business Enterprise*, Kelley, New York, p. 197 (originally published 1904).
10. See Table 11.1.
11. Clerical Medical (1966) Investigation into the Society's assets.
12. Yields adjusted to take account of the discounted value of future expected rental increases.
13. NPI (1968) Board paper 1968/28, 27 June.
14. Clerical Medical (1968) Investigation into the Society's assets.
15. Clerical Medical (1969) Investigation into the Society's assets.
16. Clerical Medical (1970) Investment Forum notes, 10 Aug.
17. Clerical Medical (1970) Investment Forum notes, 3 April. These figures were set for gross pension funds and would not apply to insurance companies without modification due to their different tax position.
18. Legal & General (31 Dec. 1970) Memorandum on investment policy for 1971.
19. Legal & General (1971) Investment policy: Second half 1971, Board Paper, 5 July.
20. Clerical Medical (1972) Investment Forum minutes, 15 Sept.
21. Clerical Medical (1972) Investment Forum minutes, 1 Dec.
22. Clerical Medical (1972) Investigation into the Society's assets.
23. R. Redden (1984) *The Pension Fund Property Unit Trust: A History*. Privately published, London, p. 34.
24. R. Redden (1984) *The Pension Fund Property Unit Trust: A History*. Privately published, London, p.94.
25. R. Redden (1984) *The Pension Fund Property Unit Trust: A History*. Privately published, London, p. 95.
26. Property unit trusts for pension funds and charities (1969) *Bank of England Quarterly Bulletin*, **ix**, Sept., 294–6.
27. Panmure Gordon & Co. (1976) *The Property Sector 1976–1977*, Panmure Gordon & Co., London, p. 29.
28. C.J. Baker (1970) The changing property scene – unit trusts, bonds and finance. *Estates Gazette*, **214**, 27 June.
29. P. Ambrose and R. Colenutt (1975) *The Property Machine*, Penguin, London, p. 54.
30. Panmure Gordon & Co. (1976) *The Property Sector 1976–1977*, Panmure Gordon & Co., London, p. 30.

REFERENCES AND NOTES

31. Church Commissioners (1974) Annual Report, 31 March, pp. 22–3.
32. N. Moor (1979) The contribution and influence of office developers and their companies on the location and growth of offices, in *Spacial Patterns of Office Growth and Location* (ed. P.W. Daniels), Wiley, Chichester, p. 195.
33. N. Moor (1979) The contribution and influence of office developers and their companies on the location and growth of offices, in *Spacial Patterns of Office Growth and Location* (ed. P.W. Daniels), Wiley, Chichester, p. 204.
34. I. Alexander (1979) *Office Location and Public Policy*, Longman, London, p. 65.
35. I. Alexander (1979) *Office Location and Public Policy*, Longman, London, p. 63.
36. Legal & General (1955) Annual report on freehold and leasehold properties, to Board of Directors, 2 March.
37. The Brown Ban was introduced in the 1964 White Paper *Offices: a statement by Her Majesty's Government*.
38. J.B. Cullingworth (1988) *Town and Country Planning in Britain*, 10th edn, Unwin Hyman, London, p. 43.
39. I. Alexander (1979) *Office Location and Public Policy*, Longman, London, p. 66.
40. I. Alexander (1979) *Office Location and Public Policy*, Longman, London, p. 68. Based on Department of the Environment figures.
41. I. Alexander (1979) *Office Location and Public Policy*, Longman, London, p. 70.
42. A.W. Evans (1967) Myths about employment in Central London. *Journal of Transport Economics and Policy*, **1**, 216.
43. A.W. Evans (1967) Myths about employment in Central London. *Journal of Transport Economics and Policy*, **1**, 215.
44. Developers overdeveloped (1965) *The Economist*, 27 Nov., 980.
45. H. Mason (1989) The twentieth-century economy, in *The Portsmouth Region* (eds B. Stapleton and J.H. Thomas), Alan Sutton, London, p. 175.
46. W.G. McClelland (1963) *Studies in Retailing*, Blackwell, Oxford, p. 19.
47. D. Cadman and A. Catalano (1983) *Property Development in the UK – Evolution and Change*, E & FN Spon, London, p. 5.
48. O. Marriott (1967) *The Property Boom*, Hamish Hamilton, London, p. 61.
49. O. Marriott (1967) *The Property Boom*, Hamish Hamilton, London, p. 121.
50. R. Schiller (1985) Land use controls on UK shopping centres, in *Shopping Centre Development: Policies and Prospects* (eds J.A. Dawson and J.D. Lord), Croom Helm, London, p. 49.
51. The number of shops included in the Bull Ring and Elephant and Castle projects were 140 and 120, respectively.
52. O. Marriott (1967) *The Property Boom*, Hamish Hamilton, London, p. 214.
53. O. Marriott (1967) *The Property Boom*, Hamish Hamilton, London, Ch. 14.

54. Including schemes with a gross floorspace of 50 000 sq. ft. or over.
55. Hillier, Parker, May and Rowden (1983) *British Shopping Developments 1965–1982*, Hillier, Parker Research, London, Table D.
56. D.J. Bennison and R.L. Davies (1980) The impact of town centre shopping schemes in Britain: Their impact on traditional retail environments. *Progress in Planning*, **14**(1), 9.
57. R. Schiller (1985) Land use controls on UK shopping centres, in *Shopping Centre Development: Policies and Prospects* (eds J.A. Dawson and J.D. Lord), Croom Helm, London, p. 50.
58. R. Schiller (1985) Land use controls on UK shopping centres, in *Shopping Centre Development: Policies and Prospects* (eds J.A. Dawson and J.D. Lord), Croom Helm, London, p. 46.
59. D.J. Bennison and R.L. Davies (1980) The impact of town centre shopping schemes in Britain: Their impact on traditional retail environments. *Progress in Planning*, **14**(1), 11.
60. R. Schiller (1985) Land use controls on UK shopping centres, in *Shopping Centre Development: Policies and Prospects* (eds J.A. Dawson and J.D. Lord), Croom Helm, London, pp. 54–5.
61. D.J. Bennison and R.L. Davies (1980) The impact of town centre shopping schemes in Britain: Their impact on traditional retail environments. *Progress in Planning*, **14**(1), 25. The survey covered only shopping centre schemes of over 50 000 sq. ft. gross retail floorspace.
62. Legal & General, list of ordinary shares held by Life and General funds as at 31 Dec. 1969, appended to Board minutes.
63. NPI (1972) Property Committee paper, 27 July.
64. Standard (1972) Asset Investigation Report.
65. Eagle Star (1973) General Purposes and Finance Committee minutes, 2 Oct.
66. Eagle Star eventually acquired the Bernard Sunley Investment Trust in 1979.
67. *The Stock Exchange Gazette* (1965) 10 Sept., 879.
68. H. Smyth (1985) *Property Companies and the Construction Industry in Britain*, Cambridge University Press, Cambridge, p. 166.
69. D. Cadman and A. Catalano (1983) *Property Development in the UK – Evolution and Change*, E & FN Spon, London, p. 7.
70. J. Plender (1982) *That's the Way the Money Goes*, Andre Deutsch, London, p. 99.
71. J. Plender (1982) *That's the Way the Money Goes*, Andre Deutsch, London, pp. 120–65.
72. J. Plender (1982) *That's the Way the Money Goes*, Andre Deutsch, London, p. 141.
73. J. Plender (1982) *That's the Way the Money Goes*, Andre Deutsch, London, p. 143.
74. J. Plender (1982) *That's the Way the Money Goes*, Andre Deutsch, London, p. 144.

REFERENCES AND NOTES

75. T. Veblen (1965) *The Theory of Business Enterprise*, Kelley, New York, pp. 197–8 (originally published 1904).
76. Panmure Gordon & Co. (1976) *The Property Sector 1976–1977*, Panmure Gordon & Co., London, p. 44.
77. M. Reid (1982) *The Secondary Banking Crisis, 1973–75: Its causes and course*, Macmillan, London, p. 62.
78. Source: Department of the Environment (1978) *Commercial and Industrial Property Statistics 1978*, HMSO, London, p. 26.
79. Based on the sources used for Table 10.7.
80. S. Jenkins (1975) *Landlords to London*, Book Club Associates, London, p. 236.
81. Department of the Environment (1975) *Commercial and Industrial Property Facts & Figures*, HMSO, London, Table 15.
82. D. Cadman and A. Catalano (1983) *Property Development in the UK – Evolution and Change*, E & FN Spon, London, p. 9.
83. E. Reade (1987) *British Town and Country Planning*, Open University Press, Milton Keynes, p. 62.
84. Source: CSO (various issues) *Annual Abstract of Statistics*.
85. J. Plender (1982) *That's the Way the Money Goes*, Andre Deutsch, London, p. 95.
86. J. Plender (1982) *That's the Way the Money Goes*, Andre Deutsch, London, p. 96.
87. M. Reid (1982) *The Secondary Banking Crisis, 1973–75: Its causes and course*, Macmillan, London, p. 66.
88. C. Hamnett and W. Randolph (1988) *Cities, Housing and Profits*, Hutchinson, London, p. 107.
89. J. Rose (1981) Institutional investment: Have the funds got it wrong? *Estates Gazette*, **260**, 21 Nov., 779.
90. J. Plender (1982) *That's the Way the Money Goes*, Andre Deutsch, London, pp. 96–7.
91. M. Reid (1982) *The Secondary Banking Crisis, 1973–75: Its causes and course*, Macmillan, London, p. 63.
92. N.H. Dimsdale (1991) British monetary policy since 1945, in (eds N.F.R. Crafts and N. Woodward) *The British Economy Since 1945*, Clarendon, Oxford, p. 118.
93. M. Reid (1982) *The Secondary Banking Crisis, 1973–75: Its causes and course*, Macmillan, London, p. 49.
94. Yes at last, revolution for the City (1971) *The Economist*, **239**, 22 May, 70–5.
95. N.H. Dimsdale (1991) British monetary policy since 1945, in (eds N.F.R. Crafts and N. Woodward) *The British Economy Since 1945*, Clarendon, Oxford, pp. 118–19.
96. J. Plender (1982) *That's the Way the Money Goes*, Andre Deutsch, London, p. 48.

97. N.H. Dimsdale (1991) British monetary policy since 1945, in (eds N.F.R. Crafts and N. Woodward) *The British Economy Since 1945*, Clarendon, Oxford, p. 120.
98. William Keegan (1972) Banks are asked to exercise restraint. *Financial Times*, 9 Aug., 1.

The property crash, aftermath and recovery 1974–1980   8

8.1 INTRODUCTION

The 1974 property crash marked the end of a boom in property investment which had lasted, almost without interruption, since the end of the Second World War. A market which had offered high returns, with little risk, to investors and developers during the 1950s had become so overcrowded with institutional funds by the early 1970s that investment could be conducted only on the basis of very optimistic assumptions regarding future income and capital growth. Events during 1974 shattered those assumptions, and marked a new phase in the history of the property market, investors realizing that prices could go down as well as up.

The crash had a severe impact on the property company sector, which did not regain its former importance in the property investment and development markets until the mid-1980s. It also led to a severe contraction in the secondary banking sector. However, the faith of the financial institutions in property soon recovered, and after a couple of years of depression and relative inactivity the property investment market began to experience another boom.

While the real value of net institutional investment in the sector never exceeded its 1974 peak, investment during the 1974–1980 period remained well above that for the years prior to 1973. This relatively high level of investment occurred despite initial yields which fell to, or below, 1973 levels during the late 1970s.[1] Negative differentials between initial property yields and yields on other assets widened during these years; the gap between shop and gilt yields grew from an average of −1.74% during 1965–1973 to −8.47% during 1974–1980. During 1974 the differential between property and equities also became negative, for the first time, falling further during the late 1970s.[1] This occurred despite unattractive returns to investment in property. From 1974 to 1980 property offered an average return of 12.3%. For the first time since the

Second World War both shares and gilts out-performed property, with average returns of 19.3 and 13.6%, respectively.

This chapter examines the impact of the property crash on property companies, banks and institutional investors. It also discusses why, despite low yields and unattractive investment returns, the property investment sector made such a strong recovery during the late 1970s.

8.2 THE PROPERTY CRASH

Veblen stated that the price rises experienced during a commercial boom induce a massive expansion in credit, based on the value of the commodity which has experienced the price rise as the collateral security, as occurred with bank lending to the property market during the early 1970s. However, should the price of the commodity in question begin to fall the collateral will decline in value to a point where it will not support the outstanding credit which it has secured.[2] This results in liquidations, as particular agents in the market find themselves unable to meet their debts.

The 1973/74 property crash was remarkable not only in its severity but also in the speed with which conditions turned from buoyant to catastrophic. In July 1973 Legal & General considered market conditions to be extremely favourable, its property committee's *Property Market Report* noting that:

> Yields on prime properties, especially in the industrial sector, have hardened further and this, coupled with increasing rental levels for new buildings has produced a substantial rise in the capital values of prime investments. The property market yields are almost back to 'pre-freeze'. The shortage of suitable properties is still acute.[3]

Conditions in the retailing sector appeared prosperous, the annual reports of major retailers showing substantial increases in pre-tax profits. A high level of demand for office space was also reported, it being stated that the demand from large space-users was pushing smaller users out of the City and that the 3 million sq. ft. of office space which was available in December 1972 had dropped to 1.2 million sq. ft.[3]

As late as August 1973 the property market showed few signs of an impending crash. As a note from Clerical Medical's investment forum during that month stated 'There seems to be a flight from money into bricks and mortar at any price'.[4] The impact of rent control had not yet begun to bite into property values, though a month later it was suggested that investment in property would be unwise at the present time, since prices were about to peak.[5] This view was not shared by Legal & General's property committee, which still viewed conditions as buoyant, the *Financial Times* of 31st August being quoted as stating that there was a continuing shortage of property in the market, with prime yields below 4%.[6]

However, the storm clouds were already beginning to gather over the property market. The major cause was the failure of the Heath–Barber 'dash for growth', which resulted in a sudden and dramatic rise in interest rates. On July 20th 1973 minimum lending rate was increased from 7.5 to 9%; a week later it was increased again, to 11.5%. This marked the beginning of the end for the property boom.[7] The quadrupling of oil prices during the autumn of that year led to further rises in interest rates, minimum lending rate ending the year at 13%.

Further downward pressure on property prices resulted from the threat of anti-property legislation, which the government felt obliged to introduce to appease public outrage at the activities of some developers. The 1970–1973 property boom had witnessed a number of well-publicized scandals in the property sector. Many of these concerned the flat break-up market, an area not directly connected with commercial property development. More important, however, were claims that developers were deliberately leaving completed offices empty, the most notable example being the activities of Mr Harry Hyams.

By 1967 Harry Hyams had already attracted a certain notoriety for his use of this practice; in *The Property Boom* Marriott described him as 'a past master of the art'.[8] His properties were kept empty not by refusing to put them on the market, but by the simple expedient of asking for an unrealistically high rent. There was a simple commercial logic to this; rents were rising so rapidly during the late 1960s and early 1970s that leaving a property empty, rather than tying it down to a rent that would be fixed for a number of years, provided sufficient capital appreciation to more than compensate for the loss of rental income. Furthermore, no rates had to be paid on empty property.

The building which brought Harry Hyams into the spotlight was Centre Point; completed in 1964 at a cost of £5.5 million, its estimated value had soared to almost £20 million by May 1973, despite being vacant since completion.[9] He also owned a number of other large office blocks in Central London which remained unlet despite high demand for office space, though it was Centre Point which attracted the public attention. The building dominated the skyline in a part of London which lacked other tall buildings, which, together with its distinctive appearance, made it a prominent local landmark. Furthermore, it contained 36 flats, which were also left vacant at a time of chronic housing shortage in London.

In November 1972 Camden Council lodged an application for the compulsory purchase of these flats.[9] This started a political storm in which Centre Point became the focal point of popular discontent at developers who were perceived to be making huge profits at society's expense. At a demonstration on the 24th June 1973 (attended by Mr Reg Freeson, Labour's housing spokesman), Mrs Lena Jeger, the Labour MP in whose constituency Centre Point stood, stated 'Centre Point is a symbol of a society in which those who make money are more blessed than those who earn money'.[10]

As early as June 1972 the Environment Minister, Peter Walker, had threatened to introduce legislation to deal with the 'incredible scandal' of those who built large properties and then left them empty.[11] He informed Parliament that within a few months he would be ready to introduce legislation to guarantee that these office blocks were occupied, even if this involved their use by public bodies. These threats were not followed up by action, however, and the public outcry continued to the growing embarrassment of the government. Eventually, in November 1973, in the run up to an election, Walker's successor, Geoffrey Rippon, felt obliged to act, stating in a television interview that it was a matter of public concern that 'no one section of the community makes excessive profits at the expense of the rest. This is largely a matter of taxation.'[11]

Taxation measures followed in December, the government announcing proposals for Capital Gains Tax on the first letting of commercial property, plus a new tax, at Corporation Tax rates, on new commercial property sales. Together with a sudden rise in interest rates the new taxes threatened the liquidity of the property sector's precarious finances. As R. Einstein noted in a *Sunday Times* article shortly after the government's announcement 'The real problem for the property companies will be to find the money to pay the tax. Most of them are long in assets and short on cash.'[12]

These tax changes were estimated by the government to involve annual taxation of approximately £80 million. However, commentators in the property industry put the likely tax burden far higher; Sidney Mason, chairman of one of Britain's largest property companies, Hammersons, stated that at least one nought should be added to the £80 million figure mentioned by the Chancellor.[12] New legislation to give the government powers to manage and let office blocks which had been empty for two years or more, and to allow local authorities to impose punitive rates on such buildings, was also announced during the same month.

The final weeks of 1973 did see one piece of government action which, in different circumstances, might have been a substantial Christmas present for the property industry: the announcement by the Department of the Environment on the 18th December of an immediate ban on new office building in London and the South East.[13] According to the Minister, Geoffrey Rippon, the measure was not aimed at the commercial property sector directly, but was instituted in order to channel all available building resources into house-building programmes at a time of crisis in the sector.

However, this measure proved insufficient, in itself, to stem the tide of a property crash in which government legislation reinforced the adverse effects of rising interest rates. By refraining from taking action against abuses in the commercial property market, then announcing legislation as an election loomed, at the very time when a reversal of government monetary policy threatened the stability of the market, the government had played a significant role in bringing about the property crash.

THE PROPERTY CRASH

Veblen stated that the immediate cause of a business crisis is a practical discrepancy between the earlier effective capitalization on which collateral had been accepted by creditors, and the subsequent effective capitalization of the same collateral, revealed by quotations and market transactions.[14] The above conditions were sufficient to bring about such a discrepancy. Veblen added that:

> The point of departure for the ensuing sequence of liquidation is not infrequently the failure of some banking house, but when this is the case it is pretty sure to be a bank whose funds have been 'tied up' in 'unwise' loans to industrial enterprises of the class spoken of above.[15]

The November 1973 failure of such a bank with considerable 'unwise' commitments to the property sector, the fringe bank London and County Securities, marked the start of the property and secondary banking crisis. Such a liquidation would, according to Veblen, trigger an avalanche of bankruptcies, as forced sales resulting from the initial liquidation lower the price of the commodity in question, reducing the profits of competing firms and leading to further bankruptcies which in turn lower prices still further.[16]

During the following months several other secondary banks with heavy loans to the property sector faced bankruptcy, including the Cornhill Consolidated Group, Cedar Holdings and Twentieth Century Banking. A crisis of confidence in the secondary banks forced them to call in loans to the property companies, which were soon caught up in the expanding wave of defaults. A number of property companies, including Amalgamated Investment and Property, Town and Commercial Properties, Guardian Properties (Holdings) and the private Stern and Lyon groups, failed during the crisis, while some of the largest companies in the sector, including Town and City and MEPC, faced severe difficulties. Many property companies were forced to sell properties for whatever price they could get in order to generate income to repay loans and fears grew that a flood of property assets entering the market as a result of these forced sales would send prices plummeting still further.

The secondary banking crisis also raised fears that the failure of a large number of secondary banks, which had borrowed heavily from the clearing banks (often using short-term borrowing to pay for long-term lending to property companies), could lead to a liquidity crisis for the clearing banks. The clearing banks had lent considerable funds to the property sector directly, in addition to their indirect lending via the secondary banks, and were thus extremely vulnerable to any severe downturn in property market conditions.

Any crisis of confidence in the clearing banks, the most stable of British financial institutions, threatened the entire banking and financial system. As one senior banker remarked after the crisis 'This would have been worse than the Wall Street Crash ...'.[17] Veblen noted the potential role of government, and of creditors to the sector, in mitigating the effects of such a crisis:

The abruptness of the recapitalization and of the redistribution of ownership involved in a period of liquidation may be greatly mitigated, and the incidence of the shrinkage of values may be more equally distributed, by a judicious leniency on the part of the creditors or by a well-advised and discreetly weighted extension of credit by the government to certain sections of the business community.[15]

However, Veblen noted that despite such efforts a large and prevalent discrepancy between earning capacity and capitalization makes a drastic readjustment of values unavoidable, leading to a redistribution of ownership of the commodity in question from solvent debtors to creditors.[18] While creditors would gain the commodity at a lower price by pushing for immediate payment of debts, Veblen argued that they generally do not do so, since they are more interested in maintaining the money value of the collateral concerned rather than gaining the maximum physical amount of that collateral.[19] Furthermore, Veblen points out that creditors often have their own debt relationships involving this commodity and would therefore suffer following a severe collapse in prices.

In line with Veblen's predictions the government, via the Bank of England, together with the major creditors to the property sector – the UK clearing banks – launched a rescue bid to prevent a wider banking crisis. This became known as the 'lifeboat operation', a fund of £2000–3000 million being established to increase the liquidity of the banking system. Bankers were persuaded not to call in loans to property companies and secondary banks, while the financial institutions were encouraged and cajoled into purchasing properties from the troubled property companies.[20]

By organizing an orderly transfer of assets from the property companies to the financial institutions a flood of sales by property companies that were desperate for funds was avoided. The financial institutions acquired the assets of these companies over a period of several years, at prices which were often well below those prevailing during 1973, but markedly higher than would have been the case in the absence of the Bank of England's intervention.

By organizing an orderly transfer of assets from the property companies to the financial institutions a flood of sales by property companies that were desperate for funds was avoided. This intervention was clearly in the interests of the overall financial system. The prevention of a fall in property prices to levels reflecting market conditions was also of some direct advantage to the insurance companies. General insurance companies are required by law to maintain an adequate margin of assets over liabilities, known as the margin of solvency, expressed as a percentage of annual premium income.[21] Insurance companies generally aim to keep this margin above 40%. In 1974 the average solvency margin for the eight largest UK insurance companies which conducted general business fell to 25%, compared to 48% in the previous year. This level was still substantially in excess of the legal minimum, but given the collapse in share

THE PROPERTY CRASH

prices, which was the major cause of falling solvency margins, insurance companies had a strong incentive to prevent the value of their property assets from falling to too low a level.

The insurance companies were somewhat taken by surprise by the crash, there being no liquidity crisis in their own financial relations with the property sector. On the 6th December 1973 Abbey Life announced that it was bailing out of the property investment market until uncertainties regarding the effects of possible legislation were clarified,[22] though the insurance company sector was relatively slow to accept that the new market conditions marked the start of a severe and protracted property crash. For example, a review of the prospects for the property investment market in 1974, undertaken for Legal & General's property committee at the end of 1973, did not forecast any dramatic downturn in the market, though rising yields were predicted, as property companies were expected to dispose of some assets as a result of high interest rates.

As late as February 1974 many insurance companies remained uncertain whether the property market was experiencing a short period of uncertainty or a major downturn. The market for property was virtually non-existent during the early months of the year. By the start of February, however, there was enough activity to provide at least an indication as to yields and valuation rates, suggesting that property values were now 15–20% below November 1973 levels, with yields 1–1.5% higher.[23] By March auctions were being cancelled and the few property deals that were taking place were generally being conducted by private treaty, resulting in a dearth of market information regarding the basis on which transactions were being undertaken. Even in April 1974 the situation was still sufficiently confused for one major property to be sold at a yield of 4.75%.[24] By the end of May, however, it had become clear that a severe fall in property values had taken place and that conditions were unlikely to improve for some time.

The property crash does not, therefore, appear to have been triggered by any sudden crisis of confidence in the sector on the part of institutional investors. Instead, the financial institutions maintained a high degree of confidence in property until the end of 1973, and even in the opening months of 1974 the first signs of the crash were interpreted largely as a temporary phenomenon of rising yields rather than a long-term market depression. This reinforces the picture of the crash as a secondary banking and property company crisis rather than a crisis of the investment market, investors responding to conditions rather than initiating them.

The long-term effects of the crash were, as Veblen predicted, a considerable transfer of assets from solvent debtors to creditors. The property crash led to a considerable reduction in the importance of property companies in both the property investment and development markets. Rowe and Pitman estimated that committed and authorized expenditure by property companies declined from a peak of £700 million in 1974 to £158 million in 1977, recovering slightly to £251 million in 1979.[25] By 1980 property companies owned assets worth just over £7 billion, a third less, in real terms, than their holdings in 1970.[26] Meanwhile, the late 1970s saw a considerable transfer of property from property companies to financial institutions as companies liquidated their holdings in order to reduce debt burdens; between 1974 and the early months of 1978 British property companies disposed of over £2 billion of assets, mainly to the financial institutions.[27]

The immediate impact of the failure of the Lyon, Stern and Guardian property companies in 1974 was the existence of a potential £500 million worth of properties overhanging the market, in addition to any disposals by other property companies which were forced to sell for liquidity purposes. Rumours also circulated that several of the large quoted property companies were in financial difficulties and the potential future supply of investment properties was such that attempts by sellers to hold out for yields in the region of 6.5% were unsuccessful.[28] By the middle of the year yields demanded by sellers and those offered by buyers were separated by 0.5–1.5%. However, largely as a result of the Bank of England's lifeboat operation, property companies were able to hold on to their assets in the immediate aftermath of the crash and most of the transfer of assets from the property companies to the financial institutions did not take place until 1976–1978, by which time the market had recovered considerably.

The secondary banking sector contracted substantially in the aftermath of the crash, as a number of exposed fringe banks were forced into asset sales or absorbed into larger banking organizations. Of the 25 banking groups assisted by the lifeboat operation, eight collapsed as a result of the crisis. A further 11 had been taken over, wholly or largely, by major banking groups by 1981. Only a handful remained as independent entities in the banking sector.[29] A number of other banking and financial organizations not included in the lifeboat operation were also absorbed into larger banking groups as a result of difficulties experienced during the crisis.

The property unit trusts and property bonds also proved substantial long-term victims of the crisis. The property crash hit the property bond market particularly severely; from early 1974 to early 1975 funds invested in this sector fell from about £570 million to £380 million.[30] Property unit trusts were also badly hit, the real value of net investment in property unit trusts falling from £161 million in 1973 to only £36 million in 1974.[31] These new entrants to the property market were never to fully recover the importance they had assumed in the early 1970s; in 1973 net investment by the property unit trusts had amounted to 8.6% of all direct institutional investment in property, and by 1980 this figure had fallen to 5.1%.

This was partly due to severe practical difficulties which became apparent in the operation of property unit trusts from the time of the crash. Even under buoyant property market conditions they had been obliged to keep a substantial amount of investors' funds in the form of cash to meet occasional redemptions from unit holders. When the market weakened and redemptions became much more substantial they were forced to sell properties, and often found that only their best property assets were marketable under such conditions.[32]

8.3 RECOVERY AND BOOM, 1975–1980

It was announced on the 19th December 1974 that an order would be laid before Parliament on February 1st 1975 to the effect that the business rents freeze would end on that date and it was understood that rents would be allowed to rise to their contractual levels from the 19th March 1975. This news did not boost market confidence, however, and at the end of 1974 there was an air of stagnation in the property market, with purchasers anticipating a further fall in prices.[33] The after-shocks of the initial crash were still being felt and many observers expected a substantial amount of property to enter the market during 1975 as a result of the financial difficulties experienced by some property-holders. In an atmosphere of such uncertainty the safety of gilts yielding 15% was preferred by many investors to property, even at 8%. Despite such conditions, Clerical Medical's 1974 asset report concluded that the long term future of property as an investment medium was 'probably good'.[33]

Not all investors had responded to the crash by withdrawing from the property market. Some foreign investors took advantage of low UK property prices during 1974, the most spectacular overseas purchase being the Kuwait Investment Office's acquisition of the St Martins property group for £107 million. A few UK institutional investors also took the opportunity to purchase properties at what were to prove very attractive prices. The NCB pension funds, which had been investing £30–40 million in property each year, increased investment to £100 million in the year from June 1974, a strategy which was to prove very lucrative as property prices recovered during the following years.[34] In early 1974 lack of confidence in the gilt-edged and stock markets had led Standard Life to increase its property investment target from £22 million to £32.7 million. In addition to development funding, where schemes had already been commenced or planned, a number of completed investments were also purchased. With regard to this heavy investment, made during a collapsing market, it was stated in a review of Standard's property investments that:

> It will be another two years before a final assessment can be made as to whether this policy has been fully justified. Certainly the present building costs of replacing these developments already substantially exceeds our expenditure to date and providing there is not a substantial recession in rent levels it seems likely that this decision will prove to have been of considerable value in future years.[35]

Standard did, however, decide to curtail new development funding, in the light of building cost inflation of 25–30% per annum and the proposed new development taxes discussed below. It was decided to limit development commitments to the existing programme until conditions became more certain, with the exception of pre-let schemes or the redevelopment of Standard's own properties.

History was to prove Standard and the NCB pension funds right, the late 1970s witnessing a return to the property market conditions of 1968–1973. Net institutional investment in the sector did not return to its 1973 level until 1980, though the intensity of demand in relation to supply is shown by initial yields, which fell below 1973 levels both in absolute terms and relative to yields on gilts and ordinary shares.

Accounts of the property crash have placed much emphasis on the Bank of England's efforts to persuade the financial institutions to take on the assets of the troubled property companies during the mid to late 1970s. However, the company records of those institutions examined in this study reveal no evidence that investment departments acquired property mainly at the behest of the Bank of England, or even that the Bank's recovery plan entered into their investment thinking at all. While the Bank of England's intervention may have been a decisive factor behind the acquisition of properties by particular financial institutions from property companies with which they were closely associated, the substantial switch of property assets from the property companies to the institutions occurred due to a belief by the institutions that property represented an attractive long-term outlet for their funds.

Supply conditions returned to those prevailing in the late 1960s. The property crash was parallelled by a slump in the construction sector; the value of new orders obtained by contractors for commercial property fell to only 43% of the 1973 level, in real terms, in 1975, and for the 1974–1980 period as a whole the average real value of new orders for commercial property amounted to only 63% of its 1971–1973 level.[36] As there is a gap of several years between the decision to develop a property and the completion of the development, this downturn in orders in the mid-1970s resulted in a fall in the volume of completed buildings in the late 1970s, by which time many of the factors which had led to the downturn in development activity no longer applied. This time-lag between the decision to develop and the completion of development builds a marked cyclical element into the property market, which was masked to some extent during the 1950s and 1960s by government development controls, credit restrictions and other measures, the introduction or removal of which distorted the development cycle.

Legislation introduced in the mid-1970s reduced the perceived profitability of property development, adding to the downturn in supply. The return to power of the Labour party in 1974 led to a renewed examination of betterment taxation, while the Conservatives had also been considering measures to tax development gains prior to their defeat. The Finance Act, 1974 introduced Development Gains Tax. This was levied at income tax rates on the development value released by the sale, the granting of a lease at a premium and, in certain cases, on the first letting of a property following a development project but not on the carrying out of such a project.[37] Development Gains Tax was replaced two years later by Development Land Tax.

In the same year the government published the White Paper *Land*, which resulted in the introduction of the Community Land Act of 1975 and the Development Land Tax Act of 1976. The Community Land Act gave local authorities wide powers of land acquisition and placed on them the obligation to consider the desirability of using these powers to acquire land for development by themselves or others wherever a planning application for development covered by the legislation was made. The Secretary of State was able to designate certain types of development covered by the Act and thus oblige local authorities to acquire all land needed for such development.[38] The long-term aim of the Act was that at a certain time in the future, called Second Appointment Day, there would be a general duty on the part of local authorities to acquire all land for relevant development, the ultimate objective of the Act being to take all land (with certain exceptions) into public ownership.

The Development Land Tax Act introduced Development Land Tax at rates of 66.66 and 80%, with phased increases up to 100% of realized development value. The Act failed to achieve its objectives, very little tax being collected. No tax was payable for developments which had been granted planning consent prior to the introduction of the Act, and by the time the supply of such land had begun to run short the legislation was abolished following the election of a Conservative government in 1979. However, the introduction of this Act, together with the Community Land Act, added substantially to uncertainty in the development market during the mid-1970s, further depressing a market that was already facing conditions of rising costs and falling user-demand.

These years saw a divergence in demand for commercial property between the user and investment markets, a trend which had begun during the 1960s and early 1970s. While user-demand stagnated due to adverse economic conditions the financial institutions still looked to property as an outlet for their rapidly growing funds, encouraged by a rent review pattern that now allowed the upward-only adjustment of rents to market levels at five-yearly intervals. The supply restrictions mentioned above served to mask a continuation of the trend towards a reduction in the rate of growth of user-demand for commercial property, demand appearing buoyant as a result of very limited supply, as had been the case in the mid to late 1960s.

The property crash had hit the office sector particularly severely; the shop market fell far less during the crash and recovered more quickly. During the 1970s the number and size of new shopping centre schemes grew substantially.[39] The proportion of fully enclosed, air-conditioned centres among the new schemes increased, and there was a general improvement in quality, as the enclosed shopping centre reached maturity. This decade saw the construction of some of Britain's largest shopping centres, such as Brent Cross and the Arndale Centres in Luton and Manchester. In contrast to the 1960s there were virtually no failures among the shopping centre developments launched from 1972 to 1980, developers and local authorities having learned from their earlier mistakes.[39]

The balance of power between the developer and the local authority tilted in the latter's favour during this decade; local authorities had learned that their interests were best served by using compulsory purchase powers to acquire land in their own right and, having decided the basic specifications and location of the centre, invite tenders for development according to the authority's own plan.[40] This placed the local authority in the position of landlord and property investor, bringing about a potential conflict of interest between the authority's financial and regulatory interests. As Russell Schiller, the head of research at Hillier, Parker, stated:

> A larger scheme might be held to be more viable commercially and whereas before the local authority might argue that this was of little concern to them, now they took care to see that any scheme in which they were intimately involved should succeed.[40]

Despite growing pressure from some sections of the retail industry, a significant conflict between town centre and out-of-town shopping centre locations was not allowed to occur during these years. Local authorities used their considerable planning powers to encourage shopping centre development in existing centres and vigorously opposed out of town schemes. Other forces also acted to defend traditional shopping centres against pressure for out-of-town developments. While leading supermarket retailers pressed for development outside town centres, in order to meet their requirements for greater space, access and parking facilities, there were no such calls from durable goods retailers, variety stores, department stores, or shoe and fashion chains.[41] Most leading retailers in these sectors had a large proportion of their corporate assets made up of valuable town centre properties and thus had an important incentive to maintain the *status quo*.

Out-of-town shopping development was largely restricted to supermarkets and retail warehouses prior to the 1980s. The first British out of town retailers appeared during the mid-1960s, when Woolworths developed its Woolco stores and Associated Dairies launched its ASDA superstores.[42] However, it was not until the early 1970s that planning applications for this type of retailing began to grow significantly. Local and central government initially resisted this trend, as did many property developers and retailers, fearing that the viability of traditional town centre shopping might be undermined.[43]

In addition to vested interest considerations, which also influenced local authorities via their part-ownership of many shopping centres, there were also firmly-founded social objections to out-of-town retailing. Out-of-town schemes were often inaccessible to the non-car owning section of the population. Furthermore, it was feared that the decline of town centre shopping as a result of these trends might lead to urban decay and the social stratification of retailing, with the development of suburban shopping centres to cater for middle-class neighbourhoods. However, evidence suggested that larger shopping centres reduced unit costs and proved attractive for shoppers with cars who did not have to haul heavy bags considerable distances to car-parks. These arguments were especially true of convenience goods, and while the development of hypermarkets and regional shopping centres, which included a significant element of comparison shopping, continued to be severely restricted, supermarkets were allowed to develop in out-of-town locations during the 1970s. A small number of hypermarkets and regional shopping centres did obtain planning consent, mainly as examples in order to test the validity of these forms of retailing.

The shop development market continued to be dominated by a few large companies. Although 125 developers had built large shopping centres between 1965 and 1982, 10 large developers were responsible for 46% of the floorspace of major shop developments (in excess of 50 000 sq. ft. gross) during this period, according to a survey by Hillier, Parker.[44] The three largest of these accounted for 27% of new floorspace, making shopping centre development one of the most concentrated sectors of the property development market. Local authorities and new town development corporations were also important developers, accounting for 7 and 10% of all floorspace developed, respectively, in addition to their partnership schemes with private developers.

Property development continued to receive the financial support of the insurance companies during the late 1970s, their confidence in the sector making a remarkable recovery in the years following the 1973/4 crash. The crash was perceived to be largely the result of reckless lending to property companies by the banking sector, something which was unlikely to occur again in the foreseeable future, and rent control, which was never intended to be more than a temporary measure, and was not thought likely to be reintroduced, given the consequences of its imposition in the early 1970s.

By March 1975 some recovery in the property market was being noted by Legal & General, though rents for City office blocks were still reported to be falling. Two months later further signs of recovery were evident, though property in the hands of liquidators, with an estimated value of over £1 billion, still overhung the market. This, together with fears of further liquidations, made investors unwilling to commit new funds to property.

Other factors which served to depress the property investment market during 1975 included high gilt yields, the poor state of the letting market for commercial property and fears of new legislation, particularly the Community Land Bill. By the middle of the year Clerical Medical began to notice an upturn in conditions, however, a memorandum from its Investment Forum stating that:

> Probably prompted by the large London firms of Estate Agents drumming up business for funds advised by each other, the yields on really top class properties are being forced down. Properties other than first class are not easily saleable on any yield basis.[45]

It was becoming clear that while the prime property market was undergoing a slow recovery investors had lost interest in secondary property, which had experienced the greatest falls in value during the property crash and was viewed as having inferior prospects for long-term capital appreciation. Many institutional investors embarked on a process of portfolio rationalization during the 1970s, weeding secondary property out of their portfolios and concentrating on large, valuable units of prime property. For example, the number of properties in Legal & General's long-term fund fell from 2001 in 1974 to 1548 in 1980, while the aggregate value of the portfolio rose from £488 million to £835 million over the same period.[46] Clerical Medical also sold off many of its smaller, secondary properties; at the end of 1977 the number of properties it owned with a value less than £100 000 amounted to 72, a year later it had fallen to 53.[47] This move away from secondary property led to an effective reduction in the supply of investment property, as the range of property which was acceptable to the financial institutions became progressively narrower.

During 1975 many insurance companies encountered problems with regard to funding links they had established with property companies. For example, the Amalgamated Investment & Property Co. Ltd had difficulty with the repayment of a mortgage granted to its subsidiary, Shop Investments Ltd, by several insurance companies. These agreed to grant a moratorium on interest payments on the mortgage.[48] Such action was in line with the Bank of England's lifeboat operation and was also in the interests of the financial institutions, as there was little point in liquidating companies whose assets could not be sold at anything close to their balance sheet values.

Legal & General experienced some difficulties with regard to its property company funding arrangements. It was usually able to resolve them on commercial terms, sometimes involving the acquisition of property assets at prices which reflected the current buyers market. Equity holdings in the companies affected did not force Legal & General into giving them special treatment, though the support of the Society was usually sufficient to keep them afloat during what were generally only temporary liquidity difficulties. As a Board paper regarding one such company stated:

> In practice given its near 10% shareholding and the presence of one of its own directors on the ... Board, it is not an available option for the Society to let this company fail. Neither is it desirable for it to be seen overtly as a prop.[49]

By the beginning of 1976 there were indications that the pension funds, property bonds and insurance companies were beginning to re-enter the property market.[50] By the middle of June the market had picked up considerably, so much so that it was suggested by H.N. Beetlestone of Clerical Medical that there might be a repetition of the 1972/73 boom.[51] Standard's 1976 Property Investment Report noted that a shortage of good quality investment properties was leading investors to look at development projects once again. Direct development, and development funding arrangements, accounted for the bulk of Standard's property acquisitions during the 1974–1980 period, as it was the only form of property investment which offered what the company considered attractive yields.

The property market continued its recovery during 1977, which *The Estates Gazette* described as 'the year when property as an investment medium came back from the dead'.[52] A mini-boom in property investment and development had begun, which was to last until 1981. A return to low yields on investment property during the late 1970s led many investors to switch from investment to development projects, while building contractors also entered this market, in order to secure continuity of employment for their work force. As a result developers' profit margins underwent a substantial decline; it was reported by Standard Life in the early months of 1979 that margins had fallen from the customary 20–25% to 10–15%; in some cases margins were being squeezed to zero.[53]

By the end of 1978 property yields were even lower than during the peak of the 1970–1973 property boom and there was a growing school of thought that the market was becoming overheated. However, there was no danger of a repetition of the 1973/4 crash, the re-structuring of property investment and finance following the crash having greatly increased the sector's stability. Most importantly, the banks, which loaned funds to developers but did not invest directly in property, had been displaced by the financial institutions, which provided both development finance and a market for completed developments. The impact of these changes was noted by H.N. Beetlestone of Clerical Medical:

> There is a fundamental difference between the current situation and that which obtained in the boom of the early 1970s. In the early 1970s the boom was sustained on a short term basis by short term money borrowed at high rates of interest from banks of all descriptions. All you needed was an address on the back of an envelope and the courage to ask for a sufficiently large sum of money. The present boom is being sustained by long term money invested in the market by insurance companies, pension funds and property bonds. To this extent, if the market should collapse, there will not be the same mad rush to sell, the holders will probably continue to hold and cool off their fingers under the cold water tap.[54]

By holding properties off the market the financial institutions could, collectively, prevent a downturn in property prices turning into a collapse. Rents, however, had no such insulation from user-market conditions. The late 1970s saw rapid rises in market rents, which were charted by the recently established property rental indices, such as the Investors Chronicle/Hillier, Parker Rent Index, introduced in October 1977. In 1973 independent property valuations had contributed to the property crash, by allowing companies to borrow funds on the basis of portfolio valuations which bore little relationship to the worth of those assets only a few months later. The Investors Chronicle/Hillier, Parker Rent Index had come under substantial press criticism at the time of its introduction and there were fears that it might similarly overheat the market by giving investors an unrealistic picture of potential rental growth. Despite such doubts Beetlestone still advocated the purchase of property at current low yields:

> By reference to history and by reference to first principles ... the Society should buy prime freehold properties regardless of yield, and the reason is the simple one that, at the end of the day when everything else has fallen by the wayside, you have the land and bricks and the mortar, which judging by today's prices and building costs, will be irreplaceable at anything like the price we pay for it.[54]

By March 1980 Standard Life could find little attraction in either the property investment or development markets. Building costs were reported to be increasing at an annual rate of 25%, in contrast to their relative stability during the previous few years, and rental values were not fully matching costs. Economic recession led to doubts regarding future demand, and increased competition from other institutions resulted in a deterioration in the supply situation. The report concluded 'We have therefore arrived at the position where for the first time since we started writing a policy statement we do not like what we see both in the completed and development market.'[55]

Buoyant investment market conditions coincided with adverse conditions in the user-market. Interest rates were rising, thereby discouraging buying on credit, unemployment figures showed sharp increases, eroding consumer spending power, retail sales were static and rental growth was showing signs of falling off. While user-market conditions were similar to those in 1974, with the exception of rent control legislation, the weight of money awaiting investment and the lack of alternative outlets for it kept the property market buoyant.

By the end of the 1970s institutional investors were clearly becoming increasingly dissatisfied with prevailing yields on investment property and expressed growing doubts regarding whether future rental growth would be sufficient to justify those yields. A number of unconventional avenues of property investment were investigated by some of the largest institutional investors in an effort to place funds in the sector on terms above those prevailing in the market. These included property development, refurbishment, the purchase of outstanding interests in currently owned properties in order to benefit from 'marriage values'[56], and the acquisition of properties of a value so large as to put them beyond the reach of most investors. However, by 1980 even returns on property development offered little advantage over those in the investment market, such was the pressure of institutional funds.

The institutions appear to have remained in property, despite low yields and unfavourable user-market conditions, due to their long-term faith in the desirability of property as an investment medium, almost regardless of current market conditions. A further factor behind their continued commitment to the sector concerns the nature of institutional investment in property. During the early post-war years the lack of a specialist property investment staff prevented many institutions from entering this market and gave those few which were prepared to undertake the initial expenditure and effort involved in setting up a property department access to a market in which they faced few competitors.

During the late 1960s, however, in addition to an increase in the number of institutions which had developed their own in-house expertise, several prestigious chartered surveying firms had begun to offer property portfolio management services for investors. These were not normally paid a fee based on the capital value of the property portfolio they helped to manage, but through commissions on buying and selling property for their clients. They thus had a vested interest in an active institutional investment market in commercial property. As Alistair Ross Goobey stated 'Most of them used their position responsibly but it was very difficult for them actively to try to dissuade the clients from investing since their livelihood depended on such activity.'[57]

The growth of a few prestigious surveyors, particularly Jones Lang Wootton and Richard Ellis, as managers of a large number of institutional property portfolios, reduced the costs of entry into the property market but led to a concentration of investment expertise in the hands of a relatively small number of agencies. As *The Economist* stated in 1978:

> The trouble is they [Jones Lang Wootton and Richard Ellis] advise virtually everybody ... Indeed, some people say that the market is moved by three small groups: Richard Ellis; Jones Lang Wootton; and a clutch of experienced institutions like the Prudential, Legal & General and Norwich Union which have a mind of their own.[58]

The chartered surveying firms appear to have taken an optimistic view of the property market during the late 1970s, backed up by information on movements in rents and values which they had begun to produce, indicating rapid long-term growth of sufficient magnitude to justify current low yields. The property departments of the larger institutional investors, who managed their own portfolios, generally took a more cautious view of the market, while arguing that investment was still worthwhile, albeit in areas such as development activity, in which they expected to obtain higher yields than those prevailing for conventional investments. The development of an in-house property investment department entails a large commitment of time and resources, and as long as investment managers had faith in the long-term value of investment in the property sector they had strong incentives to argue for a significant level of investment, even under temporarily adverse conditions, to maintain market contacts and staff numbers at levels which would be necessary when the expected improvement in market conditions took place. Such was the faith of the institutions in the underlying long-term growth potential of bricks and mortar that the sector was not to be abandoned lightly.

REFERENCES AND NOTES

1. See section 10.4.
2. T. Veblen (1965) *The Theory of Business Enterprise*, Kelley, New York, p. 201 (originally published in 1904).

3. Legal & General (1973) Property Committee, Property Market Report, 11 July.
4. Clerical Medical (1973) Investment Forum minutes, 3 Aug.
5. Clerical Medical (1973) Investment Forum minutes, 6 Sept.
6. Legal & General (1973) Property Committee, Property Market Report, 12 Sept.
7. M. Reid (1982) *The Secondary Banking Crisis, 1973–75: Its causes and course*, Macmillan, London, p. 78.
8. O. Marriott (1967) *The Property Boom*, Hamish Hamilton, p. 109.
9. A. Holden (1973) Rates rise plan is 'chicken feed' to Centre Point. *The Times*, 25 May, 5.
10. Labour MPs in new Centre Point protest (1973) *The Times*, 25 June, 3.
11. J. Bell (1973) How Heath can tackle the property developer. *The Sunday Times*, 2 Dec., 56.
12. R. Einstein (1973) How badly hit is property. *The Sunday Times*, 23 Dec., 38.
13. Ban on new office building in South-east (1973) *The Times*, 19 Dec., 1.
14. T. Veblen (1965) *The Theory of Business Enterprise*, Kelley, New York, p. 193 (originally published in 1904).
15. T. Veblen (1965) *The Theory of Business Enterprise*, Kelley, New York, p. 205 (originally published in 1904).
16. T. Veblen (1965) *The Theory of Business Enterprise*, Kelley, New York, pp. 204–5 (originally published in 1904).
17. M. Reid (1982) *The Secondary Banking Crisis, 1973–75: Its causes and course*, Macmillan, London, p. 13.
18. T. Veblen (1965) *The Theory of Business Enterprise*, Kelley, New York, p. 206 (originally published in 1904).
19. T. Veblen (1965) *The Theory of Business Enterprise*, Kelley, New York, pp. 206–7 (originally published in 1904).
20. J. Plender (1982) *That's the Way the Money Goes*, Andre Deutsch, London, p. 114.
21. J. Plender (1982) *That's the Way the Money Goes*, Andre Deutsch, London, p. 46.
22. Legal & General (1973) Property Committee, Property Market Report, 7 Dec.
23. Legal & General (1974) Property Committee, Property Market Report, 6 Feb.
24. Clerical Medical (1974) Investment Forum minutes, 5 April.
25. J. Rose (1985) *The Dynamics of Urban Property Development*, E & FN Spon, London p. 171.
26. J. Plender (1982) *That's the Way the Money Goes*, Andre Deutsch, London, p. 89.
27. The new Leviathans: Property and the financial institutions, a survey (1978) *The Economist*, **267**, 10 June, S3.

REFERENCES AND NOTES

28. Clerical Medical (1974) Investment Forum minutes, 25 July.
29. M. Reid (1982) *The Secondary Banking Crisis, 1973–75: Its causes and course*, Macmillan, London, p. 151.
30. Panmure Gordon & Co. (1976) *The Property Sector 1976–1977*, Panmure Gordon & Co., London, p. 30.
31. See section 10.2.
32. A.R. Goobey (1992) *Bricks and Mortals*, Century, London, p. 102.
33. Clerical Medical (1974) Investigation into the Society's assets.
34. M. Reid (1982) *The Secondary Banking Crisis, 1973–75: Its causes and course*, Macmillan, London, p. 104.
35. Standard (1974) Property Investment Report.
36. See Chapter 10.
37. D. Cadman and L. Austin Growe (1983) *Property Development*, E & FN Spon, London, p. 242.
38. D. Cadman and L. Austin Growe (1983) *Property Development*, E & FN Spon, London, p. 243.
39. R. Schiller (1985) Land use controls on UK shopping centres, in *Shopping Centre Development: Polices and prospects* (eds J.A. Dawson and J.D. Lord), Croom Helm, London, p. 50.
40. R. Schiller (1985) Land use controls on UK shopping centres, in *Shopping Centre Development: Polices and prospects* (eds J.A. Dawson and J.D. Lord), Croom Helm, London, p. 51.
41. R. Schiller (1985) Land use controls on UK shopping centres, in *Shopping Centre Development: Polices and prospects* (eds J.A. Dawson and J.D. Lord), Croom Helm, London, p. 42.
42. I. Wray (1972) *Town Centre, Shopping Policy and Decentralisation*. Unpublished M. Phil thesis, University of London, p. 30.
43. I. Wray (1972) *Town Centre, Shopping Policy and Decentralisation*. Unpublished M. Phil thesis, University of London, p. 32.
44. These were, in order of floorspace developed, Town and City, Ravenseft, Hammerson, Laing, EPC, Grosvenor, Norwich Union, ASDA, Capital & Counties and Neal House. Source: Hillier, Parker, May & Rowden (1983) *British Shopping Developments 1965–1982*, Hillier, Parker Research, London, Tables A and C.
45. Clerical Medical (1975) Investment Forum minutes, 6 June.
46. Source: Legal & General (1980) Estates Dept. Annual Report.
47. Clerical Medical (1978) Asset Investigation Report.
48. Clerical Medical (1975) Board minutes, 26 Nov.
49. Legal & General (1974) Board Paper, 11 June.
50. Clerical Medical (1976) Investment Forum minutes, 2 Jan.
51. Clerical Medical (1976) Investment Forum minutes, 4 June.
52. A.R. Goobey (1992) *Bricks and Mortals*, Century, London, p. 16.
53. Standard (1979) Property Investment Report for 1978.
54. Clerical Medical (1979) Investment Forum minutes, 29 June.

55. Standard (1980) Property Investment Report for 1979.
56. Marriage value' is a term used to indicate the excess in value of two interests in a property, such as the leasehold interest and the ground rent, when owned by the same investor, over their value when owned separately.
57. A.R. Goobey (1992) *Bricks and Mortals*, Century, London, p. 17.
58. The new leviathans: Property and the financial institutions, a survey (1978) *The Economist*, **267**, 10 June, S27.

The best of times and the worst of times – the commercial property market since 1980 | 9

9.1 INTRODUCTION

Few sectors of the economy typify the ups and downs of Britain's economic fortunes since 1980 as clearly as commercial property. While the recession of the early 1980s had led to a severe downturn in the property industry, by the middle of the decade the growing financial and service sector-based boom had initiated a spectacular burst of new development activity. In the space of half a decade a new generation of developers were to make vast – though largely ephemeral – profits from bricks and mortar. Regulations which had previously constrained developers from building the sort of property they believed the market wanted, where the market wanted it, were scrapped or toned down, while the property development industry was heralded by government as a vehicle of urban regeneration which could prove the saviour of the inner cities if freed from restriction.

However, by the end of the decade most of the new 'merchant developers' had been plunged into what were often terminal crises, facing mounting debts for properties they could not sell or let at anything near their cost price. Rising interest rates, and a glut of new properties, terminated the Big Bang property boom just as it had the Barber Boom of the early 1970s. As a result the property market was plunged into a crash that was in some ways more severe than its 1973/4 predecessor, and certainly more prolonged. This chapter examines the boom–bust cycle of the property market during this period, together with its economic consequences.

9.2 THE RETREAT OF THE INSTITUTIONAL INVESTORS 1981–1984

By 1981 the property market's enthusiasm following the return to power of a Conservative government, which initially maintained confidence in the face of deteriorating economic conditions, had begun to turn to pessimism. Rising unemployment, high inflation and the generally gloomy economic outlook sapped confidence in the property sector. Furthermore, interest rates began to increase significantly, both in nominal and real terms, in contrast to the mid to late 1970s which had seen negative real interest rates.

The recession led to a slow-down in the growth of market rents, which actually fell in real terms over this period. This was reflected in relatively poor returns to investment for property, which achieved a lower average return than either the equity or gilt sectors during these years. There were also some more fundamental, long-term reasons for the loss of faith in property as an investment medium on the part of the financial institutions. The Conservatives' abolition of exchange controls, together with the growing internationalization of investment markets, opened up a wide range of overseas securities to the insurance companies and pension funds, which began to invest substantially in overseas equities.[1]

Meanwhile, increasingly sophisticated domestic financial markets began to offer a wider range of investment opportunities than had been available during the 1970s. One of property's great historic attractions to the financial institutions was that it constituted a real asset, which was believed to be a reliable hedge against inflation. However, the Conservative government's emphasis on controlling inflation reduced the importance of this characteristic, while the early 1980s saw the development of alternative securities, such as index-linked gilts, which offered the same inflationary hedge while being much more attractive in terms of liquidity and transactions costs. Property has always had substantial disadvantages as an investment medium; it is illiquid, is generally available in only fairly large blocks (making it difficult for small and medium funds to achieve diversified portfolios), carries high management costs, and has transactions costs which, at about 2.75% of a property's value,[2] are more than 10 times those for equities.

Furthermore, technological developments appeared to be shortening the lifespan of commercial property. While in previous decades buildings could last for 20–30 years without major refurbishment, by the end of the 1980s it was believed that a total refit would often be needed within 10 years.[3] There was also the longer-term threat of the decentralization of commercial property away from town centres, the trend towards business parks and edge-of-town (or out-of-town) retailing to the primacy of central site values becoming increasingly apparent as a result of the government's liberalized planning policies.

In the face of these pressures the financial institutions steadily reduced their commitment to the sector; net institutional investment in property fell by 47% in real terms between 1980 and 1985. For many funds this represented a fundamental decision to substantially scale down their exposure to property. Even when the sector boomed during the second half of the decade the proportion of institutional funds allocated to property remained well below levels experienced during the 1970s. In 1982, when the property market was still suffering from

THE CHANGING FACE OF COMMERCIAL PROPERTY

depression, net institutional investment in real estate had amounted to about 12% of total net investment. By 1989, a year which saw the peak of the property boom, this had fallen to only 5%.[4]

History was to prove the institutions right in avoiding what appeared, during the boom years, to be lucrative investment opportunities. However, the absence of the financial institutions was to add to the instability of the 1985–1989 boom by removing the link between property development funding and long-term direct investment in the sector, twin roles which the institutions had played since the inter-war years.

While the early 1980s had been a period of relative inactivity in commercial property, it did witness important changes in the character of commercial property development, and of the agencies which undertook the major development projects. Sections 9.3 and 9.4 examine the impact of these trends, both in the recessionary climate of the first half of the decade and during the boom years that followed.

9.3 THE CHANGING FACE OF COMMERCIAL PROPERTY

During the 1980s a number of radical changes took place in the design, location and context of commercial property. Most of these constituted trends which had been evident during the 1970s, or before, but only achieved prominence during the Thatcher years. Their emergence was the result of a complex of factors including technology, socio-economic changes and the government's *laissez-faire* attitude towards property development and urban planning. Collectively they both accelerated the obsolescence of Britain's pre-1980 commercial property stock and undermined traditional central office and shopping districts, trends which contributed to both the institutional investors' retreat from the property sector and the slow recovery of the property market following the 1990 crash.

During the 1980s commercial architecture was a subject of unprecedented public attention. The Prince of Wales launched a successful attack on the modernist orthodoxy pioneered by Henri Le Corbusier, Mies van der Rohe and Walter Gropius during the inter-war years, characterized by extreme functionalism and simplicity of form.[5] His stinging criticism, which encompassed what were regarded in the architectural profession as among the best recent buildings and proposals, owed much of its impact to mining a significant vein of popular revulsion against three decades of commercial architecture which had blended the modernist aesthetic and the commercial considerations of the developer to produce what were widely regarded (outside the architectural world) as some of the ugliest buildings ever seen in Britain's cities.

The move against modernism found some support in the architectural profession, and during the 1980s the cubist geometry of Mies and Le Corbusier, more often than not translated by the developer into monotonous concrete slab buildings, was displaced in new commercial architecture by three competing styles, 'neo-Georgian', 'post-modernist' and 'high tech'. The classical style had never entirely disappeared from British office development, and following the reaction against modernism in the 1970s neo-Georgian architecture began to find considerable favour with clients. However, it had obvious drawbacks for taller buildings and was seen by many architects as too obvious a return to the past.

Figure 9.1 The *Estates Times'* 'STACKUP' cartoon of 24th May 1985, reflecting the new trend away from modernism in commercial architecture. Reproduced by kind permission of Mr Alan Bailey.

Post-modernism involved a more imaginative adaption of previous architectural styles to the needs of modern commercial property. Arising from a school of architecture that had been popular for many years in Italy and America, involving the incorporation of historical allusions to enliven the unremitting bleakness of modernism,[6] post-modernist architecture could be seen as a return to the jazz style of modernism combined with historical motifs which had characterized the work of many inter-war commercial and industrial developers (their work, though despised by modernist contemporaries, constituting the most popular and enduring images of the period's architecture). The parallels between these two architectural styles were summed up by Goobey, who described post-modernist offices as mixing 'the appearance of the old Cunard liners and Odeon cinemas'.[7]

By the late 1970s this architectural style had consolidated itself into a separate school, finding a name with the publication in 1977 of Charles Jencks' *The Language of Post-modern Architecture*. It rapidly gained popularity among office developers, taking those aspects of modernism that the public still considered acceptable, or which technical and financial considerations still deemed appropriate, and combining them with an eclectic mix of architectural styles from several eras. Examples include the TV-AM building in Camden and Embankment Place, both designed by Terry Farrell.

The high-tech school, leading proponents of which included Richard Rogers and Norman Foster, looked to functionalism, but, rather than producing stark geometric buildings, allowed usually hidden pipework and lifts to mould their external appearance. Examples of such buildings in the City include the Lloyds Building, Bush Lane House and 1 Finsbury Avenue.[8] Despite the dominance of technology in these buildings it has been argued that they actually work less well than traditional office buildings with the lifts and pipes on the inside. For example, Richard Rogers' Lloyds building was described by *The Economist* as being 'inflexible, cost far more than originally intended and is loathed by many who work in it'.[9]

These changes in the architecture of commercial property took place alongside an acceleration of the impact of technological change on office design. The computer revolution of the 1980s, together with attendant developments in telecommunications, had substantial implications for office layout. False floors, under which computer and telephone wires could be laid, together with suspended ceilings, became standard features of new office developments. As one of the leaders of this information technology revolution, the London architect Frank Duffy, stated 'The building has become part of the telecommunications budget'.[10] Flexible layout, with the extensive use of partitions rather than more permanent walls, also became popular, to take account of the changing working environment, while tenants generally became much more particular regarding the quality of more mundane aspects of office technology such as air-conditioning and energy efficiency.

The above changes acted to accelerate the obsolescence of office buildings, a process which both stimulated development and made completed offices less attractive to investors. However, other changes in the property market, influencing location as well as design, were to have more far-reaching effects on the property development and investment markets. The 1980s saw a rapid extension of commercial property development to areas which were not considered 'prime', a process which opened up a much wider range of sites to development, partly at the expense of traditional centres.

More liberal government planning policies acted to facilitate, or even encourage, these trends. One example of a government initiative which encouraged the development of commercial centres outside traditional districts stemmed from the government's approach to inner-city regeneration. Since the 1970s policymakers had become increasingly concerned at the problem of inner-city decline, a problem which was made violently clear during the early 1980s as a result of major riots in a number of Britain's most depressed inner-city areas. The Conservatives launched a number of policies aimed at urban regeneration, the most important of which were the Urban Development Corporations (UDC).

Established under the 1980 Local Government, Planning and Land Act, UDC were intended to oversee the physical regeneration of their localities. They were given a number of land acquisition and planning control powers, supplanting local government with regard to planning within their areas. The philosophy behind their establishment was that local authority planning restrictions and controls had contributed to inner-city decline. As Michael Heseltine noted, with regard to the London Docklands Development Corporation:

> We took their powers away from them [local authorities] because they were making such a mess of it. They are the people who got it all wrong ... UDCs do things and they are free from the delays of the democratic process.[11]

UDC would provide a free-enterprise, market-based alternative to previous urban regeneration initiatives, offering a property-based solution to the problems of the inner cities. Of the 13 UDC established by 1993 only one, London Docklands Development Corporation (LDDC), was in the South East. Established in 1981, in an area of severe unemployment, it was to become the flagship of the 1980s property boom, a flagship which spectacularly failed to weather the storms of the 1990 property crash.

Docklands was a mammoth project by any scale. The publicity accompanying LDDC's 10th anniversary in 1991 put the sum invested in the area at £8.4 billion of private investment, over £1 billion in LDDC expenditure and £3.5 billion in transport investment. This had produced 27 million sq. ft. of commercial floorspace, 15 000 new dwellings and 41 000 jobs.[12] However, a property-based, rather than people-based, redevelopment strategy, with minimal local consultation (at least during its early years), produced considerable opposition from local interests, who argued that the existing local community would gain little, if anything, from the development. Despite counter-arguments that the benefits would 'trickle down' to the existing population, there is considerable evidence that this did not, in fact, occur. For example, unemployment levels in Docklands were higher in 1992 than in 1981 and homelessness in the three docklands boroughs had increased by 200%, compared to 60% for London as a whole.[13] Nor was this purely the result of the recession; in February 1988 *The Economist* noted that net employment available close to home for those living in Docklands probably decreased during the first six years of the LDDC.[14]

Much early docklands development was centred on the Isle of Dogs. This was designated an 'Enterprise Zone', removing virtually all planning controls and freeing developments there from rates until 1992.[15] The results of this exercise were spectacular; the government was eventually able to attract one of the world's leading property developers to the site, who undertook to provide an office complex that looked set to rival the City as a prestigious London-based centre for leading British and international companies. The reality was to prove somewhat different, as is discussed below.

Deregulation went much wider than the UDCs. In the rest of Britain local authorities often found that when they refused developers planning permission their decisions were overturned on appeal to Whitehall, especially during Nicholas Ridley's term as Environment minister.[16] From 1987 developers who had received an 'unreasonable' refusal might also be awarded costs against the council if the appeal resulted in a public inquiry which the developer won. In practice, 'unreasonable' councils were forced to pay costs far more often than was the case with developers bringing frivolous appeals.[16] Thus the balance of the planning process was tilted in the developer's favour, contributing to the mushrooming of successful planning applications and the ensuing massive over-supply of new property.

In addition to taking a more flexible attitude within the existing framework of planning legislation, the government also acted to change that legislation. A number of changes were made in the Use Classes Order, which determined the scope of planning permission, the most important of which was introduction of the B1 Use Class in 1987. This allowed flexibility of use between office, research and light industrial functions in the same property.

The introduction of B1 enabled planning legislation to catch up with the way work environments had been changing for a number of years, and as such was an inevitable, and beneficial, adjustment of the planning process to meet the changing needs of industry. However, by allowing offices to be developed on industrial sites it displaced established 'light industrial' tenants in areas on the fringes of office districts in London and other major centres. For example, the tailors of Savile Row found themselves faced with substantial rent increases for their workshops, as landlords attempted to capitalize on their new ability to switch such property to office use without recourse to planning permission.[17] This undermining of traditional light industrial areas bore similarities to other major development trends of the 1980s, in that increased property market activity occurred at the expense of established land-use patterns.

Even prior to the introduction of B1 the distinctions between the office and industrial property sectors had already begun to blur. 'Business parks', which had originated in the United States, were introduced to Britain in around 1980. They were originally seen as a new kind of industrial estate serving high-tech industry, but gradually developed characteristics which both distinguished them from their industrial estate predecessors and provided advantages to high-tech and service industry tenants, enabling them to rival both the industrial estate and the urban office centre.

Business parks were typically located on a landscaped area in an edge-of-town or out-of-town location, with good access to motorways and airports, and with a relatively low density of buildings compared to traditional industrial estates. The character of the buildings was also different; these ranged in use from offices, research and development, storage or light industry, often consisting of two or three storey units which combined several of these functions in the same building.

Up to 1987 business parks had generally served the needs of high-tech companies, in industries such as computing and telecommunications. However, following the introduction of B1 they began to act as a direct competitor to the urban office centre, attracting a large number of business and financial services companies. The information technology revolution of the 1980s had considerably reduced the need for face-to-face contact with customers and suppliers for many businesses of this type, giving them the freedom to abandon town centre locations with their attendant problems of congestion and a property stock which often failed to meet their needs.

This widening of the scope of business park tenants led to a boom in the sector; the volume of new business park space completed in the UK rose from one million sq. ft. in 1985 to almost seven million in 1988.[18] By the end of 1988 over 450 business park schemes were being promoted in the UK, an average of four new projects being announced each week.[19] Such was the magnitude of the switch from traditional industrial and warehouse units that by the late 1980s there was a shortage of the older 'shed'-type factories and warehouses while an oversupply of business park units developed, particularly in and around London.

While business parks blurred the boundaries between the industrial and office sectors, retail warehouses had a similar effect on the boundaries between the industrial and retailing sectors. Retail warehouses are large retail outlets (generally over 10 000 sq. ft.), located in warehouse-style structures outside established shopping centres and offering substantial car-parking provision.[20] They generally sell a fairly narrow range of goods, in sectors such as DIY, furniture, carpets and electrical goods, which require a very substantial sales floor area.

The first retail warehouses began to appear in Britain during the 1970s, pioneered by companies such as Dodge City, which established its first retail warehouse in 1974 on a bankrupt industrial estate near Glasgow. The years from 1979 to 1982 saw most of the pioneering retail warehouse groups being taken over by large, cash-rich, retailers, as was the case with Dodge City, which was acquired by B&Q in 1981.[21] By 1982 a few, large, well-financed companies, such as B&Q and Sainsbury's Homebase, dominated the market. This process of concentration in the sector was probably accelerated by the lack of institutional interest in this new form of retail outlet. Institutional investors proved wary of retail warehouse buildings, largely due to the absence of the obvious re-letting possibilities and site value appreciation enjoyed by High Street stores. This made fund-raising difficult for the pioneers, leaving them vulnerable to take over by larger retailers with a stronger financial base.[21]

By the early 1980s retail warehouses were becoming a significant feature of the retail development market. This was only part of a wider trend towards edge-of-town or out-of-town retailing, which included the development of regional shopping centres such as Dudley's Merry Hill centre and Gateshead's MetroCentre. Both these were built in Enterprise Zones, thereby minimizing planning restraints on a type of development which usually proved unpopular with local authorities as it represented a challenge to established shopping centres and disadvantaged people without cars. Many of the most prominent developers in this sector were specialist companies not involved in City office development, such as Peel Holdings, Regentcrest and Cameron Hall.[22]

9.4 THE MERCHANT DEVELOPERS

Those companies which grabbed the headlines with the largest and most spectacular development projects of the 1980s were generally not those which had been founded during the 1950s property boom and had experienced the 1974 crash. A new breed of property developer, often styled the 'merchant developers', mushroomed during the 1980s, absorbing many of their older counterparts and pioneering the development of new locations, new financing mechanisms and new building techniques. Several rose to the ranks of Britain's largest property companies during the 1980s development boom; a great many were destroyed in the crash that followed.

High real interest rates during the 1980s made the initial margin of rental income over interest costs for newly completed buildings negative, a positive cash flow being achieved only after the first rent review. The developers who shot to prominence during the 1980s boom therefore found it difficult to retain their developments. As a result they were forced to largely restrict their activities to property trading and development for sale, building up only small investment portfolios, a strategy which earned them their 'merchant developer' tag.

The stories of these companies, classic examples of which include Mountleigh, Speyhawk and London & Edinburgh Trust, have been told in Alistair Ross Goobey's book *Bricks and Mortals*. While property trading often offered rapid and substantial profits during the boom years, they faced a common problem; having no substantial investment property portfolio to provide them with a regular income flow, their continued growth depended upon their ability to conduct ever bigger deals. For example, Tony Clegg's Mountleigh, one of the most important and fastest growing participants in the boom, built its success very largely on property trading rather than development.[23] Mountleigh was able to buy property assets and immediately have them independently valued at prices substantially above those paid during the bull market of the 1980s. For example, in January 1984 the company purchased Occidental Oil's European head office in Aberdeen for £6.4 million. Immediately following the purchase Clegg had the property valued by surveyors at £7.83 million. This surplus of £1.43 million increased Mountleigh's net assets per share, making it easier for the company to raise further finance.[24]

In 1985 Mountleigh repeated this exercise on a much larger scale, purchasing the private property company R. Hitchins for £28.4 million. Hitchins' portfolio consisted largely of developments and land, only one of its properties being fully developed and let (representing about £5 million of the purchase price). Despite this, and the fact that the portfolio was currently producing only £480 000 per annum, Jones Lang Wootton and Hepper Watson were prepared to value the company's assets at £49.25 million.[25] All of this depended on a continuation of the boom market, a fact which, of itself, should have alarmed the banks, which were providing the merchant developers with much of their capital.

Instead the merchant developers were able to build up massive debts in relation to their capital base via the use of lending techniques which kept those debts off their balance sheets. For example, if debt was incurred via associated companies in which the developer had a 50% stake, or less, it could be kept off the parent's consolidated balance sheet,[26] therefore not affecting the visible debt:equity ratio. Companies which used this form of finance extensively included Stanhope, Rosehaugh, Greycoat, London & Edinburgh Trust and Bredero.[27]

Another method of keeping debt off the balance sheet was the use of non-recourse and limited-recourse loans. These loans, a phenomenon of the 1980s, were typically used by property companies conducting developments via joint ventures with other parties. They allowed property companies to limit their financial exposure by making the property for which the finance was raised the only security for payment of the debt, without recourse to the company itself. As Richard Millward of Kleinwort Benson's corporate banking department put it 'Projects have become the borrowers, not the companies.'[28] Limited-recourse loans were similar to non-recourse loans in that they provided no guarantee as to the capital of the loan, though some guarantees would be given regarding the completion of the project and interest payments.[29]

Non-recourse loans typically involved a bank advancing around 70% of a development's cost, which was about 40–50% of its estimated value on completion.[29] Richard Guignard, finance director of the Greycoat Group, described limited recourse debt as 'damage-limitation which avoids the domino-like collapse ... Using the battleship analogy, all the loans are pigeon-holed in airtight compartments. If a torpedo hits one of the compartments, the whole ship doesn't sink.'[29] Thus ran the theory. However the property companies (like the Titanic, which was supposed to be made unsinkable by its air-tight compartments) proved much less secure when the iceberg of rising interest rates and economic recession hit the property industry towards the end of 1989.

Off-balance sheet funding made it extremely difficult to assess the true debt position of many major property companies. For example, Rosehaugh's 1989 annual report showed shareholders' funds of £476 million, against borrowings of £269 million. However, a detailed examination of the eight pages of notes covering Rosehaugh's 16 associate companies (providing detailed financial information on only six of them) reveals at least another £847 million of borrowings, of which about £400 million appeared attributable to Rosehaugh.[30] Rosehaugh was by no means atypical in this; Greycoat's 1989 group accounts showed eight associate companies with no separate balance sheets for any of them.[30]

The combination of deficit financing using financial techniques which removed the direct link between a company's debt and equity, and the substantial expansion in new development both within and outside traditional centres, constituted an explosive mix which threatened to shatter the commercial property sector. Section 9.5 examines the spectacular boom which resulted from these trends, while section 9.6 recounts what happened when the party ended.

9.5 THE BIG BANG BOOM 1985–1989

The 1985–1989 boom had much in common with the 'Barber Boom' of the early 1970s. Growth was primarily in the service and financial sectors, rather than in manufacturing. The concurrent property boom followed deregulation of the financial and property sectors, and saw a return to prominence of the property companies, as developers, and of the banks, as providers of finance. Unlike all the previous post-war commercial property booms the Big Bang Boom did not see a rapid inflow of funds from UK institutional investors, though, like its early 1970s predecessor, it was primarily a speculative boom in property values.

Several factors initiated the speculative spiral. The slump in the property development sector which preceded the boom had once again led to demand outstripping supply in most areas of the development market by the mid-1980s. The growing market for the new edge-of-town or out-of-town developments described above provided new opportunities for developers, while a number of factors boosted interest in office development in the City and other major centres. The rapid adoption of new information technology, such as the laser printer, in the office, together with demands for better working environments, led to a substantial growth in office floorspace per employee. Meanwhile, the decade saw rapid growth in City employment, in contrast to the 1970s, as a result of the growing internationalization of financial markets. This process was accelerated in 1985, as the City began to gear up in anticipation of Big Bang in October of the following year.

The City Corporation facilitated the office development boom which these pressures looked set to create. Fearing the loss of businesses, which had become less tied to the Square Mile as a result of telecommunications and related technological developments (together with competition from alternative centres such as Canary Wharf), it abandoned its existing development plan. In March 1986 the Corporation increased the allowable office space of the City by 11 million sq. ft., by adjusting maximum plot ratios and relaxing some conservation measures.[31] This marked a potential increase in the City's office space of about 20%.

Development activity was encouraged by interest rates which were lower than at any time since the 1974 crash, though in real terms they remained high as a result of the substantial decline in inflation. Booming economic conditions were most strongly reflected in the financial and service sectors, on which the prosperity of the mid-1980s was largely based. Between 1980 and 1989 the output of the financial and business services sectors, which provided much of the demand for office property, doubled in real terms.[32]

Buoyant economic conditions led to a dramatic fall in vacancy rates, from 8 and 5.5%, respectively, for City and West End offices in 1984 to troughs of 2.25 and 1%, respectively, in 1987.[33] The growing scarcity of prime property was reflected in rising market rents, which grew by almost 52% in real terms between 1985 and 1989. Despite this, investment in the sector by British insurance companies and pension funds, which had traditionally dominated commercial property investment, was lower during 1985–1989 than over the previous five years.

However, while the financial institutions reduced their exposure to property during the 1980s the banks, which had largely avoided the property market during the 10 years following the 1974 crash, stepped up their investment in the sector. During the mid to late 1980s banks found it difficult to get sufficient high margin business and the booming property sector offered what appeared to be an attractive outlet for their funds. They therefore stepped in to meet the long-term financing needs of the property developers. However, unlike the early 1970s property boom where lending to property companies had been concentrated in the secondary banking sector, during the 1980s boom it was the major clearing banks, together with some foreign banks, which provided the funds.

Foreign investors also showed substantial interest in the British commercial property market, especially the City, during the boom, particularly the Japanese, Scandinavians, Americans and Middle Eastern investors. Foreign purchasers were particularly active in the market for large, prestigious City offices; during the 18 months to December 1988 they accounted for roughly one-third of London property purchases over £20 million, involving a total investment of about £1 billion.[34] By 1991 foreign investors owned 23.7% of Central London's investment properties, compared with 10.6% held by 'traditional' institutions, 48.8% held by UK financial institutions and 16.9% held by property companies.[35]

The growth of overseas investment reflected a wider trend towards the growing internationalization of financial and investment markets. However, the inflow of investment from overseas purchasers proved insufficient to plug the gap left by the financial institutions; from 1985 to 1989 initial yields on investment property actually rose, despite the development boom, while yields on other financial assets were generally static or declining.

Like the property development boom of the early 1970s, the onset of the Big Bang boom produced a high level of amalgamations in the property company sector. In 1986 Haslemere Estates was purchased by the Dutch property company Rodamco. Stock Conversion was the take-over boom's next victim, falling to P&O after a six-day battle. BCPH made an agreed take-over bid for Land Investors and the newly formed City Merchant Developers successfully launched a reverse take-over bid for Marlborough Property. Other property companies snapped up during this year included Samuel Properties, Southend Estates and United Real Property. There was even speculation that bids might be launched for MEPC and Land Securities. They never materialized, but, as the *Estates Times* noted 'it was that sort of year'.[36] The property take-over boom continued during the following three years, as predators sought companies with investment portfolios which they could break up and sell at a profit; from 1988 to 1990 there were over 50 bids for, or by, quoted property companies.[37]

'Big Bang' produced a rush by British and overseas financial institutions to acquire City brokerage firms, there being a general belief that the major banks would need representation on the London securities market in order to remain competitive.[38] Many of these firms were rehoused by their new owners from cramped 1960s premises to new prestige offices. Deregulation also led to a fundamental reappraisal on the part of City institutions regarding the type of business they would conduct and the way in which they would go about it.

One consequence of this reappraisal was a boom in the acquisition of residential estate agents by banks, building societies and insurance companies. During the two years to November 1988 these organizations invested about £2 billion in estate agency chains. Even prior to the house price crash the returns on most of these investments proved very poor.[39] Despite the hopes of the Prudential, Lloyds and others, that agents' premises might also serve as High Street outlets for insurance and other financial products, most of these acquisitions soon proved an embarrassment to their purchasers and were later sold at considerable losses.

Financial deregulation also affected the commercial chartered surveyors. In 1986 the Royal Institution of Chartered Surveyors allowed its members to take in outside capital for the first time, starting a rush of stock market issues by firms such as Debenham, Tewson & Chinnocks, Savills and Baker Harris Saunders.[40] Some of the capital raised by the stock market flotations was used to help them develop property finance innovations, in the face of growing competition from American investment bankers which were beginning to muscle in on the business of London's major commercial chartered surveyors.

During the 1980s Salomon Brothers and Goldman Sachs had pioneered 'property investment banking'. Rather than going down the traditional development finance path (an institutional investor financing a development and then purchasing it from the developer at a pre-agreed price) they turned instead to the capital markets.[40] Using a mix of financing techniques, including zero-coupon bonds and interest rates swaps and caps, they not only reduced development costs, but allowed developers to retain ownership and thus keep any capital appreciation which might occur during the construction period. Such a financial package was put together for the Greycoat Group by Salomon Brothers in 1988, providing finance for a £123 million West End building using a deep-discount Eurosterling bond with an 8.25% coupon.[40] However, the 'securitization' of development finance proved easier to introduce in America than in Britain, and by the time the property crash intervened there had been few other attempts to finance developments via the capital market.

Like the property boom of the late 1960s and early 1970s, the property price spiral of the mid to late 1980s stimulated financial innovation. In addition to securitization, which promised to transform property mortgages into tradeable securities, there were attempts to similarly transform direct property investment. This was to be achieved via the 'unitization' of investment property, i.e. the trading in small units of a building. Its advocates claimed that unitization would encourage institutional investment in property, make possible the active management of property portfolios and enable smaller institutions to participate in the sector.

In May 1986 *The Economist* announced that 'Britain's commercial property market is on the verge of a change as radical as the deregulatory Big Bang in the equity market.'[41] John Barkshire, the chairman of Mercantile House, led a group proposing an exchange to trade shares in unit trusts based on single properties, while the National Westminster Bank group and chartered surveyors Richard Ellis promoted a rival scheme, for an exchange based on property income certificates (PINCS). This offered investors both a share in the company which managed the property and an income certificate, providing a share of the rent.[41] By 1987 there was talk of trading property units on the stock market, though tax problems, legal difficulties and a lack of interest from investors delayed the launch of the new unitization vehicles; in the end nothing came of these initiatives.

However, one new technique, the leveraged buy-out (which involved raising bank loans to finance company take-overs) was successfully imported from the USA during the final years of the boom. In March 1988 a company named Giltvote was created to allow a consortium led by veteran UK developer Stephen Wingate, and including Eagle Star, Mercury Asset Management, Kleinwort Benson and George Soros, to make a bid for Estates Property Investment Co. This corporation succeeded in purchasing Estates Property for £73 million, and in July 1989 Stephen Wingate was among the members of another consortium, again including Eagle Star and two overseas investors, which set up Marketchief to purchase Imry Merchant Developers for £134 million.[42] The downturn in the property market shortly afterwards marked an end to this form of take over.

Competition in the development market intensified during 1987. For example, when bids were sought for the 100 acre development of Kings Cross, Rosehaugh–Stanhope spent £6 million on its development proposal alone, allowing them to become the nominated developer despite competition from Speyhawk.[43] However, the October 1987 stock market crash caused considerable unease in all financial markets, not least in property.

Between Black Monday and December 4th 1987 property shares fell by 29% in absolute terms and 4% against the FTSE All-Share index. The shares of the most development orientated companies were hardest hit, with falls of 57% for Rosehaugh, 53% for Stanhope and Speyhawk, and 41% for London & Edinburgh Trust.[27] The crash hit trader–developers particularly hard, but had less effect on the 'asset-based warhorses', which had gone through their rapid growth phase during the 1950s and 1960s. As Sydney Mason, Chairman of Hammersons and one of the last survivors of the post-war property boom, stated at the time of the crash 'Last week I was considered staid and conservative; this week, I regard that as a compliment'.[43]

Following the crash some tenants cut back their expansion plans and short-term lettings on buildings due for demolition became the vogue as property companies adopted a 'wait and see' approach. However, rental growth continued during the year, Saatchi & Saatchi setting a British record by taking space in Legal & General's Landmark House, Berkeley Square at £62.50 per sq. ft.[43]

Banks increased their margins slightly in the aftermath of the 1987 stockmarket crash. However, the dull state of the securities market following the crash increased investors' interest in commercial property, while at the same time reducing the financial sector's demand for office space. The March 1988 budget gave a boost to the property industry as Chancellor Nigel Lawson cut Capital Gains Tax. The savings of Land Securities alone were estimated at £350 million.[44] This, like many of Mr Lawson's other measures, fuelled a boom which was soon to prompt a reverse in government fiscal and monetary policy, with catastrophic results for the property sector.

However, property's rapid recovery during 1988 led the banks to further step up their rapidly growing lending to the sector. During the two years to February 1987 outstanding banks loans to property companies rose by 30.63% per annum; this accelerated during the late 1980s to an average rate of 50.68% for the three years to February 1990. It was estimated that by Autumn 1988 about 100 banks in London wanted to lend money to property developers.[45] Most of this money went into huge syndicated deals to fund the large City developments that were currently in vogue. The pressure of supply forced down margins. In 1984 banks were lending at 2–3 percentage points above their own cost of funds, as represented by the London Interbank Offered Rate (LIBOR). Loans were generally repayable within five years and developers were usually required to put up at least 35% of the project's finance. By 1987 margins for some projects had dropped to below 1% above LIBOR, and the proportion of building costs required from the developer had fallen to well under 25%.[45]

Much of this borrowing was 'off-balance sheet', as discussed in section 9.5. In August 1989 Lloyds Bank published a report showing that outstanding bank loans to the property sector had increased by 60% in the year ending May, twice the rate of increase of total bank lending to UK residents. Despite showing evidence that the gearing of property companies was deteriorating, the report noted that this was not evident in the net debt to properties ratio of the 45 largest property companies.[46] The conflicting pictures presented by bank lending statistics and property company balance sheets were largely due to the prevalence of off-balance sheet funding, which hid the true debt positions of many large property companies. Thus, assessing and controlling the flow of funds to the property sector represented a considerable problem for the Bank of England, while financial deregulation following Big Bang compounded this problem. The stage was set for an explosion of credit to the property industry similar to that created by Competition and Credit Control in the early 1970s.

Meanwhile, a tidal wave of new developments gathered momentum. In addition to substantial redevelopment in the Square Mile new office districts were mushrooming on its fringes, pushing out the City's historic boundaries, especially to the east, into areas such as Bishopsgate, Brick Lane and Spitalfields.[47] And then there was Canary Wharf.

In 1985 a consortium composed of the Credit Suisse First Boston and Morgan Stanley banks, together with the American property developer Mr Gooch Ware Travelstead, had proposed a 12 million sq. ft. development on the Canary Wharf site, a strip of land running across the West India Docks on the Isle of Dogs.[48] Their ambitious plan proposed a new financial services centre, in competition with the City, the centrepiece of which was to be three towers, each over 800 ft. high (making them the tallest office blocks in Europe).

The consortium's plans came in for severe criticism on grounds of the project's risk, its aesthetics and the adequacy of their finances.[49] The first and third criticism appeared well-grounded by the summer of 1987, when the consortium refused to go ahead with the scheme unless tenants signed up in advance. Potential tenants, understandably, were unwilling to sign up until the consortium went ahead.[50] The two banks pulled out of the project, while maintaining their commitment to take office accommodation on the site. However, in July Canadian developers Olympia & York came to the rescue, buying the development company for about £170 million.

This international, Toronto-based group had already made a substantial reputation for itself as a successful developer. Owned by the brothers Paul, Albert and Ralph Reichmann, and their family, it had amassed interests in 100 million sq. ft. of property in North America, including 24 million sq. ft. in New York. Strictly observant Jews (the construction of Canary Wharf stopped for the Sabbath and the builders' holidays were based on the Jewish calendar) the Reichmann's had established themselves a reputation for 'An appetite for great risk, zeal for the long term, piety and a widely accepted honesty'.[51] Their developments were known for their high building quality, innovations in financing (they pioneered long-term development finance via tradeable securities in the USA) and innovative building techniques.

The arrival of Olympia & York boosted confidence in Canary Wharf. As *The Economist* stated at the time of the announcement 'It is the Reichmann's track record that now makes Canary Wharf credible. Their company has always completed developments on time and their financial muscle is considerable.'[52] The firm had the financial strength to develop the site without letting in advance, and to buy up prospective tenants' existing accommodation to facilitate their moves (tactics which had already been successfully employed by the group, most notably for their Battery Park World Financial Centre development in Manhattan).

Aesthetic objections to the scheme were met by reducing the size of two of the three towers and engaging prestigious architects, including Cesar Pelli, who was to design Canary Wharf's famous 800 ft. high pyramid-topped tower, Britain's tallest building. However, serious doubts as to the project's risk continued, even at the height of the boom. This development was equal to about 7% of the City and West End office market,[49] in a location which was not yet a recognized office centre, was considerably removed from London's business heart and had a transport system which was wholly inadequate for its projected needs. However, the offer of top quality office accommodation at £20 per sq. ft., compared to £55–60 in the City, made it look for a time as if Canary Wharf might indeed achieve its ambitious objectives.

Together with developments on the City fringes, the West End and on the new generation of business parks, the growth of the London region's office stock collectively defied all reason. Something had to give. In the end it was the commercial property market as a whole which gave, plunged into a slump from which it has yet to fully recover.

9.6 THE SECOND PROPERTY CRASH

Towards the end of 1989 the new wave of Big Bang office developments became available for letting in significant numbers. The prospects of them finding tenants at prevailing market rents appeared remote; during the first nine months of 1989 the City's vacancy rate had already almost doubled, to over 10%.[53] Meanwhile, in October, Nigel Lawson raised bank base rates to 15%, effectively killing the property boom which his previous policy of low interest rates had fuelled. On October 13th Robin Leigh-Pemberton, the Governor of the Bank of England, warned the banks to re-examine their lending to property companies, following the bank rate rise. Although he stated that the Bank's supervisory division did not see property lending as a 'major anxiety', he noted that 'Exposures to the more highly geared companies might well be singled out for special attention, as might the viability of the underlying projects in limited recourse financings.'[54]

The boom of the 1980s had been funded largely by massive consumer borrowing, much of it secured on the expectation of ever-rising house prices.

The interest rate hike of October 1989 contributed to the most severe and protracted collapse in house prices witnessed in Britain since the Second World War, as the credit-induced boom was swiftly transformed into a debt-induced slump. Like the Barber Boom before it, the Big Bang property development boom ended with the imposition of restrictive monetary policy to reduce inflationary pressures which had, at least partly, been the direct result of the government's earlier expansionist policies. This dealt a double blow to the property development market, reducing the demand for new buildings while increasing their development costs as a result of rising interest rates.

Unlike the recession of the early 1980s this downturn hit the South East and the City particularly severely. Vacancy rates for City and West End offices rose from 5.6 and 2.3%, respectively, in 1989 to 18.7 and 9.0%, respectively, by 1992.[55] There was a consequent collapse in market rents, which fell by 31% between 1989 and 1992. Meanwhile, the property boom of the mid to late 1980s had created a considerable development 'pipeline' which resulted in a large number of new developments coming on to the market during these years, accentuating the imbalance between demand and supply for commercial property. It had also financially overstretched most of the merchant developers, for whom the downturn in market conditions, at a time of sharply rising interest rates, proved fatal.

Value added tax on new buildings, introduced from April 1990, increased accommodation costs for organizations such as banks which could not claim it back, while the Uniform Business Rate added substantially to commercial property rates in the South East.[56] Meanwhile, the banks were becoming increasingly concerned at their vast loans to the property sector, and the sector's growing inability to repay them. In August 1991 property companies owed over £40 billion to domestic and overseas banks. The proportion of outstanding bank loans accounted for by property company lending was even greater than had been the case during 1974, as is shown in Figure 9.2.

However, unlike the 1973/4 crash, when loans had been concentrated in the hands of highly exposed fringe banks, the banks which held property company debt in 1990 were both larger and more diverse; 44% of the property sector's outstanding bank debt was accounted for by UK retail banks, 7% by UK merchant banks and a further 6% by other UK banks. The remainder had been loaned by overseas banks, the most important being the Japanese (11%) and the Americans (6%).[57] The more stable nature of the property sector's domestic creditors, together with the much greater internationalization of bank lending to UK property companies, actually weakened the position of the property companies compared to that following the 1973/4 crisis. The British banking system was not threatened to anywhere near the same extent as had been the case in 1974 and the Bank of England therefore felt less need for intervention that might have allowed more property companies to survive the crash.

In February 1990 J.M. Jones, a major M4 corridor-based developer, became the first major victim of the crash, going into receivership with debts of

THE SECOND PROPERTY CRASH

£50 million.[58] Many other property companies were to suffer the same fate during the next three years. During 1991 the City's vacancy rate grew to nearly 20%, the worst hit areas being those on the fringes of the Square Mile.[59] More than 230 property companies went into receivership or administration during the year, including Berish Berger's Land & Property Trust, City Gate Estates, Alpha Estates, Erostin and Sheraton. Many others faced very severe difficulties; by the end of the year the share prices of some 20 quoted property companies, including such former luminaries as Rosehaugh, Mountleigh and London & Metropolitan, had fallen to or below 10p.[59] A survey conducted by Chesterton Financial in 1991 found that 17% of borrowers were in breach of covenants and that over 70% of banks believed that the number of receiverships would increase over the next few months.[60]

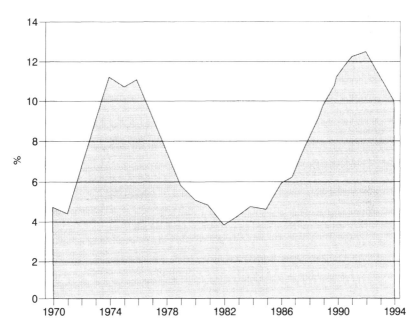

Figure 9.2 Bank property company debt as a percentage of all commercial loans, 1970–1994.
Source: information provided by DTZ Debenham Thorpe Research (based on Bank of England data).

December 1991 saw the departure of probably the most famous developer of the 1980s, Godfrey Bradman, as Chief Executive of his Rosehaugh property group. Bradman's financial expertise had built up Rosehaugh from being a virtually unknown tea company in 1978 to the best known and, while the boom lasted, most successful, British property development company. Bradman had also become well-known for his work outside the property sector, taking an active role in campaigns such as those for lead-free petrol, freedom of information, compensation for the victims of the drug Opren and environmental aspects of the building trade.

However, his business empire, which he had founded on the principle of minimizing risk, had overextended itself during the late 1980s, leaving itself vulnerable to the rise in interest rates, reduction in tenant demand and oversupply of commercial property which marked the end of the 1980s boom.[61] In the year to 30th June 1990 Rosehaugh lost £165.5 million; the following year it lost £226.6 million, the announcement of the loss, on December 6th 1991, coinciding with that of Bradman's departure as Chief Executive. Only the day before he had accompanied the Queen at the opening of one of the flagship developments of the boom, Broadgate, an office development of over 3 million sq. ft. on the site of the former Broad Street Station, developed jointly by Rosehaugh and Stuart Lipton's Stanhope.

Rosehaugh limped on for another year, until December 1st 1992, when its 26 bank creditors finally called in the receivers. By this time it owed its bankers some £350 million, of which it was expected that perhaps £100 million would be recouped from the sale of its assets.[62] By this time many of the sector's leading development companies had suffered a similar fate. In March 1992 Randsworth had gone into receivership with debts of over £350 million. In May, Mountleigh called in the receivers and in the same week Olympia & York, the world's largest developer, also reached the end of the line.

Canary Wharf, which had been heralded during the late 1980s as a monument to Thatcherism, became so in a very different sense in the harsh light of the days during and following its opening. By the time the development opened, in the middle of 1991, the property market was in deep recession. Furthermore, while the Reichmann's produced the high-quality development they had promised, the surrounding transport system, consisting only of a miniature railway and tiny roads, meant that a journey to the City could mean a half-hour taxi ride. Substantial transport improvements were promised, but remained some time away. This made Docklands very unpopular with office workers, who faced a lengthy journey to and from work on an inadequate transport system, often on top of their existing commute to a London mainline station. It was said that the cheapest method of reducing staff in the City was to move to docklands; resignations could be counted on to halve staff numbers.[63]

Despite attracting some prestigious tenants, such as the *Daily Telegraph*, Olympia & York had only managed to let 60% of Canary Wharf by 1992, many of its tenants having been attracted by rent-free periods or other expensive concessions.[64] In May 1992 the banks finally refused to lend the further funds necessary to keep the project going and Canary Wharf was placed in the hands of administrators.

The collapse of Olympia & York shook the industry; a Gallup survey of property investment and development companies, conducted shortly after the announcement, revealed that more than one-third of the companies interviewed expected the clearing banks to clamp down on their credit lines as a result and

THE SECOND PROPERTY CRASH

for a time there were real fears of a 'melt-down', in the property company sector.[65] However, as property company debt was generally held by large banking groups which were not themselves in real danger of insolvency, the banks were able to keep enough of the troubled property companies afloat for long enough to prevent such a scenario.

The older, established property companies generally fared the crash much better than the merchant developers. As Figure 9.3 shows, the property investment companies were much less highly geared than their developer–trader counterparts. This is partly an inherent characteristic of the activities of the two types of company,[66] though it does reflect the much greater risk associated with a strategy of development for sale rather than investment, which was one of the hallmarks of the property companies established during the 1980s. As the data are based on evidence from company accounts, which often disguised the true extent of debt via the off-balance sheet techniques discussed above (and excludes companies which went bankrupt prior to producing 1991 accounts) the true disparity between the gearing ratios of the property investment companies and the developer–traders is almost certain to be even greater than is indicated.

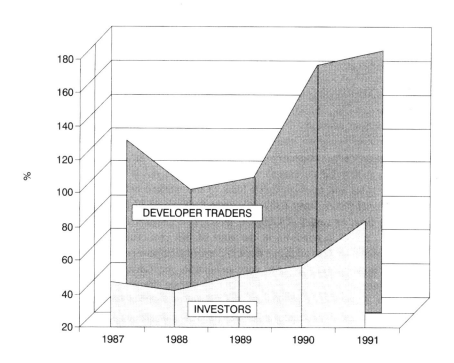

Figure 9.3 Property company gearing (debt as a percentage of net asset value), 1987–1991 (median gearing levels for each year).
Source: information supplied by R.J. Barkham.

Some well-established investment-based companies, such as Brixton Estate, were able to maintain dividend growth even during the crash. During the boom years Brixton pursued an expansionist policy, but was careful to keep its debt:equity ratio within reasonable levels. Towards the end of the decade it took steps to guard its finances against any major rise in interest rates, arranging substantial interest rate swaps, negotiating banking commitments to ensure the continued availability of finance for its development programme, and raising additional funds via a Sterling Commercial Paper issue and an £80 million issue of 10.625% first mortgage debenture stock. The company also reduced its exposure to any crash at an early stage, pulling out of the development–trading market in 1989. Thus, despite an inevitable fall in the value of its investment portfolio, pre-tax profits rose by 15.4% during 1990 and 16.3% in 1991.

Brixton was fortunate in that a large proportion of its shares were held by a few major institutional investors, their support giving it some degree of protection against the 1980s property company take-over boom. Many of Britain's other established property companies which had taken steps to limit their exposure to a downturn in the market during the boom years had been snapped up by the mushrooming merchant developers, only to collapse along with their new parents during the crash. Given a stock market which favoured rapid short-term expansion, a policy of limiting long-term risk often ran considerable short-term risks of take-over by younger, less cautious, competitors.

While the 1990 property crash did not threaten the banking system to anywhere near the same extent as its 1974 predecessor, it was certainly more severe in terms of its impact on the commercial property sector. Hillier, Parker and the London Business School estimated that the total value of commercial property in the UK, which peaked at £250 billion in 1989, had fallen by £90 billion, or about 36%, by 1992.[67] During 1974 the fall in property values had occurred in one year only, and had amounted to only 21.7%.[68]

A fall in property values of such magnitude raised initial yields substantially, despite a considerable decline in market rents. High yields attracted some insurance companies and pension funds back to the market; net institutional investment in property rose from £587 million in 1990 to £2.17 billion in 1991. Even companies such as Legal & General, which had decided to make no net additions to its property portfolio at the start of the year, found the bargains too attractive and ended up adding a net £120 million of new property to its books.[69] Institutional investors also made very substantial purchases of the equity which the beleaguered property companies issued during the early 1990s, indirectly bailing out the banks as they had done during the mid-1970s and at the same time acquiring considerable property assets at bargain prices.

While the financial institutions had pursued what (with hindsight) proved a successful contracyclical pattern of avoiding substantial property market activity during the boom years, followed by opportunistic purchases during the crash, not all institutional investors proved as fortunate. The Church Commissioners, one of the most successful institutional participants in the

1950s boom, were hit extremely hard by the crash. Their assets declined in value from nearly £2.93 billion in 1989 to £2.13 billion in 1992,[70] prompting the Archbishop of Canterbury to commission a report from accountants Coopers & Lybrand into the Commissioners' activities. This highlighted overexposure to speculative property development projects, funded by massive borrowing, and pointed to inadequate managerial control procedures and a failure to seek independent expert advice.[71]

The Commissioners had failed to keep up with the times; while their systems of managerial control and outside advice had been advanced by the standards of the 1950s, they had become archaic in the increasingly sophisticated property market of the 1980s. The Commissioners also adopted an inherently more risky strategy during the 1980s boom, borrowing from the banks in order to finance development schemes.[72] By borrowing to fund property development (something they had hitherto avoided) the Commissioners had assumed a development role more similar to that of a property company than an institutional investor, and as such exposed themselves to considerable losses, with severe consequences for the Church's funding, when conditions turned from boom to slump.

Despite a limited revival in property investment in 1991 and 1992 there were new clouds on the horizon for the property market. Government measures threatened to undermine the security of property as a long-term investment. The government announced plans to abolish privity of contract (which made a property's original tenant ultimately responsible for paying rent throughout the duration of the lease) on new leases.[73] The Department of the Environment was also conducting a review of other aspects of leases, including the all important upward-only rent review clause. The institutional lease, which accounted for much of the attraction of commercial property to long-term investors, was coming under serious legislative threat.

Even in the absence of further legislation the current buyers' market gave tenants the bargaining power to renegotiate the institutional lease in their favour. Tenants disliked the traditional 25 year lease, which restricted their mobility and tied them to buildings which might well be technically obsolete before the end of the lease term. New tenants were able to get 'break clauses' built into leases, allowing them to cancel the lease after 10 or 15 years. Others were able to reduce the term of the lease to 10 years, or less.[74] While such amendments did not mark a formal departure from the upward-only rent review they robbed it of most of its value to investors.

Tenants were also offered 'sweeteners', including rent holidays (zero rental payments for the first several months or years of a lease) and help with fitting-out costs. Such techniques proved attractive to developers as they maintained nominal rents while providing the effective rent reductions necessary to secure tenants.

In addition to its effects on the London office market, the slump had also hit provincial shopping centre development severely. In August 1990 John Parry, the managing director of Hammerson, was quoted as stating that shopping centres were currently opening with only a quarter of their space let, or none at all.[32] Ironically, after the damage was done the government belatedly moved to curb the uncontrolled expansion of retail development which had led to the oversupply. The Planning and Compensation Act, which gained royal assent in the summer of 1991, signified a major U-turn in government planning policy. The Act marked a move away from market-led planning, in favour of a more 'plan-led' approach. Local authorities were again given the power to dictate the scope, form and extent of development in their areas via a new wave of development plans.[75]

During the 1980s environmentalism had found its voice in the Tory shires when local authority planning refusals were overturned by an ultra *laissez-faire* Department of the Environment. A growing army of 'NIMBY's (Not in My Back Yard), as they were christened by Environment Secretary Nicholas Ridley, put pressure on the Conservatives to increase local control over the planning process. During the early 1990s the success rate of planning appeals against local authority decisions dropped considerably, particularly in cases where they were backed up by development plans. This change in government attitudes found further expression in 1992, the Department of the Environment issuing a new policy guidance note, PPG6, which emphasized the need to revitalise town centres, rather than develop further large out-of-town projects.

While the above analysis has focused on domestic factors, it must be remembered that there was a strong international dimension to the crash. During the early 1990s the office markets of the United States, Japan and much of continental Europe experienced severe depression. The worldwide 1980s financial services boom, itself spurred on by the growing internationalization of financial markets, had produced property booms in all the major Western economies. When the international boom turned to slump, its impact on commercial property was worldwide. However, the British property crash also encompassed sectors of the property market not directly connected with the financial services industry, such as the retail, industrial and residential sectors. This nationwide fall in property values, experienced by all major sectors, may have had an important effect in slowing down the economic recovery of the early to mid 1990s, as is discussed below.

9.7 THE GREEN SHOOTS OF RECOVERY

The fall in interest rates following sterling's departure from the Exchange Rate Mechanism (ERM) in September 1992 helped to stabilize the property market,[73] both property companies and their tenants welcoming the fall in their interest bills. Falling interest rates also stimulated investment activity, making it possible to invest in prime property using borrowed funds and achieve an immediate surplus of rental income over interest payments for the first time since the

THE GREEN SHOOTS OF RECOVERY

1950s. As Vanessa Houlder noted, by the end of 1992 entrepreneurs 'could borrow money for five years fixed at, say, 8 3/4 per cent to buy a good quality property with a yield of 10 per cent'.[76] As a result, initial yields on investment property began to decline slightly, thereby increasing values for the most sought-after properties despite continuing falls in market rents.

Overseas investors began to return to the market in significant numbers by late spring 1993, German, Middle Eastern and Far Eastern investors being particularly active. In June, George Soros, the Hungarian-born financier who had achieved fame due to his highly successful speculation against sterling prior to its exit from the ERM, announced a move into British property. His Quantum Fund was to form a partnership with British Land, involving the investment of £500 million in British commercial property. This news was sufficient to raise the share price of British property companies by 6.4%.[77]

Nineteen ninety-three was a good year for property company shares, which out-performed most other sectors during a year of booming share prices. Property companies took advantage of the booming stock market to repair their balance sheets, making about £2 billion of equity and debenture issues during the year, the highest level since 1987.[78] As a result outstanding bank loans to property companies, which had peaked at just over £41 billion in May 1991, had fallen to £33.5 billion by March 1994.[79] However, the banks continued to steer clear of the market, the lack of bank finance proving a significant constraint on the sector's recovery.

Development activity remained low during 1993; only 6.2 million sq. ft. of offices were currently under development by November of that year, compared to 15 million sq. ft. completed during the peak year, 1991.[80] Much current development activity represented pre-let property, tenants taking the opportunity to secure their future accommodation needs at prevailing low land and construction costs. Retail development activity still remained very depressed; by November 1993 only 2.54 million sq. ft. of shopping centre floorspace was currently under construction, compared to 18.9 million sq. ft. during the peak of the 1980s boom.[81]

By the early months of 1994 there was talk of a return to speculative development.[82] Economic growth strengthened tenants' covenants, while declining bond yields following Britain's withdrawal from the ERM made high-yielding property appear attractive to investors. However, rents had not yet begun to rise and a substantial surplus of un-let property still overhung the market. As Figure 9.4 shows, while vacancy rates in the City and West End had fallen considerably compared to their 1992 peak, there was still a higher proportion of vacant property in both these centres in 1994 than had been the case in 1990, when the market was already in depression. The property market, like the wider economy, was experiencing only a slow and weak recovery.

Evidence suggests that the depressed state of the property market and the slowness of economic recovery during 1993 and 1994 may be directly linked. The relationship between the collapse in house prices and depressed consumer spending is widely recognized. People who had seen the value of what was generally their most valuable asset fall by about 25% over a few years were more interested in using their income to save, or reduce, debt rather than increase consumption. A report by Douglas McWilliams in November 1992 argued that this 'wealth effect', had a counterpart in the corporate sector as a result of the commercial property crash.[83] Property makes up around 24% of the assets of non-financial companies in Britain and the 45% decline in commercial property values from the peak of the boom to the beginning of January 1993 is likely to have had a considerable effect on corporate borrowing. McWilliams estimated that (even ignoring the direct effects on corporate profits or GDP) a 10% drop in commercial property values would result in a reduction in corporate borrowing of £9.7 billion (1.5% of GDP) over the following eight quarters, while 'second round' effects would act to further depress growth.[84]

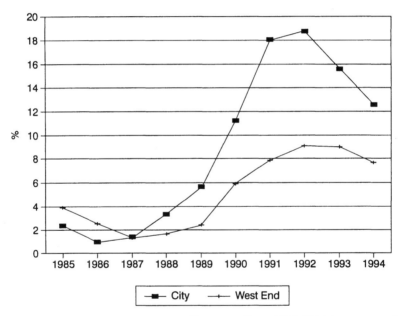

Figure 9.4 Vacancy rates for City and West End offices, 1985–1994 (percentage of stock vacant, for units over 4999 sq. ft.). Source: information supplied by Jones Lang Wootton.

According to McWilliams, commercial property price changes have both a direct effect and a monetary effect. The direct effect results from property price movements causing companies to alter their borrowing patterns and, therefore, their expenditure. The monetary effect occurs since corporate borrowing itself influences an element of the money supply; low levels of borrowing would therefore reduce monetary growth, with consequent effects on future economic activity.[84]

Another way in which the property crash might have intensified the depression of the early 1990s is via its effects on bank lending. The importance of property values to bank lending is far greater than is indicated by figures on bank loans to property companies. Julian Robins, a banking analyst at BZW, estimated in 1992 that three-quarters of bank lending was dependent on property.[85] Property's role as collateral for a large proportion of bank loans represented an unquantifiable, but potentially massive, problem for the banks, magnifying the direct effects of the property crash on their portfolios.

The collapse in property values had a particularly severe effect on loans to small businesses. This mainly involved difficulties in obtaining new loans, though there were claims that banks were even calling in existing loans as a result of the fall in property values. In April 1992 Stephen Alambritis of the Federation of Small Businesses stated that 'Bank managers say they need more security. But as soon as they have more security to make sure the loan is covered, they call it in. There is a lot of sharp practice.'[85]

Thus, the property crash had a twin effect on economic growth; it reduced companies' willingness and ability to borrow, and the banks' willingness and ability to lend. By eroding wealth and increasing debt levels the effects of a (largely avoidable) speculative boom in property values, and the inevitable slump which followed, not only resulted in catastrophe in the commercial property sector but may have had a significant impact on British economic growth during the early and mid-1990s.

REFERENCES AND NOTES

1. V. Houlder (1990) Feeling the lending pinch. *Financial Times*, 23 Nov., 1.
2. British property: New dimensions (1992) *The Economist*, 11 April, 102.
3. Pickings for the birds (1991) *The Economist*, 20 April, 97.
4. Commercial property: Hurry on down (1990) *The Economist*, 18 Aug., 19.
5. S. Connor (1989) *Postmodernist Culture*, Blackwell, Oxford, pp. 66–7.
6. L. Esher (1981) *A Broken Wave: The Rebuilding of England 1940–1980*, Allen Lane, London, p. 288.
7. A.R. Goobey (1992) *Bricks and Mortals*, Century, London, p. 26.
8. A.R. Goobey (1992) *Bricks and Mortals*, Century, London, p. 30.
9. Corporate Headquarters: The shape of things to come (1988) *The Economist*, 5 March, 22.
10. Corporate Headquarters: The shape of things to come (1988) *The Economist*, 5 March, 21.
11. Cited in R. Imrie and H. Thomas (1993) Urban policy and the Urban Development Corporations, in *British Urban Policy and the Urban Development Corporations* (eds R. Imrie and H. Thomas), PCP, London, p. 12.

12. S. Brownill (1993) The docklands experience: Locality and community in London, in *British Urban Policy and the Urban Development Corporations* (eds R. Imrie and H. Thomas), PCP, London, p. 41.
13. S. Brownill (1993) The docklands experience: Locality and community in London, in *British Urban Policy and the Urban Development Corporations* (eds R. Imrie and H. Thomas), PCP, London, pp. 46–7.
14. London Docklands (1988) *The Economist*, 13 Feb., p. 77.
15. London Docklands (1988) *The Economist*, 13 Feb., p. 71.
16. Building controls: Shaky ground (1990) *The Economist*, 27 Oct., p. 35.
17. *Estates Times Review: Twenty Years of Property* (1988) Estates Times, London, p. 75.
18. A. King and S. Bryant (1988) *UK 2000: An Overview of Business Parks*, Applied Property Research, London, p. 5.
19. A. King and S. Bryant (1988) *UK 2000: An Overview of Business Parks*, Applied Property Research, London, p. 1.
20. J.M. Wilson (1984) Retail warehousing in Greater London. *Estates Gazette*, **272**, 20 Oct., 244.
21. D. Wright (1986) DIY retailing comes of age. *Estates Gazette*, **277**, 22 Feb., 712.
22. See A.R. Goobey (1992) *Bricks and Mortals*, Century, London, Ch. 9, for a fuller discussion of the activities of provincial retail developers.
23. A.R. Goobey (1992) *Bricks and Mortals*, Century, London, p. 87.
24. A.R. Goobey (1992) *Bricks and Mortals*, Century, London, p. 89.
25. A.R. Goobey (1992) *Bricks and Mortals*, Century, London, p. 93.
26. The consolidated balance sheet would only show the parent's net equity holding in each associated company, i.e. its share of the company's estimated surplus of assets over liabilities.
27. J. Huntley (1987) Balancing act. *Estates Gazette*, **284**, 12 Dec., 1460.
28. J. Huntley (1989) More ways to limit financial risks – non-recourse lending is now widespread among newer development groups. *Financial Times*, 12 July, III.
29. M. Brett (1990) *Property and Money*, Estates Gazette, London, p. 94.
30. The Lex Column: Property accounting (1989) *Financial Times*, 4 Dec., p. 24.
31. A.R. Goobey (1992) *Bricks and Mortals*, Century, London, p. 35.
32. Commercial property: Hurry on down (1990) *The Economist*, 18 Aug., 18.
33. Source: P.C. Beverley (1992) UK business and office development cycles. MSc dissertation, University of Portsmouth, p. 54 (based on Richard Ellis data).
34. Following the global property men's star (1988) *The Economist*, 24 Dec., 89.
35. Chesterton (1991) *London: A Capital Investment*, Chesterton, London, p. 9.
36. *Estates Times Review: Twenty Years of Property* (1988) Estates Times, London, p. 71.

REFERENCES AND NOTES

37. A.R. Goobey (1992) *Bricks and Mortals*, Century, London, p. 207.
38. A.R. Goobey (1992) *Bricks and Mortals*, Century, London, pp. 202–3.
39. Old prop. nr tube. nds attn.(1988) *The Economist*, 5 Nov., 134.
40. London's great property grab (1989) *The Economist*, 30 Sept., 121.
41. Britain's property developers build themselves a market (1986) *The Economist*, 31 May, 91.
42. A.R. Goobey (1992) *Bricks and Mortals*, Century, London, p. 206.
43. *Estates Times Review: Twenty Years of Property* (1988) Estates Times, London, p. 79.
44. *Estates Times Review: Twenty Years of Property* (1988) Estates Times, London, p. 80.
45. British property finance: Holes in the ground (1988) *The Economist*, 17 Sept., 128.
46. R. Atkins (1989) Property lending may "leave banks exposed". *Financial Times*, 7 Aug., 7.
47. A. Seidl and N. Faith (1990) High-rise anxiety grips City builders. *The Independent on Sunday*, 3 June, 11.
48. London Docklands (1988) *The Economist*, 13 Feb., 71.
49. London Docklands (1988) *The Economist*, 13 Feb., 72.
50. Singing a Canadian tune (1987) *The Economist*, 25 July, 64.
51. The Reichmann Brothers: Kings of officeland (1989) *The Economist*, 22 July, 17.
52. Singing a Canadian tune (1987) *The Economist*, 25 July, 65.
53. The next crash: Unreal estate (1989) *The Economist*, 21 Oct. 138.
54. P. Norman (1989) Warning on property lending. *Financial Times*, 14 Oct., 24.
55. Information supplied by Jones Lang Wootton.
56. London property: Crash or crush? (1989) *The Economist*, 4 March, 35.
57. Chesterton (1991) *London: A Capital Investment*, Chesterton, London, pp. 9 and 15.
58. A. Seidl (1990) Pop goes the Big Bang boom. *The Independent on Sunday*, 11 Feb., 6.
59. V. Houlder (1991) The property market: A year of living on the edge. *Financial Times*, 20 Dec., 25.
60. Chesterton (1991) *London: A Capital Investment*, Chesterton, London, p. 16.
61. For a fuller account of Godfrey Bradman's career see A.R. Goobey (1992) *Bricks and Mortals*, Century, London, Ch. 3.
62. Rosehaugh: Snuffed out (1992) *The Economist*, 5 Dec., 112–17.
63. A. Seidl and N. Faith (1990) High-rise anxiety grips City builders. *The Independent on Sunday*, 3 June, 12.
64. S. Brownill (1993) The docklands experience: Locality and community in London, in *British Urban Policy and the Urban Development Corporations* (eds R. Imrie and H. Thomas), PCP, London, p. 51.

65. V. Houlder (1992) Banks shore up shaky property foundations. *Financial Times*, 15 June, 19.
66. R. Barkham (1995) The financial structure and ethos of property companies: an empirical analysis of the influence of company type. Paper presented to the International Conference on the Financial Management of Property and Construction, University of Ulster, May 1995, p. 10.
67. V. Houlder (1992) The property market: Cracks became chasms. *Financial Times*, 18 Dec., 9.
68. Source: IPD long-term property returns index.
69. British property: New dimensions (1992) *The Economist*, 11 April, 102.
70. Church Commissioners for England (1994) Report and Accounts, 1993.
71. A. Pike (1993) Church of England investment managers censured for losses. *Financial Times*, 23 July, 20.
72. F. Kane and D. Atkinson (1992) Church faces crisis as boom turns sour. *The Guardian*, 9 Nov., 2.
73. V. Houlder (1993) Optimism strengthens in spite of downbeat statistics. *Financial Times*, UK property survey, November, 3.
74. Waiting, wishing (1991) *The Economist*, 14 Dec., 107.
75. S. Robinson (1991) The property market: U-turn that surprised an industry. *Financial Times*, 1 Nov., 17.
76. V. Houlder (1992) Brighter outlook for the bargain-hunters – lower values and reduced interest rates have created opportunities for investors. *Financial Times*, 4 Dec., 16.
77. Selling up (1993) *The Economist*, 5 June, 100.
78. DTZ Debenham Thorpe Research (1994) *Money Into Property*, DTZ, London, p. 3.
79. DTZ Debenham Thorpe Research (1994) *Money Into Property*, DTZ, London, p. 12.
80. V. Houlder (1993) Institutions spur the recovery. *Financial Times*, UK property survey, Nov., 2.
81. V. Houlder (1993) High streets see most development. *Financial Times*, UK property survey, Nov., 5.
82. Market focus: Hot property – has Britain's commercial-property market risen too sharply from the dead? (1994) *The Economist*, 19 Feb.
83. Douglas McWilliams (1992) *Commercial Property and Company Borrowing*, Royal Institution of Chartered Surveyors, Paper No. 22, Nov.
84. Douglas McWilliams (1992) *Commercial Property and Company Borrowing*, Royal Institution of Chartered Surveyors, Paper No. 22, Nov., p. 3.
85. V. Houlder (1992) Why bricks are no longer bankable. *Financial Times*, 10 April, 16.

SECTION TWO

A statistical analysis of the commercial property market

A statistical overview of the property investment market | 10

10.1 INTRODUCTION

The preceding chapters have presented a number of arguments regarding the behaviour of the commercial property market, largely based on qualitative evidence. In order to ascertain whether this micro-level evidence conforms to the aggregate behaviour of the sector a long-run statistical analysis of the property investment market is presented here. The main difficulty encountered with any such analysis concerns the virtual absence of data covering most aspects of property investment prior to the very recent past. It was, therefore, necessary to construct, or extend, series covering annual net investment by the financial institutions, initial yields on investment property, the distribution of property portfolios by property type and geographical region, and returns to investment, in order to conduct this exercise.

The procedures adopted in the construction of these series, and the data obtained, are outlined in this chapter, together with a discussion of the main trends they reveal. Chapter 11 contains an analysis of the determinants of the rate of return on, and volume of, property investment, using this data to test hypotheses derived from the chronological Chapters 1–9.

10.2 NET INVESTMENT IN PROPERTY 1922–1993

One of the most important variables in the analysis of the institutional property investment market is the magnitude of the flow of institutional, and other, funds to the sector. This section provides estimates of net investment (i.e. purchases minus sales) in property by the financial institutions and other investors. The first part of the section deals with the compilation of series for annual net direct investment in property by various categories of financial institution for the years 1923–1937 and 1946–1993. The second part provides estimates of the relative

| 248 | A STATISTICAL OVERVIEW OF THE PROPERTY INVESTMENT MARKET |

importance of financial institutions and various other classes of property investor from the mid-1940s to the 1960s, together with a discussion of the growing participation of non-institutional investors in the property market since 1980.

10.2.1 Investment by financial institutions

Table 10.1 gives annual figures for net direct investment in land, property and ground rents by the financial institutions from 1923 to 1937 and 1946 to 1993. The data are derived from a number of sources, which have varying degrees of accuracy. Figures for net property investment by insurance companies are taken from annual data published by the Board of Trade, to 1961, and from *Financial Statistics* from 1962.[1] The Board of Trade figures are subject to a number of imperfections for the purposes of this research. Firstly, they refer to insurance company balance sheets which are mainly dated from the 31st December but which, in some cases, refer to March, or even June, of the following year. Secondly, the figures underestimate net investment, as they are reduced by the regular writing-down of property book costs by some institutions as an allowance for depreciation. This is acceptable for leasehold interests, which have a finite term, but was also widely applied to freeholds which were rising, rather than depreciating, in value over the years covered. However, this would not substantially distort the data.

Table 10.1 Net direct property investment by UK financial institutions (£M), nominal values 1923–1993

Year	Insurance companies	Pension funds	Property unit trusts	Investment trusts	Total
1923	0.3				0.3
1924	−1				−1
1925	−0.1				−0.1
1926	1				1
1927	1				1
1928	2				2
1929	1				1
1930	2				2
1931	3				3
1932	2				2
1933	4				4
1934	5				5
1935	7				7
1936	8				8
1937	5				5
1946	8	0.4			8
1947	17	1			18
1948	16	1			17

NET INVESTMENT IN PROPERTY 1922–1993

1949	23	1			24
1950	23	1			24
1951	32	2			34
1952	28	2			30
1953	29	2			31
1954	26	5			31
1955	65	6			71
1956	32	4			36
1957	41	10			51
1958	49	12			61
1959	54	13			67
1960	58	13			71
1961	102	12			114
1962	51	13			64
1963	63	27			90
1964	59	32			91
1965	89	39			128
1966	117	45	1		163
1967	96	82	16		194
1968	120	97	40		257
1969	186	117	43		346
1970	198	104	25		327
1971	198	85	23		306
1972	131	100	39		270
1973	307	300	57		664
1974	405	441	15		861
1975	406	323	33		762
1976	449	536	71		1 056
1977	410	472	66		948
1978	549	610	102		1 261
1979	633	548	83	–	1 264
1980	855	948	72	9	1 884
1981	1 074	847	108	0	2 029
1982	1 059	983	57	0	2 099
1983	845	680	–9	2	1 518
1984	744	997	47	1	1 789
1985	815	590	–5	2	1 402
1986	830	434	–101	–4	1 159
1987	832	197	–516	20	533
1988	1 424	272	99	12	1 807
1989	1 892	171	30	1	2 094
1990	1 298	–660	–61	10	587
1991	1 665	485	19	3	2 172
1992	678	977	–7	6	1 654
1993	274	155	92	2	523

Sources: Insurance Companies – 1923–1937, Board of Trade (1923–1938) *Annual Report on Life and Other Long-term Assurance Business,* HMSO, London; 1946, Post Magazine (1948) *Post Magazine Almanack*; 1947–1961, Board of Trade (1947–1962) *Annual Summaries of New Insurance Business*, HMSO, London; 1962–1993, CSO (various issues) *Financial Statistics*, HMSO, London. Pension funds – 1946–1953, extrapolated from the data used for Table 10.4, taking the average figure over this period for pension fund property investment as a percentage of insurance company property investment; 1954–1955, E.V. Morgan (1960) *The Structure of Property Ownership in Great Britain*, Clarendon, Oxford, pp. 116–25, extrapolated from pension fund property holdings in 1953 and 1955; 1956, E.V. Morgan (1960) *The Structure of Property Ownership in Great Britain*, Clarendon, Oxford, pp. 116–25, and *Committee on the Working of the Monetary System, Appendix Vol. 2 to the published report* (1956) HMSO, London, pp. 222–4; 1957, *Committee on the Working of the Monetary System, Appendix Vol. 2 to the published report* (1957) HMSO, London, pp. 222–4; 1958–1960, extrapolated from *Committee on the Working of the Monetary System, Appendix Vol. 2 to the published report* (1958–1960) HMSO, London, pp. 222–4 and CSO (1961–1963) *Financial Statistics*, HMSO, London; 1961–1962, CSO (1961–1963) *Financial Statistics*, HMSO, London; extrapolated from pension fund property holdings at the end of 1962 and public sector pension fund property investment during 1961 and 1962, on the assumption that the proportion of total pension fund property holdings accounted for by public sector funds remains constant during these years; 1963–1993, CSO (various issues) *Financial Statistics*, HMSO, London. Property unit trusts – 1966–1967, estimated from R. Redden (1984) *The Pension Fund Property Unit Trust: A History*. Privately published, London, p. 94 and CSO (1970) *Financial Statistics*, HMSO, London; 1968–1993, CSO (various issues) *Financial Statistics*, HMSO, London. Investment trusts – figures taken from CSO data. Horizontal lines indicate breaks in the series due to the use of different sources or methods of calculation.

The more recent data, taken from *Financial Statistics*, do not incur this problem. However, as a result of an adjustment to the way the figures were calculated in 1974, the values before and after that date are not directly comparable. In 1974 the figures were revised to include non-members of the British Insurance Association. This increased the recorded value of insurance company net property investment from £352.6 million to £405.1 million, and overall net insurance company investment from £1749 million to £1909 million, increases of 14.89 and 9.19%, respectively.

Figures for net investment in property by pension funds are published in *Financial Statistics* from 1963. Prior to that date, information on pension fund investment is extremely sparse. A survey carried out by the Radcliffe Committee estimated that the property holdings of all pension funds[2] amounted to £42 million, 3% of total pension fund assets, at the end of 1957. Net investment by pension funds in 1957 was estimated in the same survey at £10 million, 5% of total net pension fund investment during that year. These figures can only be regarded as estimates, however, as they are based on a survey covering the majority of, but not all, pension fund assets, the resulting figures being grossed up. Furthermore, the balance sheets the figures are based on do not all date from December 1957, but cover periods ending on various dates, from April 1956 to March 1958.[3]

Financial Statistics provides a figure of £104.9 million for pension fund property holdings at the end of 1962. Values for net property investment by public sector pension funds are given for 1961 and 1962, but are not available for private sector funds until 1963. The figures for pension fund property investment during 1961 and 1962 given in Table 10.1 were obtained by calculating public sector pension fund property holdings at 31st December 1962 as a percentage of total pension fund holdings, and grossing up the public sector fund figures for 1961 and 1962 on that basis. When these figures were subtracted from the difference between the 1962 and 1957 values for total pension fund property holdings £38 million was left, as net investment during the years 1958–1960. It was decided to allocate this as £12 million for 1958 and £13 million for 1959 and 1960, to allow for some increase in net investment during these years as overall funds grew.

A survey by E. Victor Morgan provides grossed up figures for pension fund holdings of real property in 1953 and 1955 of £16.75 million and £27.75 million, respectively, based on information from funds representing approximately 40% of total pension fund assets.[4] Property holdings were estimated to represent 1.69% of total pension fund assets in 1953 and 2.03% in 1955. In allocating net investment between 1954 and 1955 it was assumed that half the increase in the proportion of total pension fund assets represented by property occurred in 1954. The figure for pension fund property investment during 1956 was derived by subtracting pension fund property holdings at the end of 1955, as given by Morgan, from the Radcliffe Committee's estimate of pension fund property holdings at the end of 1957. Pension fund property investment during 1957 was then subtracted from this to obtain a figure for investment during 1956.

For 1946–1953 no published information on pension fund property holdings is available. The analysis of properties sold by Healey & Baker, discussed later in this section, indicates that pension fund property investment represented an average of 5.7% of insurance company investment over these years. In order to provide an indication of the level of pension fund investment during this period the annual figures for insurance company investment have been multiplied by this percentage to obtain estimates of pension fund investment. The resulting estimate of total pension fund property investment during the period, £10.01 million, compares with Morgan's estimate of £16.75 million for pension fund property holdings at the end of 1953, suggesting that pension fund property assets amounted to about £6.75 million by the end of 1945. This suggests that pension fund investment prior to the Second World War was on too low a level to significantly influence the volume of overall investment in property by the financial institutions, and no adjustment has been made to the inter-war figures for insurance company property investment to take account of the small amount of investment by pension funds during this period.

Figures for property unit trust investment are taken from *Financial Statistics*. The values for 1966 and 1967 have been adjusted since *Financial Statistics* only provides an aggregate figure for the period from 31st March 1966, when the first property unit trust was established, to the end of 1967. A figure of £1 million was estimated for 1966 using information provided in R. Redden's book on the Pension Fund Property Unit Trust.[5]

Table 10.2 Net direct property investment by UK financial institutions (£M), real (1990) values, 1923–1993

Year	Insurance companies	Pension funds	Property unit trusts	Investment trusts	Total
1923	7				7
1924	−25				−25
1925	−2				−2
1926	25				25
1927	26				26
1928	52				52
1929	26				26
1930	55				55
1931	88				88
1932	60				60
1933	124				124
1934	154				154
1935	213				213
1936	236				236
1937	140				140
1946	144	7			151
1947	290	17			307
1948	254	16			270
1949	357	16			372
1950	348	15			364
1951	444	28			472
1952	352	25			377
1953	355	<u>24</u>			379
1954	312	60			372
1955	747	<u>69</u>			816
1956	353	<u>44</u>			397
1957	434	<u>106</u>			540
1958	502	123			625
1959	554	133			687
1960	586	<u>131</u>			717
1961	1 000	118			1 117
1962	479	<u>122</u>			601
1963	579	248			827
1964	524	284			809
1965	756	331			1 087
1966	957	368	8		1 333
1967	766	655	<u>128</u>		1 549
1968	914	739	305		1 958
1969	1 347	848	312		2 507
1970	1 347	707	170		2 224
1971	1 233	529	143		1 906
1972	759	580	226		1 565
1973	<u>1 630</u>	1 593	303		3 526
1974	1 855	2 020	69		3 944
1975	1 499	1 193	122		2 814
1976	1 422	1 697	225		3 344

NET INVESTMENT IN PROPERTY 1922–1993					
1977	1 120	1 289	180		2 590
1978	1 386	1 540	258		3 184
1979	1 409	1 220	185	–	2 813
1980	1 612	1 787	136	17	3 552
1981	1 810	1 427	182	0	3 419
1982	1 643	1 525	88	0	3 257
1983	1 254	1 009	– 13	3	2 253
1984	1 052	1 409	66	1	2 529
1985	1 086	786	– 7	3	1 869
1986	1 070	559	– 130	– 5	1 494
1987	1 030	244	– 639	25	660
1988	1 680	321	117	14	2 132
1989	2 071	187	33	1	2 292
1990	1 298	– 660	– 61	10	587
1991	1 573	458	18	3	2 052
1992	617	890	– 6	5	1 506
1993	246	139	82	2	469

Source: Table 10.1. Price series – 1920–1947, C.H. Feinstein (1972) *Statistical Tables of National Income, Expenditure and Output of the U.K., 1855–1965*, Cambridge University Press, Cambridge, Table 65; 1948–1993, CSO (1995) *Economic Trends Annual Supplement*, HMSO, London, p. 148.

Tables 10.1 and 10.2 show annual net property investment by the financial institutions, in nominal and real values, from 1923 to 1937 and 1946 to 1993. Changes in the real level of institutional property investment in the inter-war and post-war periods are illustrated in Figures 10.1 and 10.2.

When adjusted to take account of inflation, the data for the inter-war years show a low level of investment in property during the 1920s, higher investment from 1930 to 1931, and an investment boom during 1933–1936, followed by a substantial decline in 1937.[6] The values for the post-war years show a steady increase in net property investment from 1946 to 1951, a stable level from 1952 to 1954, more rapid growth from 1955 to 1961 and an acceleration of growth from 1962 to 1969. Declining investment during the next three years was followed by a sharp increase during 1973 and 1974, though the addition of data for non-BIA members to the insurance company figures exaggerates the 1974 level somewhat, as noted above. The mid and late 1970s saw a high, but volatile, level of investment in real property, which never exceeded the 1974 peak. From the early 1980s net institutional property investment declined substantially, while remaining extremely volatile. Even during the peak of the late 1980s property boom, investment levels never returned to those experienced during the 1970s. The period since 1989 has witnessed extreme volatility in net institutional property investment, though the overall trend in investment has continued downwards.

Figure 10.3 shows real net property investment, by category of institution, from 1954 to 1993. Pension fund property investment experienced slower growth than insurance company investment prior to 1961. However, over the

254 A STATISTICAL OVERVIEW OF THE PROPERTY INVESTMENT MARKET

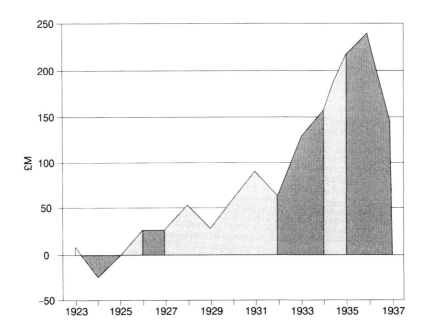

Figure 10.1 Real net institutional property investment 1923–1937. Source: Table 10.2.

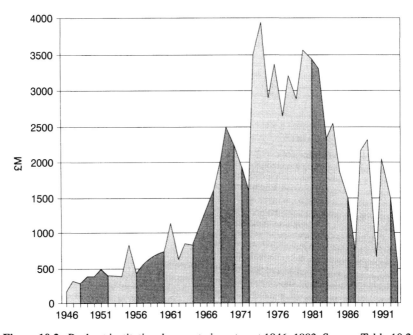

Figure 10.2 Real net institutional property investment 1946–1993. Source: Table 10.2.

following decade the growth of pension fund investment was generally more rapid. During the 1970s the level, and variation, of insurance company and pension fund property investment were broadly similar, though during the 1980s the pension funds reduced their commitment to the sector to a much greater extent than the insurance companies. The decline in insurance company and pension fund property investment is even greater in terms of property's contribution to total investment by these classes of institution, as is shown in Table 10.3. While property had absorbed an average of over 15% of insurance company and pension fund investment during the mid to late 1970s, average net investment during 1989–1993 had fallen to only 5.7% of insurance company funds and 3.1% of pension fund investment.

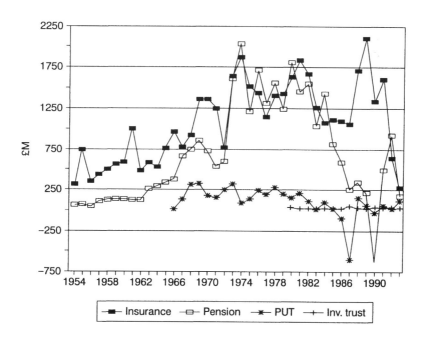

Figure 10.3 Real net property investment by category of institution 1954–1993. Source: Table 10.2.

Investment by property unit trusts peaked as early as 1969, declined substantially in the aftermath of the 1974 property crash and experienced substantial net disinvestment during the 1980s. Investment trusts have undertaken some limited investment in property during the last 15 years, though their contribution to total investment has been negligible.

10.2.2 Property investment by other categories of investor

In addition to the financial institutions other classes of investor, including property companies, individuals and 'traditional institutions' have been active purchasers of investment property. These non-institutional investors were particularly important in the British property market during the first two post-war decades. Unfortunately, there is very little evidence regarding the value of property purchases by such investors, even for the very recent past. In order to gain some indication of the relative magnitude of property purchases by non-institutional investors a register of property transactions, compiled by the chartered surveyors Healey & Baker over the years 1946–1966, was analysed. Investors were divided into six categories: individuals, insurance companies, property companies, pension funds, traditional institutions[7] and miscellaneous institutions.[8] In some cases a purchaser could not be assigned to a category of institution or, more commonly, the name of the purchaser was not given in the ledger record. This was a particular problem for the years 1956–1958, the percentage of overall investment in the 'unknown' category exceeding 60% of the total for each year over this period, compared to a maximum of 7.7% in any other year.

Table 10.3 Property investment as a proportion of total investment for insurance companies and pension funds, 1963–1993

	Insurance companies	Pension funds
1963–1966	13.13	8.07
1967–1973	16.55	13.96
1974–1979	15.01	15.24
1980–1988	11.03	7.30
1989–1993	5.70	3.11
1962–1993	12.46	9.77

Source: CSO (various issues) *Financial Statistics*, HMSO, London.

A total of 2950 transactions were recorded in the ledger over the 1946–1966 period, of which 2635 could be classified. Two hundred and twenty of these were excluded from the analysis, as they were recorded in the names of estate agents and other market intermediaries, leaving a sample of 2415 transactions, with an aggregate value of £171.14 million, on which the following analysis is based. The total number of transactions recorded varied considerably from year to year, in a manner which suggests that in some years many transactions were not recorded. This may introduce some bias in the data. Further bias might be expected, as Healey & Baker's clients, and the areas of the property market in which it specialized, were not entirely representative of the property investment market as a whole. Much of the agency's early post-war business was concentrated in shop property rather than City offices, for example.

Despite these imperfections the above data is acceptable as an indication of the broad order of magnitude of the distribution of property investment between categories of investor, as Healey & Baker's sale business constituted a very large volume of investment property, with a geographical coverage that spanned the entire country. However, in order to increase sample size the estimates provided below are grouped into several periods, each covering a number of years.

Table 10.4 provides an estimate of overall direct property investment activity from 1946 to 1966, and Table 10.5 shows the same information expressed as a percentage of total investment in each period. The figures for categories other than the financial institutions were derived by calculating their value relative to that for the insurance companies. The resulting percentage figures were then multiplied by the values for insurance company investment given in Table 10.1.

Table 10.4 An estimate of overall property investment activity (£M, annual averages for each period), 1946–1966

Year	Individuals	Property companies	Traditional institutions	Miscellaneous institutions	Financial institutions	Total
1946–1950	3.26	3.61	2.59	3.32	18.20	30.98
1951–1954	3.74	2.67	2.68	2.84	31.50	43.44
1955–1960	6.29	12.54	2.16	12.19	59.50	92.68
1961–1966	7.61	30.08	2.25	35.61	109.67	185.21

Sources: Register of property transactions compiled by Healey & Baker, 1946–1966 and Table 10.1. Figures derived by dividing insurance company property investment, as given in Table 10.1, by Healey & Baker's sales to insurance companies, calculated from the register. The figures for sales to categories of investor other than the financial institutions were then multiplied by the results of this division. Figures for investment by financial institutions are taken from Table 10.1.

Table 10.5 An estimate of the percentage distribution of property investment activity by category of investor, 1946–1966

	Individuals	Property companies	Traditional institutions	Miscellaneous institutions	Financial institutions
1946–1950	10.52	11.66	8.35	10.73	58.74
1951–1954	8.62	6.16	6.17	6.54	72.52
1955–1960	6.79	13.53	2.33	13.15	64.20
1961–1966	4.11	16.24	1.22	19.23	59.21
1946–1966	7.26	12.46	4.18	13.05	63.06

Source: As for Table 10.4.

The data have been grouped into four periods. As the figures for the years 1956–1958 are subject to serious deficiencies, as discussed above, the values for 1955–1960 are based on information for the years 1955, 1959 and 1960 only. Tables 10.4 and 10.5 may overestimate the volume of property investment to some extent, as some transactions recorded, particularly in the miscellaneous category, were purchased by organizations such as retailers for use rather than investment purposes, though the volume of such transactions is not sufficient to significantly distort the results.

The figures indicate that the proportion of total property investment activity undertaken by the financial institutions amounted to about 63% during 1946–1966. Investment by individuals amounted to about 11% of total investment during 1946–1950, but fell steadily to 4% during 1961–1966. Many of the individuals who bought property during the late 1940s went on, however, to form property companies to hold their acquisitions, and the decline in this sector may represent that transition from individual to corporate activity by property entrepreneurs. Falling profit margins on property speculation and development during the 1950s and 1960s may also be a factor behind the decline in the significance of the individual investor, who was often in search of a high short-term return on limited funds.

The growth in the property company sector, both in absolute terms and relative to the rest of the market, is shown in the figures; this category of institution accounted for 16% of all investment in 1961–1966, compared to about 12% in 1946–1950. It should be noted, however, that as the figures refer to investment, rather than development activity, they very considerably understate the overall importance of property companies in the commercial property market during this period.

The level of investment by the traditional institutions declined, both in absolute terms and relative to overall property investment, during the period covered in the tables. This might be the result of the increasing size of property transactions, which made property investment less attractive for all but the largest traditional institutions, such as the Church Commissioners. It may also be a function of the relatively static funds of many traditional institutions, which inevitably resulted in a decline in their importance relative to the rapidly expanding insurance company and pension fund sectors.

Overall investment in property appears to have risen roughly in line with institutional investment, the proportion of all investment accounted for by the financial institutions being approximately equal in 1946–1950 and 1961–1966. The financial institutions appear to have accounted for a greater proportion of total investment during the 1950s, however, particularly from 1951 to 1954.

The only category of these 'non-institutional' investors which continued to play an important role in the UK property investment market during the 1970s and 1980s was the property company sector. However, it is not possible to disentangle the flow of property company funds into the property sector for investment purposes from that for development and trading purposes, in order to provide a direct comparison of the relative importance of the financial institutions and the property companies as long-term investors in the property market.

Data is, however, available on annual capital raised by the property company sector. This is compared with net property investment by the financial institutions, over the period 1980–1993, in Figure 10.4. Figure 10.4 also shows net purchases by overseas investors; during the 1980s the growing internationalization of the property investment market led to a rising volume of investment in UK property on the part of foreign investors, as was discussed in Chapter 9.

The contrast between the period from 1980 to 1986 and the years since 1987 is striking. During the 1980–1986 period finanical institutions dominated the property investment market, their investment in the sector averaging £1697 million over these years, in current prices, compared to average levels for capital raised by property companies, and investment from overseas sources, of £273 million and £97 million, respectively.

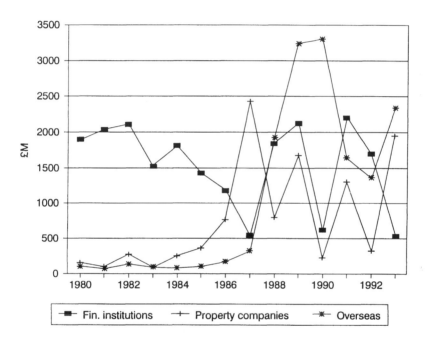

Figure 10.4 Net property investment by financial institutions, property companies and overseas investors, 1980–1993. Source: financial institutions – Table 10.1; property companies and overseas investors – information provided by DTZ Debenham Thorpe Research.

However, from 1987 property company capital issues became much larger and more volatile. Investment from overseas sources experienced an even greater increase, its pattern of growth closely following the cycle of the property market. Meanwhile, institutional property investment was generally lower than had been the case during the early 1980s, despite high investment in particular years. From 1987 to 1993 property investment by financial institutions averaged £1339 million in current prices, slightly more than average property company capital issues (£1213 million), but well below average net investment from overseas (£1988 million).

10.3 THE COMPOSITION OF INSTITUTIONAL PROPERTY PORTFOLIOS 1870–1990

In order to assess the economic significance of institutional property investment it is important to know to which sectors of the property market investment was directed. As there are no published statistics covering this area prior to the 1970s it has been necessary to derive figures from information collected in the course of this research, and other sources, as outlined below.

It was not possible to provide more than a rough indication of the distribution of rack-rented property[9] prior to 1939. This was due to the fact that insurance companies appear to have varied greatly in the composition of their property holdings, a large proportion of property acquired prior to the 1930s consisting of offices wholly or partly occupied by the institution in question, or property acquired as a result of mortgage default.[10] However, it is evident that by the late 1930s the largest single category of institutional property held for purely investment purposes (rather than for occupation) was shops. Office property also formed a substantial proportion of insurance company property assets, while investment in residential property (mainly flats) might have formed as much as 20% of all institutional property holdings. The figures presented in Table 10.7 for 1945 give some indication of the distribution of insurance company property assets by the end of the inter-war period.[11]

It is, however, possible to estimate the distribution of holdings between rack-rented property and other property interests during this period, and thus discover what proportion of the figure for holdings of 'land, property, and ground rents' shown in the Board of Trade returns actually refers to the ownership of buildings rather than land. Table 10.6 gives figures for the distribution of insurance company property holdings between rack-rented property and ground rents, feu duties, and rent charges, at five intervals between 1870 and 1937. The table is derived from the Board of Trade returns for the 10 largest life assurance companies in each year.

Table 10.6 The distribution of insurance company direct property holdings between rack-rented property and other property interests, 1870–1937

Year	Rack-rented property	Ground rents etc.
1870	51.0	49.0
1897	64.3	35.7
1914	61.4	38.6
1925	63.7	36.3
1937	66.0	34.0

Source: Derived from Board of Trade returns for the 10 largest life assurance companies in each year.

THE COMPOSITION OF INSTITUTIONAL PROPERTY PORTFOLIOS 1870–1990

The figures indicate that after a sharp increase in the proportion of property investment allocated to rack-rented property during the late nineteenth century the distribution of property between these categories of tenure remained relatively stable. The pattern of variation in investment between the higher-yielding but more risky rack rents and the more stable income provided by ground rents, etc., conforms to the general cycle of the property market; the relative magnitude of investment in rack-rented property increasing during periods of boom such as the late nineteenth century and the 1930s, and falling during the Edwardian property market depression.

Since the end of the 1930s ground rents and related securities ceased to form a significant proportion of new institutional property investment. The analysis of property holdings from 1945 therefore focuses on the distribution between different categories of rack-rented property. Table 10.7 shows estimates of the distribution of property investment by property type at five-yearly intervals from 1945 to 1990. The data from 1945 to 1970 (a and b) are based on the property portfolios of three large insurance companies. The figures are based on market values, where available, otherwise book values are used. The book value of the property holdings of these companies together formed 20.3% of all direct holdings of UK property by British insurance companies in 1950, and 19.3% in 1960, according to Board of Trade data.[12]

Table 10.7 The percentage distribution of institutional property holdings by property type, 1945–1990

Year	Offices	Shops	Industrial	Residential	Other	Total
1945	37.5	42.2	0.1	19.5	0.7	100
1950	36.2	40.1	4.3	13.4	5.9	100
1955	40.0	47.1	5.0	6.6	1.4	100
1960	41.3	46.7	5.6	6.3	0.2	100
1965	44.9	42.0	7.7	5.2	0.1	100
1970 (a)	58.9	30.4	6.1	4.0	0.5	100
1970 (b)	61.4	31.7	6.4	–	0.5	100
1970 (c)	63.8	29.7	5.7	N/A	0.8	100
1975	59.1	29.7	9.8	N/A	1.3	100
1980	56.6	27.6	14.8	N/A	1.0	100
1985	52.5	35.1	12.4	N/A	N/A	100
1990	51.9	35.3	12.8	N/A	N/A	100

Sources: 1945–1970 (a and b), based on data collected from the investment records of three insurance companies; 1970 (c)–1980, based on data provided by IPD (1980 market values). 1985–1990, based on IPD data (current market values). N/A, not available.

(a) Based on data from three insurance companies, includes residential property.
(b) Based on data from three insurance companies, excludes residential property.
(c) Based on data from Investment Property Databank database.

The values for 1970(c)–1980 are derived from the Investment Property Databank (IPD) database.[13] They are based on properties still held by the institutions on the database in 1980, but purchased prior to the end of the year in question. Distribution is calculated according to 1980 market values. The omission of properties sold prior to 1980, and the use of 1980 rather than current market values, imparts some bias to the data, though the figures should be sufficiently accurate to provide an indication of, at least, broad orders of magnitude. One advantage of using the IPD database was the very large sample size that could be drawn upon; the 1970–1980 figures are based on an average of 6395 properties with an average aggregate value of £6431 million. Data for 1985–1990 are based on the sectoral distribution of the IPD database in those years, measured at current market values.

For 1970 three estimates are provided. The first is based on the same sources as the 1945–1965 data. The second is also based on these sources, but excludes residential property (which was not included in the 1970–1990 data sample). The third estimate is derived from the IPD database. The two data sources provide broadly similar figures for the distribution of property in 1970, the largest percentage difference between the 1945–1970 source (excluding residential property) and the IPD source being 2.4%, for the office category.

The table indicates that shops were the dominant category of institutional property prior to the mid-1960s. However, by 1965 offices had become the most important sector of institutional property investment and have dominated investment ever since. Industrial property holdings expanded rapidly from 1945 to 1950, grew only slowly from 1955 to 1970, again experienced rapid expansion during the 1970s and witnessed a modest decline during the industrial depression of the early 1980s, after which they have remained relatively stable.

The proportion of residential property appears to have fallen substantially from 1945 to 1955, and to have decreased at a much slower rate over the next decade. As the IPD database does not include residential property, figures are not available from this source for 1970–1990. However, an analysis of the residential property holdings of the three institutions from which the 1945–1970 (a and b) figures are based indicates that the proportion of their portfolios made up of residential property fell from 4.0% in 1970 to 3.4% in 1975 and 0.3% in 1980. These institutions all had long histories of investment in property, and as very little residential property was purchased by financial institutions after about 1951 it is likely that their residential holdings would be substantially in excess of those of institutional investors as a whole, many of which did not enter the property market until the 1960s. These figures can, therefore, be regarded as an upper bound for institutional holdings of residential property during the 1970s, and an indication of the sharp decline in holdings of such property during this decade.

Table 10.8 provides an analysis of the geographical distribution of investment property, using standard economic regions with, in addition, separate categories for the City, the West End and the rest of Greater London. The data

THE COMPOSITION OF INSTITUTIONAL PROPERTY PORTFOLIOS 1870–1990

sources are identical to those for Table 10.7, though two of the intervals at which observations are taken are slightly different as a result of factors connected with the availability of data.

As with Table 10.7, figures for 1970 are provided using both the IPD data and the 1945–1965 data source. The figures provided by the two estimates are of similar orders of magnitude for the standard economic regions, including the South East (other than London). There are considerable differences in the London figures however, the data from the three institutions underestimating the percentage of properties in the City and West End, and overestimating the percentage in the rest of London, by a considerable margin. The data from the three institutions could be expected to be more representative of overall institutional property investment prior to the late 1960s than for 1970, however, as a large number of 'new' investors, such as the property unit trusts, entered the property market in the late 1960s, with property portfolios that contained no property purchased prior to the 1960s.

The table shows that investment property has been highly concentrated in the City, the West End, other areas of Greater London and the rest of the South East. In order to measure the relative concentration of investment property in each region, compared to the level of economic activity in that region, the percentage figures given in Table 10.8[14] were divided by the proportion of British GDP accounted for by each region for four years from 1965 to 1990. The results, shown in Table 10.9, indicate that the proportion of investment property in the South East averaged almost twice as much as would be expected from the region's contribution to GDP during these years. With the exception of the West Midlands in 1965 no other region had a proportion of national investment property in excess of 72% of its contribution to national GDP in any of the years shown.

The areas of Britain experiencing the heaviest underinvestment were Northern Ireland, Wales and the North, with average ratios for the four years of 12.8, 18.8 and 35.3%, respectively. Most other regions also conform to a general pattern of investment being concentrated in areas relatively close to London and the South East. The main exception is Scotland, with an average ratio of 58.3%, substantially in excess of that for other peripheral regions. This may be partly the result of several major insurance companies being based in Scotland, facilitating their access to information on the region's property market.

In order to further examine the causes of the uneven regional distribution of investment property the calculations made in Table 10.9 were repeated for each class of property for the years 1970, 1980 and 1990. The results are given in Table 10.10. The table shows that, while investment in shops and industrial property is significantly less concentrated in the South East, substantial regional inequalities are present for all three classes of property. The maldistribution of regional investment in shop property is particularly surprising, as the level of retailing activity is broadly determined by purchasing power, which is closely correlated with GDP.

Table 10.8 The geographical distribution of investment property (%), 1945–1990

	1945	1952	1956	1960	1965	1970(a)	1970(b)	1975	1980	1985	1990
City	14.5	5.6	9.4	9.5	8.6	14.4	19.2	16.1	14.2	16.3	15.1
West End	23.6	23.8	19.8	18.8	13.6	10.5	24.2	18.5	15.4	12.0	12.5
Other Greater London	37.7	38.0	33.6	32.2	33.0	32.0	23.1	22.5	24.9	20.6	18.4
Other South East	7.3	8.1	7.7	8.5	10.3	10.9	9.8	14.8	17.3	20.1	22.9
South West	6.6	2.4	2.5	2.1	2.3	2.3	3.2	3.8	4.2	5.0	5.1
East Anglia	0.2	0.3	0.6	0.9	1.6	1.9	1.7	1.8	2.0	2.3	2.6
East Midlands	2.0	1.2	1.8	1.6	1.9	1.6	1.3	2.4	2.4	2.6	2.9
West Midlands	0.8	4.2	5.2	8.7	11.2	8.6	5.2	5.0	4.9	4.6	4.5
North West	1.0	3.7	5.1	4.9	4.8	7.1	3.8	4.1	3.9	4.9	4.4
Yorkshire and Humberside	3.3	4.1	4.1	3.6	4.3	4.4	3.1	4.3	3.7	3.6	3.3
North	1.0	1.5	1.6	2.3	1.9	1.5	0.8	1.6	1.3	1.7	1.9
Scotland	1.8	5.5	6.9	5.5	5.5	3.9	3.5	4.2	4.6	5.3	5.2
Wales	0.2	0.7	0.8	0.8	0.7	0.6	0.8	0.7	0.9	0.8	0.9
Northern Ireland	0.0	0.9	0.6	0.4	0.3	0.3	0.5	0.3	0.3	0.2	0.2
Total	100.0	100.0	100.0	100.0	100.0	100.0	100.0	100.0	100.0	100.0	100.0

Sources: 1945–1970(a), based on data from three insurance companies; 1970(b)–1980, based on data from IPD database (1980 market values); 1985–1990, based on data from IPD database (current market values).

Table 10.9 Regional distribution of investment property as percentage of regional GDP distribution, 1965–1990

	1965	1975	1985	1990
South East	183.1	210.7	196.2	190.8
South West	38.4	54.7	66.9	66.8
East Anglia	56.9	61.2	65.9	71.4
East Midlands	33.1	36.4	38.1	43.5
West Midlands	114.0	54.6	53.8	53.7
North West	40.8	36.4	46.6	43.9
Yorkshire and Humberside	50.8	51.6	44.1	42.9
North	37.8	29.4	34.4	39.4
Scotland	64.6	46.3	60.7	61.7
Wales	16.0	16.8	20.6	21.6
Northern Ireland	17.4	13.3	9.1	11.3
Total	100.0	100.0	100.0	100.0

Sources: Table 10.8 (1970 figures based on IPD data) and CSO (various issues) *Regional Trends*, HMSO, London. Nineteen sixty-five property investment figures are divided by GDP figures for 1966, the first year for which the regional distribution of GDP is available. Figures exclude continental shelf.

Table 10.10 Regional distribution of offices, shops and industrial property as a percentage of regional GDP distribution, 1970–1990

	Offices			Shops			Industrial		
	1970	1980	1990	1970	1980	1990	1970	1980	1990
South East	244.5	242.7	233.0	149.5	147.7	126.0	175.0	179.8	198.2
South West	17.7	26.0	34.6	108.8	97.3	112.1	104.6	102.3	72.5
East Anglia	23.0	27.6	45.0	112.3	110.1	109.9	152.2	122.5	72.5
East Midlands	5.2	12.2	9.1	54.0	79.2	83.8	56.9	54.2	72.3
West Midlands	19.9	29.5	25.5	120.2	92.6	91.9	85.2	92.2	62.8
North West	23.5	26.3	22.2	46.7	47.9	82.2	64.0	46.2	26.2
Yorkshire and Humberside	20.8	24.7	19.2	77.6	89.4	79.2	14.5	45.1	39.1
North	6.9	8.2	8.3	34.5	59.0	95.7	28.0	24.4	10.9
Scotland	27.2	37.3	50.2	67.3	84.9	86.7	41.9	42.9	39.9
Wales	10.2	13.7	7.6	33.9	33.4	45.3	19.8	16.4	13.2
Northern Ireland	22.9	15.5	9.1	32.1	14.0	18.7	0.0	1.7	0.0
Total	100	100	100	100	100	100	100	100	100

Sources: Table 10.8 (1970 figures based on IPD data) and CSO (various issues) *Regional Trends*, HMSO, London.

10.4 INITIAL YIELDS ON INVESTMENT PROPERTY 1920–1993

Initial yields on investment property, i.e. the relationship between the immediate rental income produced by a property and its purchase price, provide a valuable indication of the buoyancy of the property investment market, the value placed on an equivalent immediate income from property compared to other classes of asset, and the extent to which expected future capital and income appreciation are reflected in property values.

Fortunately, while statistical data on most aspects of property investment are unavailable prior to the very recent past, a series assembled by the chartered surveyors Allsop & Co. provides details of initial yields for shop, office and residential property for 1933–1938 and 1947–1993, and for industrial property from 1959.[15] Additional yield information has been collected during the course of the present research. Table 10.11 shows the Allsop & Co. yield figures, together with additional values for the years 1920–1932 and 1946 for offices, shops and flats, and 1946–1958 for industrial property. The additional data were derived using evidence from the investment transactions of a number of financial institutions, together with information from investment property advertisements in *The Times* and *The Estates Gazette*.[16] Yields on consols and ordinary shares are also given in the table, for comparison.

Unlike most other institutional investments, property is subject to substantial deductions from gross income associated with 'outgoings', such as insurance, repairs and management costs. In order to take account of this, and make the property yield figures compatible with those for other assets, the figures in Table 10.11 are given net of outgoings but not of taxation.

Figures 10.5 and 10.6 show yields on offices, shops and flats from 1920 to 1938, and yields on offices, shops and industrial property from 1946 to 1993. The inter-war period was a time of relatively stable property yields, though yields declined substantially during the 1930s property boom, rising during the late 1930s due to deteriorating economic conditions and the threat of war. The early post-war years saw very stable yields on investment property; the standard deviation for shop yields was only 0.22% during the years 1946–1954 and fell further to 0.16% during 1955–1964. However, yields became markedly less stable during the late 1960s and 1970s, the standard deviation for shop property during 1964–1973 and 1974–1980 rising to 0.62 and 1.37%, respectively. During the 1980s property yields experienced an upward trend, and unlike the booms of the early and late 1970s, the mid to late 1980s property boom did not witness a substantial decline in yields. With the onset of the 1990 crash initial yields on investment property rose substantially, especially for the office sector.

Yield data for investment property are particularly revealing when examined relative to other assets. The yield differential between property and gilts provides an indication of the attractiveness of property relative to a risk-free, fixed-interest benchmark, while the differential between property and equities is indicative of investors' expectations regarding the relative future capital growth

INITIAL YIELDS ON INVESTMENT PROPERTY 1920–1993

prospects of the two asset classes.[17] Figure 10.7 examines differentials between shop and office yields, and the yield on gilts. During 1920–1929 the average excess of the yield on shop property, the property category with the lowest initial yield,[18] over the yield on gilts was 1.89%. During 1930–1938 the differential widened to 2.52%, despite a boom in institutional property investment. Property yields exceeded those on gilts by a similar margin during the late 1940s and early 1950s, though the introduction of the rent review in the mid-1950s built an equity element into investment property, leading to a considerable reduction in the property/gilt yield differential. A 'reverse yield gap' emerged during the early 1960s, gilt yields exceeding those on property. This gap widened during the late 1960s and 1970s (due to property's growing equity element at a time of rising inflation) averaging –8.47% from 1974 to 1980. However, this trend was reversed during the 1980s, the reverse yield gap falling to an average of only –4.59% during 1987–1993.

Table 10.11 Initial yields on investment property (%), 1920–1993

	Shops	Offices	Flats	Industrial	Gilts	Shares
1920	6.00	7.00	9.25	N/A	5.3	3.36
1921	6.50	7.50	9.00	N/A	5.2	3.33
1922	6.75	7.75	10.00	N/A	4.4	3.08
1923	6.5	7.25	9.50	N/A	4.3	2.80
1924	7.00	7.75	9.75	N/A	4.4	2.43
1925	6.25	7.50	10.00	N/A	4.4	2.43
1926	6.25	7.25	9.75	N/A	4.6	2.58
1927	6.75	7.50	9.25	N/A	4.6	2.39
1928	6.50	8.75	9.75	N/A	4.5	2.93
1929	6.75	8.00	9.25	N/A	4.6	2.64
1930	7.00	8.75	9.75	N/A	4.5	2.61
1931	7.00	8.25	9.75	N/A	4.4	2.54
1932	7.00	8.00	9.00	N/A	3.7	2.66
1933	6.50	7.00	7.50	N/A	3.4	2.29
1934	5.50	7.00	7.00	N/A	3.1	2.10
1935	4.50	6.00	6.50	N/A	2.9	2.19
1936	5.00	7.50	6.50	N/A	2.9	2.20
1937	5.50	6.50	8.00	N/A	3.3	2.49
1938	6.25	7.25	9.00	N/A	3.4	3.05
1939–1945 Insufficient data to estimate property yields
1946	5.25	5.75	7.00	7.00	2.6	2.44
1947	5.50	6.00	6.50	7.00	2.8	3.31
1948	5.00	6.00	6.00	7.00	3.2	3.83
1949	5.00	6.00	6.00	7.00	3.3	4.18
1950	5.50	6.50	7.50	7.50	3.5	4.60
1951	5.25	6.50	7.50	7.50	3.8	4.14
1952	5.50	6.50	7.50	8.25	4.2	4.87
1953	5.50	6.50	7.50	8.25	4.1	4.90
1954	5.50	7.00	7.50	8.25	3.8	4.32
1955	5.50	7.00	8.00	7.50	4.2	3.93
1956	5.50	8.00	8.00	8.75	4.7	3.99

Year						
1957	5.50	7.50	8.00	9.00	5.0	3.97
1958	5.50	7.50	8.00	8.50	5.0	4.25
1959	6.00	7.50	8.00	10.00	4.8	3.68
1960	5.50	7.00	8.00	10.00	5.4	3.54
1961	5.50	7.00	8.00	10.00	6.2	3.51
1962	5.50	6.00	7.50	10.00	6.0	3.88
1963	5.50	6.00	7.00	10.00	5.30	3.60
1964	5.50	6.50	7.00	10.00	5.80	3.72
1965	6.00	6.50	6.50	9.00	6.43	4.88
1966	6.00	6.50	6.50	9.00	6.91	5.21
1967	6.50	6.50	6.50	9.00	6.80	4.69
1968	7.00	7.00	6.50	9.00	7.54	3.38
1969	7.00	6.50	6.00	8.75	9.05	3.28
1970	7.50	7.50	6.00	9.00	9.21	4.75
1971	7.00	8.00	6.00	8.50	8.85	4.16
1972	6.00	5.00	6.00	7.50	8.90	3.25
1973	5.75	5.00	7.00	7.50	10.71	4.15
1974	7.50	7.50	9.00	10.00	14.77	8.23
1975	6.25	6.25	10.00	9.00	14.39	6.80
1976	5.75	6.25	10.00	9.00	14.43	5.96
1977	4.75	5.75	8.00	8.00	12.73	5.42
1978	4.25	4.50	8.00	7.00	12.47	5.65
1979	3.75	4.50	7.50	6.75	12.99	6.49
1980	4.00	5.50	7.00	7.25	13.78	7.62
1981	3.75	5.00	N/A	7.25	14.74	6.23
1982	3.75	5.00	N/A	7.75	12.88	5.29
1983	3.75	5.00	N/A	7.75	10.80	4.70
1984	3.75	5.00	N/A	7.50	10.69	4.66
1985	3.75	5.00	N/A	7.50	10.62	4.62
1986	4.00	6.50	N/A	8.50	9.87	4.25
1987	4.25	6.00	N/A	8.00	9.47	3.73
1988	4.50	6.00	N/A	7.50	9.36	4.62
1989	5.00	6.00	N/A	8.00	9.58	4.46
1990	5.75	8.00	N/A	9.50	11.08	5.29
1991	5.00	8.50	N/A	8.75	9.92	4.98
1992	5.25	8.75	N/A	9.00	9.12	4.68
1993	4.50	7.00	N/A	7.50	7.87	4.10

Sources: offices, shops and flats, 1933–1938 and 1947–1993, and industrial property 1959–1993, information supplied by Allsop & Co. Offices, shops and flats, 1920–1932 and 1946, and industrial property, 1946–1958, assembled from information collected on the property purchases of a number of financial institutions, plus investment property advertisements in *The Times* and the *Estates Gazette*. In some years the Allsop & Co. data gave two yield figures, such as 6/7. The mean value was taken in such cases. Ordinary shares: 1920–1969, L. Foldes and P. Watson (1978) *Quarterly Returns to U.K. Equities 1919–1970: Papers on Capital and Risk, No. 6*, RTZ, London; 1970–1993, CSO (1995) *Economic Trends Annual Supplement*, HMSO, London, p. 246. 1920–1969 data are taken from the value-weighted, rather than equally-weighted, Foldes and Watson indices. Gilts – 1920–1962, B.R. Mitchell (1988) *British Historical Statistics*, Cambridge University Press, Cambridge, p. 678 (2.5% consols); 1963–1993, CSO (1995) *Economic Trends Annual Supplement*, HMSO, London, p. 246 (long-dated British government bonds).

INITIAL YIELDS ON INVESTMENT PROPERTY 1920–1993

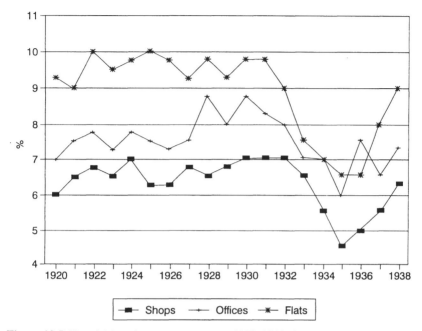

Figure 10.5 Net yields on investment property 1920–1938. Source: Table 10.11.

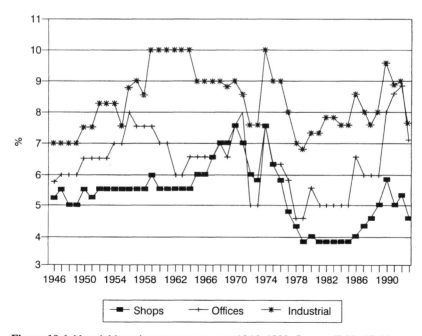

Figure 10.6 Net yields on investment property 1946–1993. Source: Table 10.11.

270 A STATISTICAL OVERVIEW OF THE PROPERTY INVESTMENT MARKET

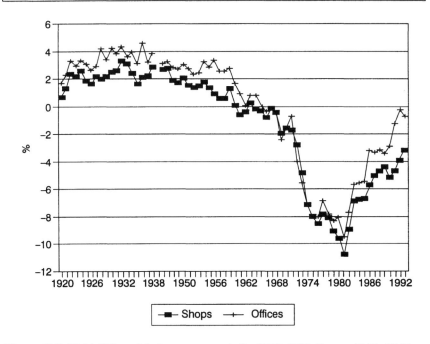

Figure 10.7 Yield differentials for property and gilts 1920–1993. Source: Table 10.11.

Figure 10.8 Yield differentials for property and equities 1920–1993. Source: Table 10.11.

| MARKET RENTS | 271 |

Figure 10.8 shows yield differentials for shop and office property with respect to UK ordinary shares. During the inter-war years differentials were high, the average excess of initial yields on shop property over those on ordinary shares during 1920–1929 and 1930–1938 being 3.73 and 3.57%, respectively. After 1945 the differential fell somewhat, with average values for the periods 1946–1954 and 1955–1964 of 1.27 and 1.74%, respectively. The excess of property over equity yields widened markedly during the late 1960s and early 1970s, averaging 2.33% from 1965 to 1973. During 1974, however, yields on shares exceeded initial shop yields for the first time, and over the mid to late 1970s unsettled stock market conditions led to a widening of the property/equity reverse yield gap, which averaged –1.42% from 1974 to 1980. From the early 1980s the differential between equity and property yields diminished considerably, roughly in parallel with the recovery in gilt yields, the reverse yield gap disappearing by the end of the decade.

10.5 MARKET RENTS

Market rental data indicate the value of newly-determined rents for investment property, i.e. the level of rents if property rentals were adjusted to market levels every year. This information is extremely useful as it reveals the underlying rental growth of commercial property, which is not fully reflected in property portfolio income growth due to there being gaps of several years between rent reviews.

Table 10.12 shows investment property market rentals, sub-divided into retail, office and industrial property, from 1962 to 1993. The series was taken from three separate published indices in order to extend the period covered as far as possible. The aggregate property index was weighted, from 1971, using the sectoral distribution of the IPD long-term property returns index, and prior to 1971 using the five-yearly estimates of the sectoral distribution of property portfolios, contained in Table 10.7, with interpolation for the intervening years.

Figure 10.9 shows the rate of growth of market rents over the years 1963–1993. The graph shows that rental growth for the three major investment property classes is strongly correlated over this period, especially since the mid-1970s. The data also has a strong cyclical pattern, its three major peaks corresponding to the property investment booms of the early 1970s, late 1970s–early 1980s and late 1980s. This cyclical trend is still apparant when the data is corrected for changes in retail prices, as is illustrated in Figure 10.10, which shows the rate of growth of aggregate investment property market rents in nominal and real terms.

The average rate of growth of market rents is shown, over four sub-periods, in Table 10.13, together with average retail price changes. The data indicate that the two most important classes of investment property – shops and offices – have experienced broadly similar rental growth over the whole period under consideration. Offices experienced very rapid rental growth prior to the 1974

A STATISTICAL OVERVIEW OF THE PROPERTY INVESTMENT MARKET

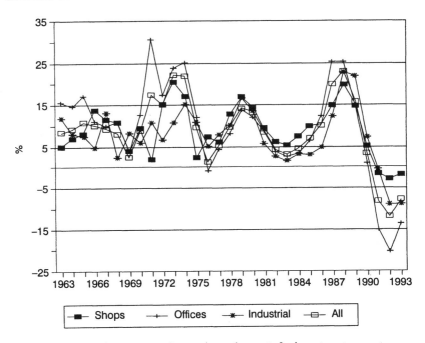

Figure 10.9 Annual percentage changes in market rents for investment property, 1963–1993. Source: Table 10.12.

Table 10.12 Market rental indices for investment property 1962–1993 (1975 = 100)

Year	Retail	Office	Industrial	All property
1962	31	16	34	24
1963	33	18	38	26
1964	35	21	41	29
1965	38	24	44	32
1966	43	27	46	35
1967	47	29	52	38
1968	53	33	53	41
1969	55	34	57	42
1970	60	38	60	46
1971	61	49	67	54
1972	70	58	71	62
1973	84	72	78	75
1974	98	89	90	92
1975	100	100	100	100
1976	107	99	105	101
1977	113	103	113	106
1978	127	111	124	116
1979	148	126	144	132
1980	169	141	163	149

MARKET RENTS				
1981	184	153	172	161
1982	195	158	176	167
1983	205	162	178	172
1984	219	166	183	179
1985	240	176	188	191
1986	263	197	196	210
1987	301	247	219	250
1988	360	308	268	307
1989	412	351	325	354
1990	431	354	347	365
1991	423	299	344	333
1992	409	237	312	293
1993	400	204	283	269

Sources: 1962–1966, Commercial property 1962–1974 (1975) *Estates Gazette*, **233**, 18 Jan., p. 216 (based on data prepared by the Economist Intelligence Unit for Michael Laurie & Partners); 1967–1974, Jones Lang Wootton (1993) *Property Index*, June (data refer to June of each year); 1975–1992, information supplied by Investment Property Databank.

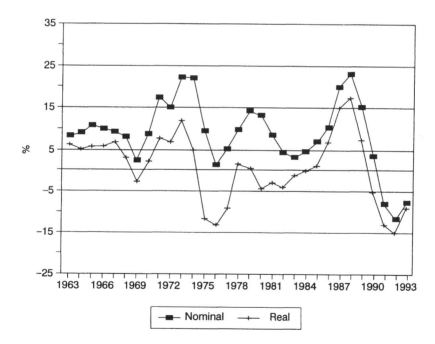

Figure 10.10 Annual percentage change in market rents for investment property, in nominal and real terms, 1963–1993. Source: Table 10.12.

property crash, but in subsequent years rental growth was more rapid for the shops sector. Industrial property experienced substantially lower rental growth over the period as a whole, but performed considerably better than the office sector during the aftermath of the 1974 and 1990 property crashes.

Table 10.13 Average rates of growth of property market rents (nominal values), 1962–1993

	Shops	Offices	Industrial	Total	RPI
1963–1973	9.45	14.84	7.98	10.84	5.31
1974–1979	9.94	9.88	10.64	9.86	15.60
1980–1988	10.38	10.43	7.13	9.85	7.31
1989–1993	2.13	−7.90	1.13	−2.65	5.64
1963–1993	8.59	8.64	7.10	8.07	7.87

Source: Table 10.12.

Over the 31 years shown in Table 10.13, property market rental growth exceeded inflation by a small margin. However, this masks an important change in property's inflationary hedge during this period. As with many aspects of the commercial property market, the 1974 property crash was a watershed; from 1963 to 1974 real market rental growth was negative in only one year; from 1975 to 1984 it was positive in only two years. The mid-1980s saw a return to real rental growth, though following the onset of the 1990 crash rents began to decline in nominal, as well as real, terms for the first time.

10.6 THE RATE OF RETURN ON INVESTMENT PROPERTY 1921–1993

In order to compare the performance of property as an investment medium over time, or in relation to other assets, some measurement of the financial gains or losses associated with the ownership of investment property over a certain time period is necessary. Such a measure is known as the rate of return and measures two components: (1) income return, the income derived from an investment during the time period under examination as a proportion of its market value at the start of that period; and (2) capital return, the extent to which the asset has appreciated (or depreciated) in value.

While income is relatively easy to measure, changes in the capital value of property assets over time are much more difficult to estimate. Property is fundamentally different from equities and gilts in that while stock exchange securities have their value set and recorded on a daily basis, via market transactions, property values cannot be measured in this way. Each property is unique, therefore, short of selling a property, it is impossible to precisely determine its value by reference to the market. Property values can, therefore, only be estimated imperfectly by comparison with recent market transactions involving similar, though not identical, properties.

This leads to several potential problems, including a tendency to smooth the data, which is thought to arise due to the use of valuation data collected over a period of several months as a basis for valuations, and as a result of conservatism among valuers when setting new valuations for properties. In a recent article S. Lee argued that despite these problems 'it is likely that the appraisal figures used do represent the 'true' underlying prices', citing evidence from various studies of appraisal values and subsequent sale values for property, which generally support this conclusion.[19] However, Macgregor and Nanthakumaran, who undertook a more recent examination of this subject, argue that the valuation basis of property indices may indeed lead to a smoothing of the time-series data, though they do concede that the unusual time-series characteristics of property returns compared to other assets may be, in part, an inherent characteristic of the property market rather than the result of data distortions.[20]

There are a number of different means of calculating the rate of return on investment property. The differences between them mainly relate to the treatment of capital injections and withdrawals during the analysis period. The formula chosen for the following returns calculations is the money-weighted rate of return[21] used by the IPD, which produces the most broadly-based index of property investment returns currently available, beginning in 1971. The formula treats capital flows during the analysis period as if they had taken place halfway through the period, a necessary assumption since the precise timing of capital injections was not recorded for much of the data on which the following series are based. The equation used by IPD, and applied here, to calculate total returns is given in Equation 10.1:

$$\text{Total return} = \left(\frac{C_t - C_{t-1} - E + I}{C_{t-1} + 0.5 E} \right) 100$$

where C is the capital value, E the capital expenditure, ie. the net injection of funds into the property during the year and I the net income.

This equation is the sum of the income and capital returns. In both cases the denominator is the capital value at the beginning of the year plus half the capital expenditure during the year ($C_{t-1} + 0.5E$). For the income return the numerator is the net income (I); for the capital return it is the change in capital value during the year minus net expenditure on the property ($C_t - C_{t-1} - E$).

The following property returns data are based, from 1971, on the IPD long-term property returns index, as this is the most broadly-based index currently available, based on properties with a December 1994 market value of £45 454 million. Prior to the early 1970s there is little published evidence on property investment returns of a character that would be suitable for comparison with other assets. An annual index of returns on investment property was, therefore, compiled, from 1921 to 1970, using data from the property records of a number of financial institutions and other sources. The basis on which this index was compiled varies, due to the availability of data, over three periods covered by the series, as outlined below.

1. 1956–1970. Information was compiled for the aggregate performance of the UK property portfolios of two large financial institutions. The market values of the property holdings of these institutions amounted to £73 976 000 in

1956 and had risen to £395 221 000 by 1970. For the period after 1958 it was possible to disaggregate data from the larger of the two institutions by property type, and this forms the basis of the pre-1971 data on returns for offices, shops, flats and industrial property given below.
2. 1949–1955. For this period it was only possible to obtain data on annual property returns from the smaller of the two institutions. This had an average of 50 properties in its portfolio over these years.
3. 1920–1948. Prior to 1949 the construction of a property returns index proved more difficult, due to the absence of annual property valuation figures. It was, however, possible to derive a returns index for the largest single category of investment property – prime shops. During this period prime shops were customarily let on very long leases, 99 or even 999 year lease terms being typical.[22] Properties were let on a fixed rent for this period, and as a shop had to be let to a well-known multiple trader for it to be regarded as 'prime' there was very little chance of the rent not being paid in full. Rent on such property therefore displayed very little annual variation in either an upwards or downwards direction. Such properties were customarily let on full repairing leases, with tenants responsible for day-to-day costs, therefore capital expenditure and outgoings on insurance, etc., were incurred by the tenant rather than the landlord.

Given these conditions it is possible to derive an approximate measure of annual returns using the index of initial yields on prime shops given in section 10.4.[23] Given a constant rental stream, capital values can be derived by multiplying income by years purchase, the inverse of the initial yield. As capital expenditure falls on the tenant it does not enter into these return calculations, which can be undertaken using Equation 10.1, from the information derived above. The figures for rentals, yields, capital values and returns used in these calculations are given in Table 10.14.

It is not, however, possible to derive similar figures for other categories of property. Office rentals varied significantly over the medium term due to the shorter leases which were common in the office sector, especially for office blocks with multiple tenancies. Furthermore, as the covenant of tenants was generally weaker, tenant default was a real danger; it was also common to temporarily lower rents during times of depression in order to avoid default. The variability of income was even greater for residential property than for offices, as indicated by the substantially higher yields on such property. The inclusion of these classes of property for the 1921–1948 period could be expected to add a greater cyclical component to the figures, though the data presented does provide a reasonably accurate reflection of returns on the most important single category of investment property held by the financial institutions during these years.

THE RATE OF RETURN ON INVESTMENT PROPERTY 1921–1993

Table 10.14 The data used to derive the 1921–1948 property returns series

Year	Yield	Income	Capital value [a]	Return [b]
1920	6.0	100	1667	
1921	6.5	100	1538	−1.69
1922	6.75	100	1481	2.80
1923	6.5	100	1538	10.60
1924	7.0	100	1429	−0.64
1925	6.25	100	1600	19.00
1926	6.25	100	1600	6.25
1927	6.75	100	1481	−1.16
1928	6.5	100	1538	10.60
1929	6.75	100	1481	2.80
1930	7.0	100	1429	3.18
1931	7.0	100	1429	7.00
1932	7.0	100	1429	7.00
1933	6.5	100	1538	14.69
1934	5.5	100	1818	24.68
1935	4.5	100	2222	27.72
1936	5.0	100	2000	−5.50
1937	5.5	100	1818	−4.09
1938	6.25	100	1600	−6.50
1946	5.25	100	1905	
1947	5.5	100	1818	0.70
1948	5.0	100	2000	15.50

Source: Table 10.11. Yield figures refer to primary shops. The income stream is assumed to be constant due to the length of prevailing leases and the strong covenants of multiple retailers.
[a] Capital value per £100 of income, equals income(1/yield).
[b] Calculated from income and capital value using Equation 10.1.

The annual series for income, capital and total returns to investment in property are given in Table 10.15, and income and capital returns over this period are illustrated in Figure 10.11. Table 10.16 gives similar returns data for each of the four classes of investment property – offices, shops, industrial property and residential property – from 1959, based on the sources indicated above. Insufficient data was available to derive returns figures for residential property after 1972.

Table 10.15 Capital, income and total returns on investment property (%), 1921–1993

Year	Capital appreciation	Income return	Total return
1921	−7.7	6.0	−1.7
1922	−3.7	6.5	2.8
1923	3.8	6.8	10.6
1924	−7.1	6.5	−0.6

Table 10.15 (contd).

1925	12.0	7.0	19.0
1926	0.0	6.3	6.3
1927	−7.4	6.3	−1.2
1928	3.8	6.8	10.6
1929	−3.7	6.5	2.8
1930	−3.6	6.8	3.2
1931	0.0	7.0	7.0
1932	0.0	7.0	7.0
1933	7.7	7.0	14.7
1934	18.2	6.5	24.7
1935	22.2	5.5	27.7
1936	−10.0	4.5	−5.5
1937	−9.1	5.0	−4.1
1938	−12.0	5.5	−6.5
1947	−4.5	5.3	0.7
1948	10.0	5.5	15.5
1949	−4.3	4.0	−0.4
1950	8.3	6.0	14.3
1951	−3.3	4.5	1.2
1952	−7.7	5.1	−2.6
1953	1.9	5.1	7.0
1954	5.0	5.5	10.6
1955	−4.7	5.4	0.7
1956	1.0	5.1	6.1
1957	−2.8	5.4	2.6
1958	2.4	5.7	8.1
1959	5.3	5.8	11.1
1960	4.5	5.3	9.8
1961	0.6	5.7	6.3
1962	1.2	5.6	6.9
1963	4.5	5.8	10.3
1964	2.6	5.7	8.3
1965	−1.1	5.8	4.7
1966	−2.8	5.9	3.2
1967	1.8	6.1	7.9
1968	15.2	6.1	21.3
1969	−0.1	5.4	5.2
1970	19.2	5.4	24.6
1971	11.7	5.0	16.6
1972	22.9	5.1	27.9
1973	23.3	4.6	27.9
1974	−21.7	4.4	−17.3
1975	4.3	5.8	10.1
1976	2.9	6.0	8.9
1977	19.3	6.7	26.0
1978	19.5	6.1	25.6
1979	16.2	5.9	22.1
1980	11.5	5.8	17.3
1981	9.5	5.6	15.1
1982	2.1	5.5	7.6
1983	1.6	5.8	7.4

1984	2.5	6.1	8.6
1985	2.3	6.3	8.6
1986	4.8	6.4	11.2
1987	18.7	6.5	25.3
1988	22.6	6.0	28.6
1989	9.5	5.5	15.1
1990	−14.0	5.7	−8.3
1991	−10.4	7.1	−3.3
1992	−9.9	8.0	−1.9
1993	10.1	8.8	18.9

Sources: 1921–1948, Table 10.14; 1949–1955, data on the property portfolio of one insurance company (average number of properties = 50); 1956–1971, data on the property portfolios of two insurance companies (combined 1956 market value = £74M, rising to £395M by 1971); 1971–1993, data supplied by Investment Property Databank.

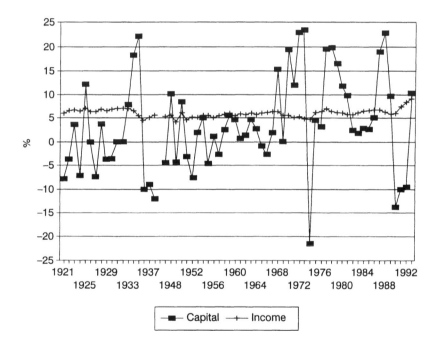

Figure 10.11 The capital and income components of property investment returns, 1921–1993. Source: Table 10.15.

Tables 10.17–10.19 provide summary statistics of the average rate of return, and standard deviation of return over time, for various classes of property and other assets. Standard deviation is included as a measure of the degree of variability of returns over time, providing an indication of the degree of risk associated with alternative investments.

Table 10.16 Annual returns on investment property (%), by sector, 1959–1993

	Shops			Offices		
Year	Capital	Income	Total	Capital	Income	Total
1959	4.7	5.7	10.4	4.2	5.7	9.9
1960	0.1	5.1	5.2	0.1	5.3	5.4
1961	−0.2	5.7	5.5	0.6	5.7	6.3
1962	−2.0	5.8	3.8	5.9	5.7	11.6
1963	2.5	6.1	8.6	2.2	5.6	7.8
1964	2.5	5.9	8.4	4.9	5.8	10.6
1965	−3.1	5.9	2.8	4.3	5.6	9.9
1966	−3.5	6.0	2.5	−2.3	5.7	3.4
1967	3.3	6.4	9.7	4.5	5.9	10.4
1968	10.5	6.3	16.8	20.9	6.0	26.9
1969	−2.6	5.8	3.2	4.8	5.0	9.8
1970	−4.5	5.9	1.4	48.6	4.7	53.3
1971	2.9	5.3	8.2	15.0	4.7	19.7
1972	12.3	5.5	17.8	27.4	4.4	31.8
1973	17.5	5.7	23.2	26.2	3.9	30.1
1974	−24.6	5.7	−18.9	−22.7	3.6	−19.1
1975	10.7	5.3	16.0	1.7	5.4	7.1
1976	7.6	5.3	12.9	0.8	5.8	6.6
1977	22.1	6.7	28.8	16.8	6.3	23.1
1978	25.0	5.9	30.9	17.1	5.9	23.0
1979	19.3	5.4	24.7	13.7	5.6	19.3
1980	14.3	5.2	19.5	11.2	5.5	16.7
1981	12.1	5.0	17.1	9.4	5.4	14.9
1982	5.3	5.0	10.3	1.4	5.3	6.7
1983	6.6	5.4	12.0	−0.4	5.6	5.2
1984	8.0	5.4	13.5	0.3	5.9	6.3
1985	6.9	5.5	12.3	1.2	6.2	7.4
1986	5.9	5.4	11.4	5.2	6.4	11.6
1987	14.5	5.6	20.0	23.0	6.5	29.5
1988	18.5	5.3	23.8	24.4	5.9	30.2
1989	4.9	5.0	9.9	10.6	5.4	16.0
1990	−13.5	5.4	−8.1	−15.1	5.5	−9.6
1991	−3.9	6.9	3.0	−17.6	6.8	−10.8
1992	−4.2	7.4	3.2	−15.2	8.2	−7.0
1993	11.8	7.8	19.6	8.8	9.3	18.1

Table 10.16 (contd).

	Industrial			Flats		
Year	Capital	Income	Total	Capital	Income	Total
1959	−0.5	6.8	6.2	29.5	6.1	35.6
1960	2.1	6.5	8.7	0.0	5.3	5.3
1961	0.0	8.1	8.1	0.4	6.2	6.6
1962	−0.9	6.5	5.6	31.6	6.4	38
1963	0.2	7.1	7.3	1.3	4.7	6
1964	0.0	7.2	7.2	0.0	4.7	4.7
1965	1.9	7.4	9.3	7.7	4.9	12.6
1966	0.0	6.9	6.9	−5	4.7	−0.2
1967	8.6	7.0	15.6	5	4.9	9.9
1968	10.8	6.5	17.4	20.1	4.5	24.6
1969	−6.2	6.1	−0.1	0.9	2.9	3.8
1970	−3.8	6.1	2.3	6.4	4	10.4
1971	3.2	8.2	11.4	14.7	3.8	18.5
1972	12.6	8.9	21.5	43.9	3.2	47.2
1973	11.5	8.1	19.6	N/A	N/A	N/A
1974	−12.0	7.5	−4.5	N/A	N/A	N/A
1975	6.6	9.0	15.6	N/A	N/A	N/A
1976	5.5	8.3	13.8	N/A	N/A	N/A
1977	25.3	8.3	33.6	N/A	N/A	N/A
1978	20.8	7.2	28.0	N/A	N/A	N/A
1979	19.3	7.3	26.6	N/A	N/A	N/A
1980	9.1	7.3	16.4	N/A	N/A	N/A
1981	5.0	7.2	12.2	N/A	N/A	N/A
1982	−1.5	7.2	5.7	N/A	N/A	N/A
1983	−1.8	7.7	5.9	N/A	N/A	N/A
1984	−2.2	8.2	6.1	N/A	N/A	N/A
1985	−5.3	8.7	3.4	N/A	N/A	N/A
1986	−1.2	9.5	8.4	N/A	N/A	N/A
1987	13.7	9.9	23.6	N/A	N/A	N/A
1988	28.2	9.2	37.4	N/A	N/A	N/A
1989	20.4	7.9	28.2	N/A	N/A	N/A
1990	−10.8	7.2	−3.6	N/A	N/A	N/A
1991	−0.4	8.8	8.3	N/A	N/A	N/A
1992	−8.4	9.1	0.7	N/A	N/A	N/A
1993	9.3	10.0	19.3	N/A	N/A	N/A

Sources: 1956–1971 (1956–1972 for flats) – data on the property holdings of a large insurance company; 1972–1993, data supplied by Investment Property Databank. N/A, not available.

Table 10.17 Average return and standard deviation of return on investment property, 1921–1993

Year	Average return			Standard deviation		
	Capital	Income	Total	Capital	Income	Total
1921–1929	−1.3	6.5	5.2	6.3	0.3	6.5
1930–1938	0.9	6.1	6.9	11.6	0.9	11.9
1947–1954	0.5	5.1	5.6	6.2	0.6	6.6
1955–1964	1.4	5.5	7.0	3.0	0.2	3.1
1965–1973	9.5	5.5	15.1	10.1	0.5	9.8
1974–1980	6.5	5.8	12.3	13.4	0.6	14.0
1981–1988	7.7	6.0	13.8	7.7	0.3	7.8
1989–1993	−3.5	7.0	3.5	10.5	1.3	10.8

Source: Table 10.15.

Table 10.17 reveals the relative stability of the income return on property over time, and the much greater variability of capital returns, as is illustrated in Figure 10.11. Prior to 1954 the long-run capital return on investment property was virtually zero, due to the very long leases on which properties were customarily let. During 1954–1964 property offered a small, but significant, long-run capital return, which became the most important component of total returns from the mid-1960s to the late 1980s. However the five years from 1989 witnessed negative capital returns to investment in property.

Table 10.18 Average rate of return for property by sector (%), 1959–1993

	Shops	Offices	Industrial	Residential	Total	Inflation
1959–1964	7.0	8.6	7.2	15.1	8.8	2.4
1965–1973	9.3	20.9	11.3	15.1*	15.1	5.9
1974–1980	15.1	10.0	17.9	N/A	12.3	15.9
1981–1988	15.0	13.6	12.3	N/A	13.8	6.0
1989–1993	5.1	0.5	10.0	N/A	3.5	5.6
1959–1993	10.7	11.8	11.9	N/A	11.4	7.2

Sources: Tables 10.15 and 10.16. N/A, not available.

* 1965–1972.

Table 10.18 shows similar information for each major class of property. Of the three property classes for which returns data are available for the whole of the period covered in the table, industrial property produced the highest average return, largely due to its superior performance during the years following the 1974 and 1990 property crashes. Office property produced the highest average return prior to the 1974 crash, but performed less well than shop property during subsequent periods. Shop property produced a significantly lower return than offices or industrial property over the 1959–1993 period. However, the volatility of returns for shop property, 9.87%, was lower than that for offices and industrial property, 13.52 and 10.04%, respectively. As volatility of return is often used as a proxy measure of risk, this suggests that while shop property produced a lower long-run average return than other property classes this was compensated for to some extent by a lower risk.

Over the period for which returns data for residential property are available, 1959–1972, residential property performed well compared to other property classes, with an annual average return to investment of 15.1%. However, high returns on residential property during the latter part of this period were the result of capital values which reflected sale values outside the rented sector, rather than values based on expectations of future investment performance. As such, the residential property figures are not directly comparable with those for other property classes.

Table 10.19 compares the average return, and standard deviation of return over time, for property, equities and gilts. Property exhibited a lower rate of return than consols and shares during 1921–1929, but out-performed both these asset classes by a small margin from 1930 to 1938. During the post-war period property out-performed gilts by a wide margin until 1973, but has performed less well during the following years. Shares provided a higher average return than property during all periods other than the 1965–1973 property investment boom. However, property investment returns displayed much less volatility than those for equities, as is indicated by their standard deviation of return, which is substantially lower than that for equities during each sub-period shown, other than 1981–1988. Property returns were also less volatile than gilt returns during every sub-period other than 1965–1973 and 1989–1993. These figures must be treated with some caution, however, as it is not clear to what extent the low standard deviation of property returns is due to problems of data smoothing, as noted above, or is an inherent characteristic of property as an investment medium.

Table 10.19 Average return and standard deviation of return for property, shares and gilts (%), 1921–1993

	Average return			Standard deviation		
	Property	Shares	Gilts	Property	Shares	Gilts
1921–1929	5.2	14.3	6.3	6.5	16.5	5.6
1930–1938	6.9	6.6	6.7	11.9	18.5	14.9
1947–1954	5.6	8.7	−1.3	6.6	16.5	8.5
1955–1964	7.0	10.2	0.7	3.1	20.2	10.3
1965–1973	15.1	9.0	3.2	9.8	24.3	8.7
1974–1980	12.3	19.3	13.6	14.0	53.6	17.1
1981–1988	13.8	20.1	14.4	7.8	7.8	12.7
1989–1993	3.5	17.1	14.3	10.8	13.3	7.3
1965–1993	12.0	15.8	10.6	11.4	31.1	13.3

Sources: Property – Table 10.15. Equities and gilts – 1920–1992, Barclays de Zoete Wedd (1993) *BZW Equity-Gilt Study*, BZW, London; 1993 – Investment Property Databank (1995) *IPD Annual Review*.

For the whole period for which rent reviews have given property a significant equity element (1965–1993), property has performed better than gilts though less favourably than equities. Meanwhile, property returns have been subject to lower volatility than both equity and gilt returns, indicating that property carries less risk than these other asset classes (if property's low standard deviation of return is not assumed to arise from the problems of data smoothing). However, property's relative performance is much less favourable if only the years since the 1974 crash are taken into account. Any assessment of the relative merits of property as an investment medium therefore depends on whether the 1974 property crash represents a fundamental discontinuity in the performance of property as an asset, from which any assessment of its relative merits should begin, or whether a longer-run comparison better reflects the true long-term performance of investment property.

A further notable feature of returns to investment in property are their low correlation with investment returns for other asset classes. Over the period 1963–1993 the correlation coefficients between returns on property and those on equities and gilts were 0.197 and 0.071, respectively, compared to a correlation between equities and gilts during this period of 0.623. This suggests that the factors influencing the property market have little relationship to those influencing these other asset markets. The factors which have determined property returns during the last 30 years are examined in Chapter 11, together with an analysis of the determinants of the volume of property investment during this period.

REFERENCES AND NOTES

1. Sources: Board of Trade (1923–1938) *Annual Report on Life and Other Long-Term Assurance Business*, HMSO, London; Post Magazine (1948) *Post Magazine Almanack*, Post Magazine, London; Board of Trade (1947–1961) *Annual Summaries of New Insurance Business*, HMSO, London; CSO (various issues) *Financial Statistics*, HMSO, London.
2. The figures exclude local authority pension funds, which had virtually no direct property holdings, and those reassured with insurance companies, which are covered in the insurance company figures.
3. Source: *Committee on the Working of the Monetary System* (1960) HMSO, London, Appendix, Vol. 2, pp. 222–4.
4. E.V. Morgan (1960) *The Structure of Property Ownership in Great Britain*, Clarendon, Oxford, pp. 116–25.
5. R. Redden (1984) *The Pension Fund Property Unit Trust: A History*. Privately published, London, p. 94. PFPUT, the only property unit trust in existence during 1966, had property holdings of £2.1 million as at 31st March 1967, the end of its first accounting year.

REFERENCES AND NOTES

6. The pattern of net property investment by the insurance companies during the inter-war years is discussed in more detail in Chapter 4.
7. This category included Oxford and Cambridge Colleges, religious organizations, livery companies and similar institutions.
8. This included retailers, industrial and commercial companies, banks, etc.
9. Property in which the institution has a freehold or leasehold interest in the building, rather than the land. It may, or may not, also own the ground rent.
10. See Chapter 2.
11. There was virtually no investment in property by insurance companies during the Second World War, as is discussed in section 5.2.
12. Source: Board of Trade (1951 and 1961) *Annual Summaries of New Insurance Business*, HMSO, London.
13. See section 10.6.
14. The 1970 figure was based on the IPD data only.
15. Source: Information supplied by Allsop & Co.
16. Properties were customarily sold at a yield calculated to the nearest quarter of one percent of the purchase price. The additional yield values in Table 10.11 have therefore been rounded to the nearest quarter of one percent to reflect this, and make them compatible with the Allsop & Co. data.
17. Other factors, such as taxation, would also influence yield differentials to some extent.
18. Excluding ground rents.
19. S.L. Lee (1988/89) Property returns in a portfolio context. *Journal of Valuation*, **7**, 252.
20. B.D. MacGregor and N. Nanthakumaran (1992) The allocation to property in the multi-asset portfolio: the evidence and theory reconsidered. *Journal of Property Research*, **9**, 25.
21. 'Money weighted' refers to the fact that this measure does not take into account the timing of capital injections during each analysis period, summing the net value of these injections and treating them as if they occurred halfway through the period over which the rate of return is calculated.
22. See Chapter 3.
23. The yield figures for shops during these years are based on properties let at the start of long leases, typically 99 or 999 years, without rent review. This was by far the most common lease structure for property entering institutional portfolios during this period.

11 An econometric analysis of property returns and the volume of institutional property investment

11.1 THE LONG-TERM DETERMINANTS OF RATES OF RETURN AND NET INVESTMENT IN UK PROPERTY

Chapters 1–9 of this study identified a number of factors which have determined rates of return, and levels of investment, in commercial property. This chapter will outline the main long-term factors which have influenced the property investment market, the importance of which will then be analysed using simple econometric models. Four major long-term determinants of property investment returns can be identified from the evidence presented in Chapters 6–9.

11.1.1 The decline in the interval between rent reviews

Prior to the mid-1950s investment property was typically let on long leases, of 99 or 999 years, at fixed rents. The persistence of inflation during the 1950s, and a growing appreciation among institutional investors of the potential which property offered as an equity investment, led to the introduction of the rent review, an upward-only adjustment of rent paid for a property to market levels, at intervals stated within the lease. The first rent review clauses began to appear around 1955; the next 20 years saw a steady reduction in the interval between reviews, until the present pattern of five yearly reviews became established in the early 1970s.

Table 11.1 and Figure 11.1 show typical rent review periods for new leases, and average times to rent review for institutional property portfolios, from 1955 to 1993. For years in which there was a transition from one current rent review period to another the average of the two is given.

Table 11.1 Current and average times to rent review for investment property, 1955–1993

	Average time to rent review	Current rent review period[a] (years)
1955	71.46	50
1956	66.02	40
1957	59.80	33
1958	53.18	33
1959	45.92	25
1960	38.90	21
1961	33.90	21
1962	30.78	21
1963	27.91	21
1964	25.41	18
1965	22.19	14
1966	20.19	14
1967	17.84	10
1968	15.48	7
1969	13.76	7
1970	11.66	7
1971	9.52	6
1972	7.67	5
1973	6.00	5
1974	4.58	5
1975	3.89	5
1976	2.80	5
1977	2.50	5
1978–1993	2.50	5

Source: Estimated from the archival records of a number of financial institutions.

[a] In some years when rent review intervals are in the process of being revised the current rent review figure is an average of prevailing review periods.

The effect of the reduction in intervals between rent reviews over the 20 years since the mid-1950s was to increase the long-run rate of return on investment property. This occurred since, given a constant rate of rental growth, a reduction in the time between adjustments of rents within leases to market levels increases the net present value of a property asset. Such a long-term increase in property investment returns can be observed over the years during which review periods were being reduced, as is demonstrated in Figure 11.2, which shows returns on property, equities and gilts over the period 1955–1977, expressed in terms of nine year moving averages. Unlike the other asset classes, property displays a steady upward trend in returns, interrupted only by the 1974 property crash.

AN ECONOMETRIC ANALYSIS OF PROPERTY RETURNS

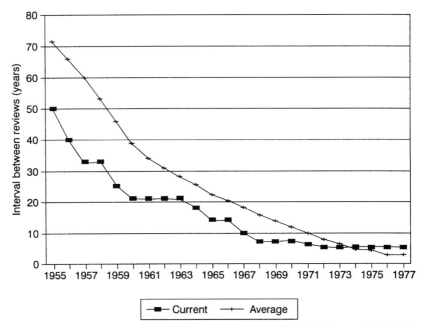

Figure 11.1 Current and average times to rent review, 1955–1977. Source: Table 11.1.

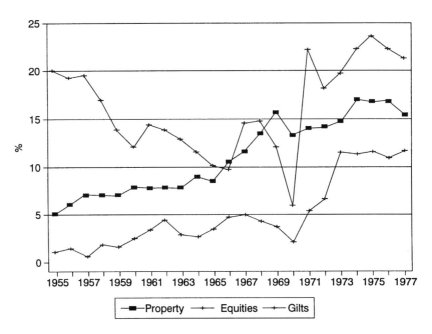

Figure 11.2 Returns on property, equities and gilts, 1955–1977, expressed in terms of nine-year moving averages. Source: as for Table 10.19.

11.1.2 Changes in market rents

Property returns are determined by the discounted value of future market rents (i.e. rents on newly-let property), the time to rent review determining the period over which rental growth has to be discounted to arrive at a property's value. Market rents are in turn determined by demand and supply in the user-market for property, each of which is influenced by a complex series of factors including general economic growth, the relative expansion of the service sector, planning legislation and other government-imposed development restrictions. Undertaking an exhaustive examination of these factors would be beyond the scope of the present analysis. Their short-term impact on the property market since 1945 has been discussed in Chapters 1–9, and they are incorporated in the following econometric analysis via their effects on market rents.

11.1.3 The property development cycle

One of the most important long-term factors influencing property investment returns is the property development cycle. This cycle occurs as a result of the lagged relationship between demand and supply for commercial property. The British property development market is largely 'speculative', i.e. most occupiers do not commission new buildings, but rely on a pool of new developments which have been constructed without pre-arranged tenants to supply their property needs.[1] Developers therefore have to anticipate potential demand for the buildings they are erecting, a demand which will not materialize for several years due to the lag between initiating and completing a development project. This time lag is typically two to three years, or longer.[1]

There is little information regarding the volume of new developments in the 'pipeline'. The developer therefore has great difficulty in predicting what the supply situation will be when the developments currently being initiated are completed. During periods of booming property values oversupply can easily build up, leading to a drastic cut-back in developments when rents begin to fall. This, in turn, creates undersupply, which will lead to another phase of rapid property value growth.

During upswings in the property market, property companies become more highly geared in order to finance their development programmes. A rising ratio of debt:equity for the property company sector reduces its ability to withstand a fall in property prices and, once the slump does set in, bankruptcies and forced sales on the part of developers accentuate the slump by increasing the flow of properties on to the market.

The development cycle produces a very strong, and regular, cyclical pattern in property investment returns, with intervals from troughs to peaks of between four and five years. Prior to the mid to late 1960s the long periods between rent reviews insulated property investment returns from this cycle to some extent, as property was subject to such infrequent changes in rental income that it was regarded by investors as a largely fixed-interest security. However, following the reduction in rent review periods, property returns began to display a very clear cyclical trend. This is illustrated in Figure 11.3, which shows property returns over the period 1963–1993, expressed as a five-year moving average in order to smooth out short-term fluctuations. A similar, but less regular, cyclical pattern can also be seen in market rental growth, as was illustrated in Figure 10.10.

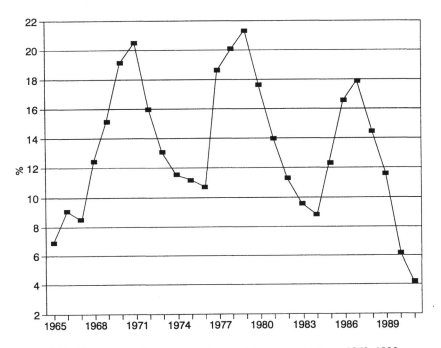

Figure 11.3 Five-year moving average of property investment returns 1963–1993. Source: Table 10.15.

Writers on business cycles have noted that business cycles are irregular in frequency,[2] something which does not appear to be true of the property development cycle. This may be due to the fact that the relatively stable time-lag between initiating a development project and completing the development has given the property development cycle a high degree of regularity, not found in other sectors of the economy. The shape of the cyclical variation in returns shown in Figure 11.3 does correspond to that often noted for business cycles, returns growth being slowest towards the peak of the cycle, while the decline in returns is greatest immediately after the cyclical peak.

The existence of a property development cycle has been proposed by a number of writers, though until recently the absence of long-run data has made it difficult to test for the presence of a cyclical pattern to property returns. Recent studies have indicated a high correlation between current and lagged property returns. Evidence for such a relationship was provided by the findings of an econometric analysis undertaken by John Whitley at the London Business School,[3] which found lagged property values to be an important determinant of current values. This was explained as being due to the importance of expectations of future price movements of commercial property.[4] Other studies have suggested that this relationship may result from smoothing due to problems of data composition, while not ruling out the possibility that it is an inherent characteristic of the data, as was noted in Chapter 10. While data composition problems might lead to smoothing in the short term, the long-term cyclical pattern outlined above could not be accounted for in this way.

11.1.4 The long-term impact of the 1973/4 and 1990 property crashes

The evidence presented in Chapters 7–9 suggests that the property crashes of 1973/4 and 1990 had a significant long-term impact on investors' expectations regarding the future growth of property values and the risk associated with investment property. While investors were generally happy to estimate rental growth on the basis of a simple linear projection of recent trends during the early 1970s, paying little attention to the possibility of oversupply and a severe downturn in market conditions, in the years following the 1973/4 crash investors payed greater attention to the risk of current boom conditions turning to slump. The 1990 property crash appears to have similarly increased the perceived risk associated with property investment.

In the following analysis the property crashes are treated as structural shifts in the property investment market, reducing the expected future rate of income growth for property assets for any given level of current market rents. As property values, and the volume of institutional property investment, are largely determined by current income plus the discounted value of future income, the crashes could therefore be expected to reduce the growth of property values and investment in the property sector.

11.2 THE DETERMINANTS OF INVESTMENT PROPERTY CAPITAL VALUES

Having outlined the main long-term factors which appear to have influenced property investment returns, it is now necessary to discover to what extent they can explain the variation in annual returns to investment in property. Developing a comprehensive model of the factors influencing the property market, both in the short and the long term, is beyond the scope of this study. The analysis will, instead, seek to determine to what extent the long-term influences outlined above can be shown to be significant determinants of changes in investment property values during the period under examination.

AN ECONOMETRIC ANALYSIS OF PROPERTY RETURNS

It was decided to model capital values for investment property rather than rates of return. The variation in property values over time approximates to that of total returns to investment in property, as the income component of property returns has shown a high degree of stability over the last 30 years (as was illustrated in Figure 10.11), almost all the annual variation in returns during this period being accounted for by changes in capital values. Furthermore, as income is adjusted only once every five years, or less, capital returns are of more interest with respect to the current analysis; indeed, capital appreciation during a period of static income would actually depress income returns.

The following variables were used in the analysis:

CAP. An index of investment property real capital values, based on the data given in Table 10.15 (set at 1964 = 100), was used as the dependent variable.

REV. This series shows the average time interval to the next rent review for property in institutional investment portfolios, as outlined in Table 11.1. It was included in order to examine the existence, and importance, of the above hypothesis regarding the relationship between declining rent review intervals and rising property returns over the period up to the late 1970s.

RENT. The index of investment property market rental values shown in Table 10.12 was used to examine the impact of changes in market rents on investment property values. Market rents are influenced by a variety of short- and medium-term factors which affect the changing balance between demand and supply in the property letting market. The inclusion of this series therefore compensates, to some extent, for the omission of these factors from the model. Like the CAP variable, market rents are expressed in real (1990) values.

Two dummy variables, **B1** and **B2**, were included in the model, to examine the long-term impact of the 1973/4 and 1990 property crashes. These were given values of 0 prior to 1974 and 1990, respectively, and 1 thereafter. The lagged value of CAP was also included as an independent variable, to account for the property development cycle. The cyclical nature of the property market could be expected to result in a positive association between current changes in property values and changes during the previous year, as was discussed in section 11.1.

The variables (other than B1 and B2) were transformed into log form, as the relationship between the time to rent review for an investment property and the rate of income growth for that property, for a given rate of growth of market rents, is non-linear. In the following equations the prefix **L** before a variable has been used to represent the log value of that variable. The analysis was conducted over the period 1965–1993, in order to avoid the need to include an extra dummy variable to account for the possible effects of the introduction of the 1964 Brown Ban on property values. Equation 11.1 gives the results of this analysis and the fitted values for LCAP produced by Equation 11.1 are illustrated, together with the actual values, in Figure 11.4.

THE DETERMINANTS OF INVESTMENT PROPERTY CAPITAL VALUES

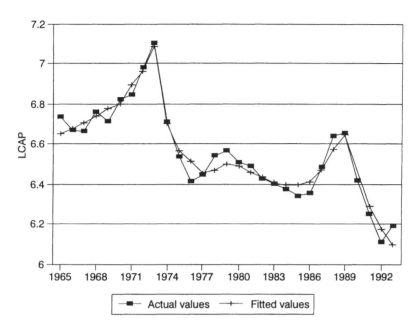

Figure 11.4 The actual and fitted values of LCAP, 1965–1993. Source: Equation 11.1.

Equation 11.1 suggests that almost all the variation in property values over these years can be explained in terms of the factors included in the model, as is indicated by the high value for R^2. The coefficients for all variables display the expected signs, and all have t-values which are significant at, or close to, the 1% confidence interval. Both the 1973/4 and 1990 property crashes are shown to have reduced the subsequent growth of property capital values, the impact of the first crash being substantially greater than that of the second. The greater importance of the first crash is to be expected, since the 1990 crash was essentially the repetition of an earlier event and would, therefore, be likely to have a weaker long-term effect on investors' expectations than the original event.

Equation 11.1. Modelling LCAP by OLS over the period 1965–1993

Variable	Coefficient	Std. error	t-value	t-prob	Part R^2
Constant	2.9473	0.43786	6.731	0.0000	0.6633
LREV	−0.1656	0.04002	−4.138	0.0004	0.4268
LRENT	0.4105	0.11930	3.441	0.0022	0.3398
B1	−0.4856	0.06257	−7.761	0.0000	0.7236
B2	−0.2083	0.03320	−6.273	0.0000	0.6311
LCAP_1	0.2828	0.10642	2.657	0.0141	0.2349

$R^2 = 0.955478$, $F(5, 23) = 98.72$ [0.0000], $\tilde{O} = 0.053616$, $DW = 1.57$,
$RSS = 0.06611752608$ for six variables and 29 observations

11.3 THE DETERMINANTS OF THE VOLUME OF INSTITUTIONAL PROPERTY INVESTMENT

The evidence presented in Chapters 7–9 suggested that the volume of institutional property investment during the last 30 years was largely determined by investors' perceptions regarding the growth of property values, together with the progressive decline in the interval between rent reviews (which increased the extent to which this underlying capital appreciation was reflected in the market value of let property). Market rents and intervals between rent reviews should, therefore, be important determinants of the volume of institutional property investment.

In the following model the dependent variable, **LINV**, shows the log value of real net institutional investment in commercial property over the period 1965–1993, based on the data given in Table 10.2. The independent variables are the same as those used in Equation 11.1 (excluding lagged capital values), plus the additional variable **LSHARE**, an index of real UK equity prices.[5]

This was included since property and shares are viewed as substitutes by the financial institutions, as they constitute the two main classes of institutional 'equity' investment. A movement in share values could, therefore, be expected to lead to an opposite movement in the volume of property investment, assuming that institutional investors intrepreted the share price movement as part of a longer-term trend. The results of this analysis are given in Equation 11.2 and the actual and fitted values of LINV are shown in Figure 11.5.

Equation 11.2 shows that the factors included in the model do appear to be significant determinants of the volume of institutional property investment. All variables had t-values which were significant at, or close to, the 5% confidence interval, and displayed the expected relationship with the volume of investment. The coefficient of the dummy variable for the impact of the first property crash, B1, was found to be greater than that for the second, B2, as was the case with the analysis of property values.

Equation 11.2. Modelling LINV by OLS over the period 1965–1993

Variable	Coefficient	Std. error	t-value	t-prob	Part R^2
Constant	10.509	4.2573	2.468	0.0214	0.2094
LREV_1	−0.554	0.2534	−2.184	0.0394	0.1718
LRENT_1	1.007	0.4887	2.061	0.0508	0.1559
LShare	−0.809	0.2510	−3.223	0.0038	0.3112
B1	−0.949	0.4910	−1.933	0.0657	0.1397
B2	−0.649	0.2455	−2.644	0.0145	0.2331

$R^2 = 0.65826$, $F(5, 23) = 8.8605$ [0.0001], $\tilde{O} = 0.348249$, $DW = 2.16$, $RSS = 2.789381734$ for six variables and 29 observations

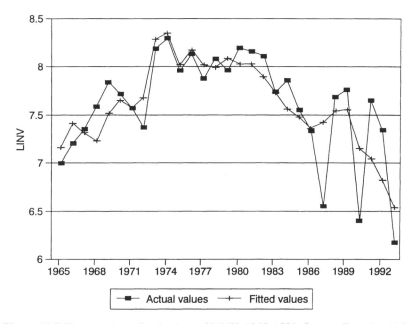

Figure 11.5 The actual and fitted values of LINV, 1965–1993. Source: Equation 11.2.

Lagged, rather than current, values of the LREV and LRENT variables were found to be significant. This may be due to the practice of institutional investors setting target levels for property investment during each year at the beginning of the year, based on information then available. It may also reflect the imperfect nature of property market information during much of this period. The R^2 value for this equation was substantially lower than that for Equation 11.1, indicating that factors not included in the model played a greater role in determining the volume of property investment than was the case with capital values. This is hardly surprising since the volume of institutional property investment has been influenced by legislative changes, such as the abolition of exchange controls in 1979, and the growth of alternative investments, such as index-linked gilts, which were not available throughout the period under consideration. Furthermore, financial innovation has also acted to increase the volume of institutional property investment, via investment vehicles such as property unit trusts. However, Equation 11.1 does indicate that as much as 65.8% of the variation in institutional investment over this period can be explained due to the factors included in the model.

Both Equations 11.1 and 11.2 reinforce the evidence presented in Chapters 7–9, indicating that the reduction in rent review intervals over the period up to the late 1970s was an important determinant of property value growth and the volume of net institutional property investment. Underlying property market conditions (as expressed by market rental growth) are also shown to be a major determinant of property values and the level of investment. The two property market crashes appear to have had a significant long-term effect in reducing subsequent property value growth and investment levels, the first crash having a greater long-term impact than the second. Finally, the buoyancy of the UK equity market is found to display an inverse relationship with the volume of property investment, in line with the hypothesis that property and equities are regarded as substitutes by investors.

REFERENCES AND NOTES

1. D. Cadman and A. Catalano (1983) *Property Development in the UK – Evolution and Change*, E & FN Spon, London, p. 1.
2. J.M. Blatt (1980) On the Frisch model of the business cycle. *Oxford Economic Papers*, **32**(3), 467–79; J. Tinbergen and J.J. Polak (1950) *The Dynamics of Business Cycles*, Routledge & Kegan Paul, London, p. 69.
3. J. Whitley (1992) Forecasting capital values in the commercial property sector. Unpublished research paper.
4. Commercial property values: forecasts from the London Business School–Royal Institution of Chartered Surveyors model (1992) *Economic Outlook*, October, 28.
5. Sources: 1964–1992, Barclays de Zoete Wedd (1993) *BZW Equity–Gilt Study,* Barclays de Zoete Wedd, London; 1993, CSO (1995) *Economic Trends Annual Supplement*, p. 246. Values are expressed in log form.

Conclusions 12

Over the last century Britain has developed one of the most sophisticated commercial property markets in the world. However, during the last 25 years a variety of factors have resulted in that market becoming subject to cycles of escalating property values followed by price collapses and depression. The boom–bust cycle of the property development industry has clearly had a detrimental effect on the British economy. While rapid inflation during the mid-1970s removed much of the long-term impact of the first property crash, by reducing the real value of the indebtedness which followed it, the consequences of the 1990 crash are still being felt in the property market and the wider economy. As Chapter 9 demonstrated, the wealth effect of a collapse in commercial property values has a substantial impact on corporate borrowing. Meanwhile, the direct effects of the crash on the property development industry have been catastrophic, leading to a sharp reduction in new development, and substantial unemployment in the construction industry and property-related professions such as surveying and architecture.

Furthermore, there is substantial evidence to suggest that the boom phases of the property cycle may also have negative economic effects. During periods of rapid monetary expansion, such as the early 1970s and mid-1980s, bank lending to the property industry expanded much more rapidly than lending to other sectors of the economy. This has resulted in speculative spirals in property values, much of the profits made from property during those years arising from take-overs, trading and other similar activities which take advantage of escalating property prices without actually creating anything of material value.

Meanwhile, channelling money into the property sector, rather than industry, has itself undermined the economic booms created by the credit expansions. By preventing industry from growing sufficiently rapidly to either supply the goods and services required to meet the expanded consumer demand, or provide sufficient user-demand to justify the volume of new accommodation being built, the funnelling of expanded credit to the property sector has itself hastened the economic slumps which followed these booms.

If a repetition of the boom–bust cycle of the commercial property sector is to be avoided, action is necessary on the part of property companies, the banks, government and institutional investors. The property companies have experienced the most rapid growth of any of the players in the property market during its upswings, and the most catastrophic collapses during its slumps. At times of easy monetary conditions many property companies have been happy to conduct business on a basis that would lead to insolvency should interest rates rise substantially, or property values fall. Indeed, during both the early 1970s and mid to late 1980s property booms some property companies were not only prepared to undertake individual projects on which interest costs were expected to exceed initial estimated rents, but allowed the interest bill on their overall portfolio to exceed their income. This willingness to take high risks is partly due to the ease of entry into the property development market, which encourages speculative activity during upswings in the property cycle. As Barkham has noted 'The barriers to entry into the property development industry are extremely low and in situations where finance is easily available and liability is limited, marginal and extremely risky schemes are initiated.'[1]

As such schemes are typically financed almost entirely with other people's money, the property company sector faces a classic principal agent problem. Developers have considerable incentives to pursue high-risk strategies which will, if successful, provide them with very considerable financial rewards, while failure will merely result in the loss of funds provided by shareholders and the banks. Both these parties appear happy to accept the risk during the boom phases of the property development cycle (though during the mid to late 1980s boom the use of off-balance sheet finance and other accountancy techniques often served to hide the full extent of that risk from those who provided the finance).

Older, more well-established, property companies generally took a more cautious attitude towards risk than the new entrants to the property sector, especially during the 1980s boom. However, pursuing policies which limit the risk of insolvency carries its own dangers; during boom periods predators generally find little difficulty in raising finance to take over property companies that the stock market considers to be growing too slowly. Thus, companies are forced to either increase their gearing, and their consequent vulnerability to any downturn in market conditions, or face absorption by their more rapidly expanding counterparts. All of this is, of course, only made possible by a stock market which focuses on short-term profits rather than long-term growth, a more cautious strategy generally being priced substantially lower than one of rapid expansion with little regard to downside risk.

A further factor contributing to the crisis-prone nature of the property company sector is the absence of research by property companies into the risks they are facing. British property companies control property assets with an aggregate value of over £24 billion,[2] while their development activities magnify their financial importance, as is evident from the sector's bank debt, which exceeded £40 billion in 1991.[3] However, despite the massive capitalization of the property sector, property companies put very little money into forecasting future demand and supply for commercial property, basing development decisions on techniques which are, to say the least, rudimentary. A questionnaire survey of 50 public property companies (representing almost 55% of the listed property company sector), undertaken in 1992 at the University of Reading, revealed an extremely low level of expenditure on market research and forecasting. The most important basis for strategic decision-making was found to be 'a reliance on common sense and judgement', while only 12% of respondents said they spent more than £20 000 a year on formal research.[4]

Given the long time-lag between initiating and completing a development project, this lack of attention to forecasting future market conditions almost guarantees the periodic undersupply, followed by massive oversupply, of commercial property, which has been a prime cause of the boom–bust cycle in the sector. While basing investment decisions on formal research may go against the grain of the entrepreneurial nature of the property industry, the use of more sophisticated forecasting techniques is vital given the high-risk, cyclical market in which the property developer operates.

Property companies face incentives to undertake high-risk strategies; however, their ability to pursue such strategies is constrained by the willingness of the banks to lend them the necessary finance. The banks' attitude towards the property company sector during the boom phases of the property cycle has earned them a bad press. For example, in his account of the 1980s property boom, and the following crash, Goobey noted that 'The ability of banks to read investment cycles makes institutional investment managers appear to be relatively clairvoyant, and the banks' enthusiasm to lend to exactly the same areas as their rivals makes investment managers look like free-thinkers; by comparison to both lemmings seem positively rational'.[5]

The evidence examined in the course of this study suggests that the bad press the banks have earned as a result of their property lending activities during the 1980s boom is fully justified. What makes the banks' attitude all the more remarkable is that they had undergone a very similar episode during the early 1970s, culminating in a severe banking crisis and a Bank of England rescue operation to bail out the banking sector, partly at public expense. Having once burned their fingers in a speculative property boom the banks might have been expected to pay more attention to the danger signals. However, the banks have their own principal agent problem; the Bank of England is almost certain to step in to avert any systemic crisis of the banking system, thus making the true risk of 'following the herd' considerably less than that of undertaking an investment of apparently equal risk in isolation.

While the property companies have only limited information regarding the development plans of their rivals and the volume of development projects in the 'pipeline', the highly concentrated nature of the banking system should have provided the banks with a clearer picture of the glut of property which was to enter the market at the end of the decade. However, they do not appear to have felt the need to collate such information, or assess its implications for the profitability of the developments they were funding.

Capie and Collins have argued that while the quasi-monopolistic structure of the British banking system may have disadvantages due to the resulting lack of competition in bank finance, this is more than compensated for by the much greater stability which it has provided.[6] Evidence of such stability is noticeably absent with regard to the banks' property lending activities; while the secondary banks were the main culprits behind the 1974 crash the clearing banks played a leading role in channelling funds to the sector during the mid to late 1980s. The banks' key position in the property development process, and the massive scale of their lending activities, provides them with a wealth of information on which to base estimates of future property market supply. This, together with their considerable economic expertise, offers the potential for a fruitful partnership between the banking sector and the property companies, should the banks base their aggregate property lending policy on an informed analysis of the likely future course of the property development market. However, the banks have chosen, instead, to be guided by the current profitability of the sector and the actions of their competitors, resulting in substantial losses which must almost certainly have been passed on to their personal, and non-property corporate, clients in the form of higher charges.

In recent years it has been suggested by some commentators that property finance should bypass the banks entirely, relying instead on the securitization of property assets and property company debt (i.e. dividing properties and mortgages on commercial property into tradeable securities). Advocates of this method of funding property development argue that as the market is far more sensitive to changes in conditions influencing future profitability than the banks, any indication of a likely future downturn in conditions would lead to a fall in the price of property securities, which would reduce the magnitude of any slump by providing an early warning of the changing conditions. As *The Economist* argued in 1993 'By adjusting quickly, prices set in a traded market would have alerted investors to a looming new-property surplus. Investors who were quick to spot this glut would have sold some, or all, of their holdings more easily, sending a much-needed signal to other investors and the construction industry.'[7]

Securitization certainly offers a number of potential benefits. In addition to providing a means of finance which would be more sensitive to changing property market conditions, it would also make the total volume of corporate debt more transparent, and would weaken the link between property price collapses and declining profits for the banking sector (thus reducing the wider economic impact of any future property crash). While previous attempts at securitization have proved unsuccessful, the recent experience of the United States with the securitization of commercial mortgages suggests that a substantial move in this direction (at least with regard to property company debt, less certainly with the securitization of property assets) may now be practicable. However, if such a move merely resulted in property companies facing a wider range of choice in raising long-term finance, either turning to the banks or the market, there is a potential danger that this would lead to an even greater flow of funds to the property company sector during times of monetary expansion.

While much of the blame for the crisis-prone nature of the British commercial property market lies with the banks, government policy has also made an important contribution to the volatility of the sector. Labour governments have typically restricted property development, via development gains taxation and/or restrictions on the volume of new development (such as Office Development Permits). Such controls have generally resulted in a decrease in property development activity, a rise in property values (resulting from the reduction in the supply of new buildings), and a consequent boost to institutional investment in the sector, as was noted in Chapters 5, 7 and 8. They have also, inevitably, brought about an increase in rents for occupiers of property affected by the legislation in question. Thus, by artificially inflating the price of commercial property, development taxation and restrictions have acted to distort the property market, leading to rising property values and market rents, which artificially raised the expectations of investors and developers regarding the long-term profitability of the sector.

The destabilizing effects of such restrictions pale into insignificance, however, alongside the actions of Conservative governments during the early 1970s and the 1980s. During the Barber Boom of the early 1970s deregulation of the financial sector, together with the relaxation of property development controls, led to a massive channelling of funds into speculative property development, the government's expansionist policies producing a boom in property rather than manufacturing. Thus, while a great deal of the blame for the property sector's reckless expansion during these years must lie with the property companies and the banks, the government provided the framework which allowed them to pursue their high-risk strategies.

After allowing the property boom to gather steam for several years (steadily increasing the property company sector's vulnerability to any downturn in conditions) the government then introduced a range of measures which collectively precipitated the 1973/74 crash and magnified its severity. Following the imposition of rent control legislation, the government made a sudden reversal of its expansionist monetary policy, while announcing a range of anti-property legislation (largely for electoral reasons). These actions were sufficient to initiate the secondary banking and property crisis which marked the end of the long post-war boom for the commercial property market, and might have had very severe economic consequences had it not been for the Bank of England's successful lifeboat operation and the effects of inflation on property values and debt.

Government policy towards the property sector during the mid to late 1980s had strong similarities with that of the early 1970s. Planning and other restrictions on development were eased via Enterprise Zones, a more liberal attitude towards edge-of-town and out-of-town developments, and a range of other measures, as outlined in Chapter 9. Meanwhile, financial deregulation and the government's relaxed attitude towards monetary policy during much of the latter half of the decade fuelled a boom in bank lending to property companies and a spectacular, but transitory, financial services boom, which appeared for a time to create a demand for the rapidly growing stock of new developments. The reversal of the government's monetary policy at the end of the decade pulled the rug out from under the property development industry, as it had in 1973, initiating a property crash from which the sector has yet to fully recover.

If a recurrence of these episodes is to be avoided, it is important that government takes greater account of the impact of its economic and planning policies on the commercial property sector. Three areas of government policy need particular attention.

Firstly, government must take its regulatory role far more seriously than has been the case during previous economic booms. During both the Barber and Big Bang property booms the Bank of England proved ineffective in restraining the rapid flow of bank funds to the property development sector. During the late 1980s, for example, it refrained from publicly announcing its concern at the growth of lending to property companies until October 1989, when interest rates had already risen substantially and the damage was done. Earlier effective action might have limited the extent to which property companies made themselves vulnerable to any rise in interest rates, or fall in tenant demand, and thereby reduced the devastating impact of the reversal of government monetary policy on the wider economy.

Property company accounting standards is another area in which insufficient regulation contributed to the magnitude of the mid to late 1980s boom, and the following slump. Property companies found that accounting legislation enabled them to produce accounts which were profoundly misleading to their shareholders but were still perfectly legal. Hence, the true extent of their indebtedness often only came to light following the property crash, when they became faced with bankruptcy. As Barkham and Purdy noted 'A number of failed companies had previously looked relatively healthy when judged by their annual accounts.'[8] Concern has been expressed regarding the valuation basis of property assets in company accounts, the capitalization of interest on development finance (instead of charging it to the profit and loss account), and various artificial means of raising the level of property company profits.[8] However, the greatest abuses of accounting procedures concerned the various means by which debts were kept off the balance sheets of property companies.[9] The regulations governing off-balance sheet finance have since been tightened up. However, many commentators believe that property companies still have scope for concealing the true extent of their debt under the new rules.

A second area in which government has considerable power to influence the volume of new property development is its approach to planning policy. In the

1980s the government's extremely *laissez-faire* attitude towards planning issues led to a substantial decentralization of commercial property development out of central urban districts. Meanwhile, its attempts at a property-based approach to inner-city regeneration resulted in a further substantial increase in property development outside recognized commercial centres. Such policies served to erode the physical limit which the scarcity of central sites placed on new office and retail development. This both reduced the long-run capital appreciation of central site values (and the properties they contained) and enabled a much greater volume of new development to be initiated than had hitherto been possible.

As property projects are not ready for letting until several years after their commencement, the sudden liberalization of planning policy contributed to the glut of new developments which entered the market towards the end of the decade. Allowing a proliferation of new business parks and out-of-town retail developments may also have reduced institutional investors' confidence in the property market, by threatening the future capital growth of the city centre shops and offices which formed the bulk of their property holdings. Again, following the property crash, there has been a partial reversal of government planning policy, as outlined in Chapter 9, though action has only been taken after the market has become depressed, with few developers seeking new sites for speculative projects.

If the government wishes to reduce the extreme volatility which the property development sector has experienced over the last 25 years, and its consequent effects on the wider economy, it must therefore adopt a much more sensitive approach to planning policy. Any repetition of the dramatic expansion in the physical area over which commercial development can be undertaken, as occurred during the 1980s, must be avoided. The merits of the trend towards decentralized commercial activities are open to question, as was noted in Chapters 8 and 9. However, even if this move of commercial property to edge-of-town or out-of-town locations is viewed by policy-makers as desirable, or at least irresistible, it must be managed by urban planners in such a way as to prevent a sudden mushrooming of competing new developments, no market for which may exist by the time they reach completion. Similarly, government would be wise to avoid any repetition of 'Brown Ban'-type restrictions, which inflate property values in the areas where development is prevented and thus lead to a boom in property investment, which will in turn lead to a glut of new development activity once the restrictions are relaxed or removed.

Thirdly, government would be well advised to pay greater attention to the influence of commercial property values on corporate borrowing and investment, together with the other effects of the property development cycle on the wider economy. If, as the evidence presented in Chapter 9 suggests, property values do have a considerable impact on corporate investment, and consequently on economic growth, any such relationship should be taken into account by its incorporation in the macroeconomic models on which governments base their economic forecasts. The commercial property market has hitherto been largely neglected in government economic calculations, as is evident from the paucity of official data on the property investment and development industries. However, given the key role of property in corporate finance, this neglect cannot be allowed to continue.

As for the property investing financial institutions, the evidence presented in this study has suggested that their property market activities have generally exerted a stabilizing influence on the sector. This is largely due to their twin roles as suppliers of development finance and direct investors in property; unlike the banks they provide a long-term home for the properties they finance.

The importance of this is illustrated by contrasting the consequences of the downturn in property market conditions during the early 1980s with the crashes of the mid-1970s and early 1990s. During the late 1970s the financial institutions played a leading role in both property investment and development finance, the relative importance of the banks and the property companies in the property development sector having been reduced substantially in the aftermath of the 1973/4 property crash. The financial institutions generally adopted a more cautious attitude to their property financing deals than the banks had at the start of the decade (or during the mid to late 1980s), and while a significant oversupply of property had built up by the early 1980s, the volume of projects in the development pipeline was substantially less than had been the case in 1974.

When the oversupply became apparent, the financial institutions responded by reducing their investment in the sector, though they were already committed to acquiring a large proportion of the developments they were financing. Thus, rather than falling precipitously following the peak of the boom, property values instead experienced a long period of very slow nominal growth (though they did decline in real terms) until the market once again entered a period of undersupply during the middle of the decade. The financial institutions were faced with relatively low returns on one of their major areas of investment for several years, but the downturn had much less effect on the wider economy than was the case in 1974 and 1990. Rather than disposing of their properties, the financial strength of the institutions allowed them to retain their poorly-performing property assets, preventing any collapse in property values with its attendant adverse economic consequences.

The recent instability of the property investment market has resulted in a substantial scaling-down of the financial institutions' commitment of funds to the property sector. While over 15% of net insurance company and pension fund investment was allocated to property in the 1970s, during recent years this has fallen to about 5%. However, property could, potentially, still hold considerable attractions for institutional investors. As property returns have a low correlation with those on equities and gilts, the allocation of a proportion of institutional funds to property provides substantial opportunities for diversification to minimize the overall volatility of portfolio returns. Furthermore, property constitutes an 'equity' security, that is subject to a much lower variability of returns over time than the institutions' main alternative medium of equity investment, ordinary shares. Though property may never again command as high a proportion of institutional funds as was the case during the inflationary era of the 1970s it could, therefore, play a significant role in institutional investment, if the recurrence of speculatively-driven property market crashes can be avoided in the future and long-run returns to investment in the sector are sufficient to make it competitive with equities and gilts.

There is a general need for greater research into the factors influencing the commercial property market, and the long-term nature of the property cycle. While the aftermath of the first property crash did see a considerable improvement in the techniques used to analyse property performance, quantitative analysis partially replacing the intuitive 'rule of thumb' approach to determining market conditions, the availability of long-run data is still much poorer for the property sector than for many other industries of similar economic importance. The fact that much research is undertaken by the large chartered surveying practices, which rely on a buoyant property sector for their income and therefore have a considerable incentive to 'talk up the market' has also led to criticism from a number of commentators.[10] If the property market is to learn from the lessons of the past a much greater volume of truly independent research into the factors influencing the sector is necessary, both the academic world and independent commercial organizations having an important role to play in this process.

However, better property market research can only benefit the sector if investors, developers and the providers of finance are prepared to make use of it. Investors generally prefer to base their decision-making on present trends rather than evidence from the past. This tendency was noted by J.M. Keynes, who stated that people deal with uncertainty in their decision-making according to the following general rules:

(1) We assume that the present is a much more serviceable guide to the future than a candid examination of past experience would show it to have been hitherto. In other words we largely ignore the prospect of future changes about the actual character of which we know nothing.
(2) We assume that the *existing* state of opinion as expressed in prices and the character of existing output is based on a *correct* summing up of future prospects, so that we can accept it as such unless and until something new and relevant comes into the picture.
(3) Knowing that our own individual judgement is worthless, we endeavour to fall back on the judgement of the rest of the world which is perhaps better informed. That is, we endeavour to conform with the behaviour of the majority or the average. The psychology of a society of individuals each of whom is endeavouring to copy the others leads to what we may strictly term a *conventional* judgement.[11]

Keynes's comments regarding a reliance on current price and output levels, and the 'herd instinct' psychology of the market, are as relevant today as in 1937. If banks are to continue to base their lending policies on the actions of their competitors and the current buoyancy of the market, and developers continue to eschew research in favour of their own assessment of current market conditions, a repetition of the boom–bust property cycle is inevitable, the next crash being due around the turn of the century. However, if the property companies, the banks and government can learn from their mistakes and take a more sophisticated approach to analysing the market, the property sector has the potential to make a positive and significant contribution to the British economy. Marx noted that history repeats itself, first as tragedy, then as farce.[12] It is to be hoped that the tragedy of the 1990 property crash may prevent the farce of a further crash during the next decade.

REFERENCES AND NOTES

1. R.J. Barkham (1995) The financial structure and ethos of property companies: An empirical analysis of the influence of company type. Paper presented to the International Conference on the Financial Management of Property and Construction, University of Ulster, May 1995, p. 10.
2. R.J. Barkham (1995) The financial structure and ethos of property companies: An empirical analysis of the influence of company type. Paper presented to the International Conference on the Financial Management of Property and Construction, University of Ulster, May 1995, p. 2.
3. See section 9.6.
4. R. Barkham, A. Baum and A. Ackrill (1994) The performance of property companies: an entrepreneurial analysis. Paper presented to the European Real Estate Society Inaugural Conference, Amsterdam, 1994, p. 11.
5. A.R. Goobey (1992) *Bricks and Mortals*, Century, London, p. 21.
6. F. Capie and M. Collins (1992) *Have the Banks Failed British Industry?*, IEA, London, p. 69.
7. Liquid property (1993) *The Economist*, 20 Feb., 17. There is some empirical evidence to suggest that securitized real estate markets are more informationally efficient than unsecuritized markets; for example, R.J. Barkham and D. Geltner (1995) Price discovery in American and British property markets. *Real Estate Economics*, **23**.
8. R.J. Barkham and D.E. Purdy (1992/93) Property company financial reporting: Potential weaknesses. *Journal of Property Valuation and Investment*, **11**, 134.
9. See section 9.4.
10. D. Scarrett (ed.) (1988) *Sources of Property Market Information*, Gower, Aldershot, p. viii.

11. J.M. Keynes (1937) The general theory of employment. *Quarterly Journal of Economics*, February; reprinted in D. Moggridge (ed.) (1973) *The Collected Writings of John Maynard Keynes, Vol. XIV*, Macmillan, London, p. 114. (Emphasis in original.)
12. K. Marx (1963) *The Eighteenth Brumaire of Louis Bonaparte*, International Publications, New York, p. 1 (originally published in 1852).

Bibliography

ARCHIVAL SOURCES

1. The papers of the Church Commissioners.

    (a) Papers at the Church of England Record Centre, 15 Galleywall Road, South Bermondsey, London SE16 3PD.
    Annual Reports (1948–1980).
    Estates and Finance Committee papers, filed under C.C. 95055 (1954–1964).
    (b) Papers held by the Church Commissioners at 1 Millbank, London SW1P 3JZ.
    Board of Governors minutes (1948–1964).
    The Church Commissioners Index: Historical Review of Investment Policy (*c.* 1965).

2. The papers of City Offices Ltd, at the Guildhall Library, London.
    Annual and half-yearly reports (1864–1939).
    Board minutes (1864–1939).

3. The papers of the Clerical Medical Investment Group, held by the company at Narrow Plain, Bristol BS2 OJH.
    An appreciation of the society's investment position (1947) Memorandum by Sir Andrew Rowell (Aug.).
    Asset Investigation Reports (1915–1980).
    Board minutes (1900–1980).
    Finance Committee minutes (1896–1939).
    Investment Forum minutes (1969–1980).
    Memorandum by A.D. Besant (12 Oct. 1942).
    Note regarding the Society's investment policy (7 Oct. 1942).
    Notes on investment policy (1949) Memorandum by Sir Andrew Rowell (14 Dec.).
    J. Pegler (1981) Unpublished memoirs.

4. The papers of Eagle Star, held by the company at Eagle Star House, Bath Road, Cheltenham, GL53 7LQ.

BIBLIOGRAPHY

General Purposes and Finance Committee minutes (1951 and 1964–1980).
Report on General Fund Properties (undated ledger).

5. The papers of Healey & Baker, held by the company at 29 St George Street, Hanover Square, London WIA 3BG.

    Aubrey Orchard-Lisle (1983) Paper on recollections of Sir Montague Burton.

    Typescript of interview by Kerrie Walkingshaw with Mervyn and Aubrey Orchard-Lisle (1 July 1987).

    Register of property transactions compiled by Healey & Baker (1946–1966).

6. The papers of Hillier, Parker, May & Rowden, held by the company at 77 Grosvenor Street, London W1A 2BT.

    Minutes of Partners Meetings (10 April 1917–1 March 1926; 31 Dec. 1928–10 Jan. 1972).

    Superannuation Fund minutes (12 Dec. 1922–27 June 1960).

7. The papers of Legal & General Group plc, held by the company at Temple Court, 11 Queen Victoria Street, London EC4N 4TP.

    Actuarial report on the quinquennial investigation into the Society's assets for the five years to 31st December 1936 (1937).

    Annual Report On Freehold And Leasehold Properties To Board Of Directors (2 March 1955).

    Board minutes (1933–1980).

    Estate Department Annual Reports (1934–1980).

    Notes on the first 50 years of the Estate Department, written by an employee of the company (1985).

    Property Committee, property market reports (1973–1980).

8. The papers of Marks & Spencer plc, held by the company at Michael House, Baker Street, London W1A 1DN.

    Annual Reports (1927–1980).

    Memo of Interview Ledger, D54.

    Report and Valuation of Marks & Spencer's properties conducted by Hillier, Parker, May & Rowden (April 1926).

9. The Montague Burton papers, at the West Yorkshire Archive Service, Chapeltown Road, Leeds.

    Box 114, E.B. Berhens (1931) *Critical Survey of Montague Burton Ltd* (Sept.).

    Box 129, Correspondence with Healey & Baker (Sept.–Dec. 1938).

    Box 130, letter from M. Burton to R.J. Pearson (21 Aug. 1939) and Pearson's reply (23 Aug. 1939).

10. The papers of the National Coal Board, at the Public Record Office, Kew.
    PRO: Coal 23.3.
    PRO: Coal 23.347.
    PRO: Coal 33.83.

11. The papers of the National Provident Institution.

(a) At the Guildhall Library, London.
   Assets reports (1898, 1917, 1933–1939).
   Board minutes (1890-1927).
(b) Held by the company at 48 Gracechurch Street, London EC3P 3HH.
   Assets Reports (1942–1978).
   Board minutes (1928–1980).
   Board Resolutions (1868–1898).
   Investment Committee minutes (1955–1980).
   Letter from Deputy Manager to St Quintin, Son & Stanley (12 Jan. 1976).
   Property Committee paper (27 July 1972).
   Report of Investment Committee (13 Dec. 1938).
12. The papers of the Prudential, held by the company at 142 Holborn Bars, London EC1N 2NH.
   Annual Reports (1900–1938).
   Board minutes (1918–1939).
13. The papers of Standard Life, held by the company at 3 George Street, Edinburgh EH2 2XZ.
   Asset Investigation Reports (1971–1980).
   Board minutes (1958–1980).
   Private Minute Book A1.
   Committee minute books, A1/6/5 and A1/6/6.
   Investment Committee minutes (1960–1964), minute book A1/6/9.
   Property Investment Reports (1973–1980).

OTHER UNPUBLISHED SOURCES

R. Barkham (1995) The financial structure and ethos of property companies: an empirical analysis of the influence of company type. Paper presented to the International Conference on the Financial Management of Property and Construction, University of Ulster, May 1995.

R. Barkham, A. Baum and A. Ackrill (1994) The performance of property companies: an entrepreneurial analysis. Paper presented to the European Real Estate Society Inaugural Conference, Amsterdam, 1994.

P.C. Beverley (1992) UK business and office development cycles. MSc dissertation, University of Portsmouth.

Edward L. Erdman (1990) *Edward Erdman: Surveyors – A Brief History of the Practice*. Unfinished draft copy, held by Mr Edward Erdman.

T.W. LaPier (1995) Real estate advisory services: Growth and competition in Japan, Europe, and the United States, 1960–1990. PhD thesis, University of London.

John Lawrence (1994) From counting house to office: the transformation of London's central financial district, 1693–1871. Unpublished paper.

D.J. Richardson (1970) The history of the catering industry, with special reference to the development of J. Lyons and Co. Ltd., to 1939. PhD thesis, University of Kent.

J.H. Treble and J. Butt (*c.* 1980) Unpublished typescript history of Standard Life, held by Standard Life at 3 George Street, Edinburgh EH2 2XZ.

R. Turvey (1994) City of London office rents: 1864–1914. Unpublished paper.

J. Whitley (1992) Forecasting capital values in the commercial property sector. Unpublished research paper.

I. Wray (1972) Town centres, shopping policy and decentralisation. MPhil thesis, University of London.

OFFICIAL SOURCES

Board of Trade (1923–1938) *Annual Report On Life And Other Long-Term Assurance Business*, HMSO, London.

Board of Trade (1947–1961) *Annual Summaries of New Insurance Business*, HMSO, London.

CSO (various issues) *Annual Abstract of Statistics*, HMSO, London.

CSO (1993 and 1995) *Economic Trends Annual Supplement*, HMSO, London.

CSO (various issues) *Financial Statistics*, HMSO, London.

CSO (various issues) *National Income and Expenditure*, HMSO, London.

CSO (various issues) *Regional Trends,* HMSO, London.

Committee on the Working of the Monetary System, Report (Aug. 1959) Cmnd 827 of 1959, HMSO, London.

Committee on the Working of the Monetary System, Appendix Vol. 2 to the published report (1960) HMSO, London.

Committee to Review the Functioning of the Financial Institutions, Report (1980) Cmnd 7937 of 1980, HMSO, London.

Department of the Environment (1975) *Commercial and Industrial Property Facts & Figures*, HMSO, London.

Department of the Environment (1978) *Commercial and Industrial Property Statistics*, HMSO, London.

M.C. Fleming (1980) *Statistics Collected by the Ministry of Works 1941–56*, Department of the Environment, London.

OTHER PRINTED SOURCES

I. Alexander (1979) *Office Location and Public Policy*, Longman, London.

P. Ambrose and R. Colenutt (1975) *The Property Machine*, Penguin, London.

S. Aris (1970) *The Jews in Business*, Cape, London.

W. Ashworth (1986) *The History of the British Coal Industry*, Vol. 5, Clarendon, Oxford.

BIBLIOGRAPHY

R. Atkins (1989) Property lending may 'leave banks exposed'. *Financial Times*, 7 Aug., 7.

C.J. Baker ((1970) The changing property scene – unit trusts, bonds and finance. *Estates Gazette*, **214**, 27 June.

Bank of England (1932–1939) *Bank of England Statistical Summary*.

Property unit trusts for pension funds and charities (1969) *Bank of England Quarterly Bulletin*, **ix**, Sept.

Barclays de Zoete Wedd (1993) *BZW Equity–Gilt Study*, BZW, London.

R.J. Barkham and D. Geltner (1995) Price discovery in American and British property markets. *Real Estate Economics*, **23**.

R.J. Barkham and D.E. Purdy (1992/93) Property company financial reporting: Potential weaknesses. *Journal of Property Valuation and Investment*, **11**.

C. Barnett (1986) *The Audit of War*, Macmillan, London.

R. Barras (1979) *The Returns from Office Development and Investment: Centre For Environmental Studies Research Series, No. 35*, Centre for Environmental Studies, London.

M. Bateman (1985) *Office Development: A Geographical Analysis*, Croom Helm, London.

A. Baum and N. Crosby (1988) *Property Investment Appraisal*, Routledge, London.

D.J. Bennison and R.L. Davies (1980) The impact of town centre shopping schemes in Britain: Their impact on traditional retail environments. *Progress in Planning*, **14**(1).

G.F.A. Best (1964) *Temporal Pillars*, Cambridge University Press, Cambridge.

J.M. Blatt (1980) On the Frisch model of the business cycle. *Oxford Economic Papers*, **32**(3).

M. Brett (1990) *Property and Money*, Estates Gazette, London.

George Bridge (1960) Investment of insurance funds. *The Chartered Surveyor*, **92**, March.

S. Brittan (1964) *The Treasury Under The Tories 1951–1964*, Penguin, Harmondsworth.

G. Bull and A. Vice (1961) *Bid For Power*, 3rd edn, Elek, London.

W. Burns (1959) *British Shopping Centres*, Leonard Hill, London.

D. Cadman and L. Austin-Growe (1983) *Property Development*, E & FN Spon, London.

D. Cadman and A. Catalano (1983) *Property Development in the UK – Evolution and Change*, E & FN Spon, London.

M. Campbell-Kelly (1992) Large-scale data processing in the Prudential, 1850–1930. *Accounting, Business and Financial History*, **2**.

D. Cannadine (1980) *Lords and Landlords: The Aristocracy and the Towns 1774–1967*, Leicester University Press, Leicester.

D. Cannadine and D. Reeder (eds) (1982) *Exploring the Urban Past: Essays in urban history by H.J. Dyos*, Cambridge University Press, Cambridge.

F. Capie and M. Collins (1992) *Have the Banks Failed British Industry?*, IEA, London.

M. Cassell (1991) *Long Lease!: The story of Slough Estates 1920–1991*, Pencorp, London.

M. Casson (1982) *The Entrepreneur*, Robertson, Oxford.

Chesterton (1991) *London: A Capital Investment*, Chesterton, London.

Church Commissioners for England (1994) *Report and Accounts, 1993*.

City of London Real Property Co. Ltd (*c.* 1964) The City of London Real Property Co. Ltd: 1864–1964. Privately published, London.

G. Clayton and W.T. Osborn (1965) *Insurance Company Investment*, Allen & Unwin, London.

D. Clutterbuck and M. Devine (1987) *Clore: The Man and his Millions*, Weidenfeld & Nicolson, London.

S. Connor (1989) *Postmodernist Culture*, Blackwell, Oxford.

T.A.B. Corley (1993) The entrepreneur: the central issue in Business history?, in *Entrepreneurship, Networks and Modern Business* (eds J. Brown and M.B. Rose), Manchester University Press, Manchester.

P. Cowen *et al.*(1969) *The Office: A facet of urban growth*, Heinemann, London.

George Cross (1939) *Suffolk Punch: A Business Man's Autobiography*, Faber & Faber, London.

J.B. Cullingworth (1988) *Town and Country Planning in Britain*, 10th edn, Unwin Hyman, London.

The Daily Telegraph (1992) 4 Jan., 2.

N.H. Dimsdale (1991) British monetary policy since 1945, in *The British Economy Since 1945* (eds N.F.R. Crafts and N. Woodward), Clarendon, Oxford.

DTZ Debenham Thorpe Research (1994) *Money Into Property*, DTZ, London.

H.J. Dyos (1961) *Victorian Suburb: A study of the growth of Camberwell*, Leicester University Press, Leicester.

Commercial property values: forecasts from the London Business School–Royal Institution of Chartered Surveyors model (1992) *Economic Outlook*, Oct.

Real property shares (1933) *The Economist*, **cxvii**, 25 Nov., 1022.

Full stop for London offices (1964) *The Economist*, **ccxiii**, 7 Nov., 616–17.

Developers overdeveloped (1965) *The Economist*, 27 Nov., 980.

Yes at last, revolution for the City (1971) *The Economist*, **239**, 22 May, 70–5.

The new leviathans: Property and the financial institutions, a survey (1978) *The Economist*, **267**, 10 June.

Britain's property developers build themselves a market (1986) *The Economist*, 31 May, 91–3.

Singing a Canadian tune (1987) *The Economist*, 25 July, 64–6.

London Docklands (1988) *The Economist*, 13 Feb., 71–7.

Corporate headquarters: The shape of things to come (1988) *The Economist*, 5 March, 21–4.
British property finance: Holes in the ground (1988) *The Economist*, 17 Sept., 128.
Old prop. nr tube. nds attn. (1988) *The Economist*, 5 Nov., 134.
Following the global property men's star (1988) *The Economist*, 24 Dec., 89–90.
London Property: Crash or crush? (1989) *The Economist*, 4 March, 35.
The Reichmann Brothers: Kings of officeland (1989) *The Economist*, 22 July, 17–20.
London's great property grab (1989) *The Economist*, 30 Sept., 121–2.
The next crash: Unreal estate (1989) *The Economist*, 21 Oct., 138.
Commercial property: Hurry on down (1990) *The Economist*, 18 Aug., 18–19.
Building controls: Shaky ground (1990) *The Economist*, 27 Oct., 35.
Pickings for the birds (1991) *The Economist*, 20 April, 97.
Waiting, wishing (1991) *The Economist*, 14 Dec., 107.
British property: New dimensions (1992) *The Economist*, 11 April, 102–7.
If only property slumped the way it used to (1992) *The Economist*, 23 May, 111.
Rosehaugh: Snuffed out (1992) *The Economist*, 5 Dec., 112–17.
Liquid property (1993) *The Economist*, 20 Feb., 17–18.
Selling up (1993) *The Economist*, 5 June, 100–2.
Market focus: Hot property – has Britain's commercial-property market risen too sharply from the dead? (1994) *The Economist*, 19 Feb.
R. Einstein (1973) How badly hit is property? *The Sunday Times*, 23 Dec., 38.
E.L. Erdman (1982) *People & Property*, Batsford, London.
L. Esher (1981) *A Broken Wave: The Rebuilding of England 1940–1980*, Allen Lane, London.
Estates Gazette (1932) 24 Sept.
Commercial property 1962–1974 (1975) *Estates Gazette*, **233**, 18 Jan.
Estates Times Review: Twenty Years of Property (1988) Estates Times, London.
A.W. Evans (1967) Myths about employment in Central London. *Journal of Transport Economics and Policy*, **1**.
C.H. Feinstein (1972) *Statistical Tables of National Income, Expenditure and Output of the U.K., 1855–1965*, Cambridge University Press, Cambridge.
The Lex Column: Property accounting (1989) *Financial Times*, 4 Dec., 24.
M.C. Fleming (1980) Construction and the related professions, in *Reviews of United Kingdom Statistical Sources, Vol. 12* (ed. W.F. Maunder), Pergamon Press, Oxford.
M.C. Fleming (1980) *Statistics Collected by the Ministry of Works 1941–56*, London.
L. Foldes and P. Watson (1978) *Quarterly Returns to U.K. Equities 1919–1970; Papers on Capital and Risk, No. 6*, RTZ, London.

Michael Foster (1986) Hammerson – From Brent Cross to Buffalo. *Estates Gazette*, 278, 7 June.

A.R. Goobey (1992) *Bricks and Mortals*, Century, London.

C. Gordon (1985) *The Two Tycoons*, Hamish Hamilton, London.

C. Gordon and G. Arnott (1962) Some observations on property financing. *Investors Chronicle*, **215**, Property Supplement, 23 Feb.

D.P. Hager and D.J. Lord (1985) The property market, property valuations and property performance measurement. *Journal of the Institute of Actuaries*, **112**.

C. Hamnett and W. Randolph (1988) *Cities, Housing and Profits*, Hutchinson, London.

L. Hannah (1976) *The Rise of the Corporate Economy*, Methuen, London.

A.C. Harvey (1981) *The Econometric Analysis of Time Series*, Philip Allen, Oxford.

J. Harvey (1981) *The Economics of Real Property*, Macmillan, Basingstoke.

J. Hasegawa (1992) *Replanning the Blitzed City Centre*, Open University Press, Buckingham.

Healey & Baker (*c.* 1970) *Healey & Baker: 1820–1970*. Privately published.

Carol E. Heim (1990) The Treasurer as developer–capitalist? British new town building in the 1950s. *Journal of Economic History*, **50**.

Hillier, Parker, May & Rowden (1983) *British Shopping Developments 1965–1982*, Hillier, Parker Research, London.

A. Holden (1973) Rates rise plan is 'chicken feed' to Centre Point. *The Times*, 25 May, 5.

K. Honeyman (1993) Montague Burton Ltd: The creators of well-dressed men, in *Leeds City Business, 1893–1993: Essays Celebrating the Centenary* (eds J. Chartres and K. Honeyman), Leeds University Press, Leeds.

V. Houlder (1990) Feeling the lending pinch. *Financial Times*, 23 Nov., 1.

V. Houlder (1991) The property market: A year of living on the edge. *Financial Times*, 20 Dec., 25.

V. Houlder (1992) Why bricks are no longer bankable. *Financial Times*, 10 April, 16.

V.Houlder, 'Banks shore up shaky property foundations,' *Financial Times* (15 June 1992), 19.

V. Houlder (1992) Brighter outlook for the bargain-hunters – lower values and reduced interest rates have created opportunities for investors. *Financial Times*, 4 Dec., 16.

V. Houlder (1992) The property market: Cracks became chasms. *Financial Times*, 18 Dec., 9.

V. Houlder (1993) Institutions spur the recovery. *Financial Times*, UK property survey, Nov., 2.

V. Houlder (1993) Optimism strengthens in spite of downbeat statistics. *Financial Times*, UK property survey, Nov., 3.

V. Houlder (1993) High streets see most development. *Financial Times*, UK property survey, Nov., 5.

J. Humphries (1987) Inter-war house building, cheap money and building societies: The housing boom revisited. *Business History*, **29**.

J. Huntley (1987) Balancing act. *Estates Gazette*, **284**, 12 Dec., 1460.

J. Huntley (1989) More ways to limit financial risks – non-recourse lending is now widespread among newer development groups. *Financial Times*, 12 July, III.

R. Imrie and H. Thomas (eds) (1993) *British Urban Policy and the Urban Development Corporations*, PCP, London.

Investment Property Databank (1995) *IPD Annual Review*, Investment Property Databank, London.

Finance for property companies (1961) *Investors Chronicle*, **213**, 15 Sept., 915.

Investors Chronicle/Hillier Parker (1983) *Investors Chronicle Hillier Parker Rent Index, Report No. 12*, London, May.

A.A. Jackson (1991) *Semi-detached London: Suburban Development, Life and Transport 1900–39*, 2nd edn, Allen & Unwin, Didcot.

S. Jenkins (1975) *Landlords to London*, Book Club Associates, London.

J.B. Jefferys (1954) *Retail Trading in Britain 1900–1950*, Cambridge University Press, Cambridge.

J. Johnson and G.W. Murphy (1957) The growth of life assurance in UK since 1880. *Transactions of the Manchester Statistical Society*, 1956–57.

Jones Lang Wootton (1989) *The Glossary of Property Terms*, Estates Gazette, London.

F. Kane and D. Atkinson (1992) Church faces crisis as boom turns sour. *The Guardian*, 9 Nov., 2.

William Keegan (1972) Banks are asked to exercise restraint. *Financial Times*, 9 Aug., 1.

J.M. Keynes (1937) The general theory of employment. *Quarterly Journal of Economics*, February; reprinted in D. Moggridge (ed.) (1973) *The Collected Writings of John Maynard Keynes, Vol. XIV*, Macmillan, London.

J.M. Keyworth (1990) *Cabbages and Things*. Privately published.

C.P. Kindleberger (1964) *Economic Growth in France and Britain 1851–1950*, Oxford University Press, Oxford.

A. King and S. Bryant (1988) *UK 2000: An Overview of Business Parks*, Applied Property Research, London.

I.M. Kirzner (1980) The primacy of entrepreneurial discovery, in *The Prime Mover of Progress: The Entrepreneur in Capitalism and Socialism. Papers on the Role of the Entrepreneur* (eds I.M. Kirzner *et al.*), IEA, London.

F.H. Knight (1921) *Risk, Uncertainty and Profit*, Houghton Mifflin, Boston.

Edward L'Anson (1864) Some notice of office buildings in the City of London. *Royal Institute of British Architects, Transactions*.

S.L. Lee (1988/89) Property returns in a portfolio context. *Journal of Valuation*, **7**.

R.M. Lester (1937) *Property Investment*, Pitman, London.

B.D. MacGregor and N. Nanthakumaran (1992) The allocation to property in the multi-asset portfolio: the evidence and theory reconsidered. *Journal of Property Research*, **9**.

W.G. McClelland (1963) *Studies in Retailing*, Blackwell, Oxford.

D. McWilliams (1992) *Commercial Property and Company Borrowing, Royal Institution of Chartered Surveyors, Paper No. 22*, Nov.

O. Marriott (1967) *The Property Boom*, Hamish Hamilton, London.

R. Marris (1971) An introduction to theories of corporate growth, in *The Corporate Economy: Growth, Competition, and Innovative Potential* (eds R. Marris and A. Wood), Macmillan, London.

K. Marx (1963) *The Eighteenth Brumaire of Louis Bonaparte*, International Publications, New York (originally published in 1852).

H. Mason (1989) The twentieth-century economy, in *The Portsmouth Region* (eds B. Stapleton and J.H. Thomas), Alan Sutton, London.

D. Massey and A. Catalano (1978) *Capital and Land: Landownership by Capital in Great Britain*, Edward Arnold, London.

G.E. May (1912) The investment of life assurance funds. *Journal of the Institute of Actuaries*, **46**.

H. Mercer, N. Rollings and J.D. Tomlinson (eds) (1991) *Labour Governments and Private Industry: The Experience of 1945–1951*, Edinburgh University Press, Edinburgh.

B.R. Mitchell (1988) *British Historical Statistics*, Cambridge University Press, Cambridge.

N. Moor (1979) The contribution and influence of office developers and their companies on the location and growth of offices, in *Spacial Patterns of Office Growth and Location* (ed. P.W. Daniels), Wiley, Chichester.

E.V. Morgan (1960) *The Structure of Property Ownership in Great Britain*, Clarendon, Oxford.

S.J. Murphy (1984) *Continuity and Change: Building in the City of London 1834–1984*, Corporation of London, London.

NCB Staff Superannuation Schemes (1953) Annual Report, 1952/53.

E. Nevin (1955) *The Mechanism of Cheap Money*, University of Wales Press, Cardiff.

P. Norman (1989) Warning on property lending. *Financial Times*, 14 Oct., 24.

A. Offer (1981) *Property and Politics 1870–1914*, Cambridge University Press, Cambridge.

D.J. Olsen (1976) *The Growth of Victorian London*, Batsford, London.

F.J. Osborn and A. Whittick (1977) *New Towns: Their Origins, Achievements, and Progress*, Leonard Hill, London.

Panmure Gordon & Co. (1976) *The Property Sector 1976–1977*, Panmure Gordon & Co., London.

A. Pike (1993) Church of England investment managers censured for losses. *Financial Times*, 23 July, 20.

J. Plender (1982) *That's the Way the Money Goes*, Andre Deutsch, London.

S. Pollard and D.W. Crossley (1968) *The Wealth of Britain 1085–1966*, Batsford, London.

Post Magazine (1948) *Post Magazine Almanack*, Post Magazine, London.

C.G. Powell (1980) *An Economic History of the British Building Industry 1815–1979*, Methuen, London.

E. Reade (1987) *British Town and Country Planning*, Open University Press, Milton Keynes.

R. Redden (1984) *The Pension Fund Property Unit Trust: A History*. Privately published, London.

G. Rees (1969) *St Michael: A History of Marks & Spencer*, Pan, London.

M. Reid (1982) *The Secondary Banking Crisis, 1973–75: Its causes and course*, Macmillan, London.

H.W. Richardson and D.H. Aldcroft (1968) *Building In The British Economy Between The Wars*, George Allen & Unwin, London.

S. Robinson (1991) The property market: U-turn that surprised an industry. *Financial Times*, 1 Nov., 17.

J. Rose (1981) Institutional investment: Have the funds got it wrong? *Estates Gazette*, **260**, 21 Nov.

J. Rose (1985) *The Dynamics of Urban Property Development*, E & FN Spon, London.

J. Rose (1993) *Square Feet*, RICS Books, London

St. Quintin (c. 1981) *The History of St. Quintin Chartered Surveyors 1831–1981*. Privately published, London.

D. Scarrett (ed.) (1988) *Sources of Property Market Information*, Gower, Aldershot.

R. Schiller (1985) Land use controls on UK shopping centres, in *Shopping Centre Development: Policies and Prospects* (eds J.A. Dawson and J.D. Lord), Croom Helm, London.

J.A. Schumpeter (1934) *The Theory of Economic Development*, Harvard University Press, Cambridge, MA.

A. Seidl (1990) Pop goes the Big Bang boom. *The Independent on Sunday*, 11 Feb., 6.

A. Seidl and N. Faith (1990) High-rise anxiety grips City builders. *The Independent on Sunday*, 3 June, 10–12.

E.M. Sigsworth (1990) *Montague Burton: The Tailor of Taste*, Manchester University Press, Manchester.

H. Smyth (1985) *Property Companies and the Construction Industry in Britain*, Cambridge University Press, Cambridge.

Spon's Handbook of Construction Cost and Price Indices (1991) E & FN Spon, London.

D.K. Stenhouse (1984) Liverpool's office district, 1875–1905. *Historical Society of Lancashire and Cheshire*, **133**.

The Stock Exchange (1965) *The Stock Exchange Gazette*, 10 Sept., 879.

Douglas Sun (1991) Legal & General Group PLC, in *The St James Press International Directory of Company Histories, Vol. III*, St James Press, London.

R.B. Sunnucks (1935) *Investment in Property*, 2nd edn, Banbury Publishing Co., London .

B. Supple (1970) *The Royal Exchange Assurance*, Cambridge University Press, Cambridge.

D. Sutherland (1968) *The Landowners*, Blond, London.

S. Taylor (1966) A study of post-war office developments. *Journal of the Town Planning Institute*, **52**.

W.A. Thomas (1978) *The Finance of British Industry,1918–1976*, Methuen, London.

Thomas Skinner & Co. (1950) *Skinner's Property Share Annual 1950–51*, Thomas Skinner, London.

F.M.L. Thompson (1963) *English Landed Society in the Nineteenth Century*, Routledge & Kegan Paul, London.

F.M.L. Thompson (1968) The land market in the nineteenth century, in *Essays in Agrarian History, Vol. 2* (ed. W.E. Michinton), David & Charles, Newton Abbot.

The Times (1924) 1 Jan., 31.

The Times (1930) 25 March, 13.

The Times (1932) 4 Jan., 21.

The Times (1932) 31 Dec., 4.

The Times (1935) 2 Jan., 5.

Labour MP's in new Centre Point protest (1973) *The Times*, 25 June, 3.

Ban on new office building in South-east (1973) *The Times*, 19 Dec., 1.

J. Tinbergen and J.J. Polak (1950) *The Dynamics of Business Cycles*, Routledge & Kegan Paul, London.

N. Tiratsoo (1990) *Reconstruction, Affluence and Labour Politics: Coventry 1945–60*, Routledge, London.

N. Toulson (1985) *The Squirrel and the Clock: National Provident Institution 1835–1985*, H. Melland, London.

R. Turvey (1993–94) London lifts and hydraulic power. *The Newcomen Society, Transactions*, **65**.

T. Veblen (1965) *The Theory of Business Enterprise*, Kelley, New York (originally published 1904).

O. Westall (1993) Entrepreneurship and product innovation in British general insurance, 1840–1914, in *Entrepreneurship, Networks and Modern Business* (eds J. Brown and M.B. Rose), Manchester University Press, Manchester.

B. Whitehouse (1964) *Partners in Property*, Birn, Shaw, London.

B. Williams (1994) *The Best Butter in the World: A History of Sainsbury's*, Ebury, London.

J.M. Wilson (1984) Retail warehousing in Greater London. *Estates Gazette*, **272**, 20 Oct., 244–6.

John D. Wood & Co. (1937) *Report on Business for the Year 1936*. Privately published.

D. Wright (1986) DIY retailing comes of age. *Estates Gazette*, **277**, 22 Feb., 712–13.

Index

References in **bold** refer to tables; those in *italics* refer to figures.

Abbey Life 199
Accountants 108–9
Actuarial theory 28–9, 90–1
Actuaries 154
Agricultural land 4, 13, 114, 138–40
Alambritis, Stephen 239
Alexander, I. 175
Alliance Assurance Co. 59
Allnatt Ltd 84–5, 86
Allsop & Co. 266
Alpha Estates 231
Amalgamated Investment and Property Co. Ltd 197, 206
Architects 108–9, 145–6
Architecture, competing styles during 1980s 216–17
Argyle Securities 182–3
Arlington Property Co. 88–9
Arndale Property Trust Ltd 126, 151, 177, 203
Arnott, Garry 160
Associated Dairies, ASDA superstores 204
Atlas Insurance Co. 133

B&Q 220
B1 Use Class 219–20
Bailey, A.H. 28, 29, 90
Bailey, Alan 216
Baker Harris Saunders 225
Bank of England
    banking regulation 186–7, 228, 229
    lifeboat operation 198–9, 200, 202, 206
Banks 83–4

    equity participation arrangements 157
    fringe 186–7
    property company debt 183, 186–7, 224, 227–8, 230, 231, 237
    and the property cycle 299–300
    overseas 230
Baranquilla Investments 153
Barber, Anthony 187
Barkham, R.J. 233, 298, 302
Barlow Report (1940) 104
Barnet, Correlli 103–4
Battery Park World Financial Centre 229
BCPH 225
Bedford, Duke of 12, 74, 84
Beecham, Joseph and Thomas 74
Beecham Estate and Pill Corporation 74
Beetlestone, H.N. 206, 207–8
Bell Property Trust 81–2
Bennison, D.J. 179
Berger, Berish 231
Bernard Sunley Investment Trust Ltd 181
Bernstein, W. 124–5, 126
Berwin Estates (Bahamas Ltd) 152
'Big Bang', 1986 223
Bilton, Percy 83–4, 126
'Black Monday', influence on property market 227
Book value, defined xiv
Bowie, Norman 172–3
Bradman, Godfrey 231–2
    see also Rosehaugh
Bredero 222
Brent Cross Shopping Centre 203
Bridge, George 134,135

INDEX

Bridgeland, Aynsley 123
Bristol 103, 184
British Empire Mutual Life Assurance Co. 16–17
British Foreign and Colonial Corporation 59
British Home Stores 53, 90
British Land 237
B.P. pension fund 182
Brixton Development Ltd, *see* Brixton Estate Ltd
Brixton Estate Ltd 84–5, 86, 123–4, 234
Broadgate development 232
Brown ban 166, 174–7
Brown, George 166
Builders, speculative 15
Building lease 11, 12–13
 defined xiv
Building licences 104, 132
Building societies 15, 91
Bull Ring Centre, Birmingham 178
Burton, Montague, *see* Montague Burton Ltd
Bush House 81
Bush Lane House 217
Business Parks 219–20
Bute, Lord 89
Butler, R.A. 147

Calthorpe, Lord 13
Cambridge College investments 14
Cameron Hall 221
Canary Wharf 228–9, 232–3
Cannadine, D. 12,13
Cantling 123
Capie, Forrest 300
Capital and Counties 154
Capital conversion factor 160–1
Capital gains taxation 167, 196, 227
Capital Issues Committee 121, 122, 128, 147–8
Capital security multiplier 54, 160
Cassell, Michael 5
Casson, Mark 117, 120
Catalano, A. 138
Cavendish-Bentick Estate 12
Cecil Estate 12
Cedar Holdings 197

Centre Point 143, 195
Charles, Prince of Wales 215
Chartered Surveyors 19, 39–47, 108, 118, 305
 and insurance companies 39
 and key plans 45–7
 and local authorities 179
 liberalization of property valuation standards during early 1970s 185
 portfolio advisory services 209
Cheap money 27, 69–72
 and construction 76, **76**
Chesterton Financial 231
Cheyne Investments Ltd 79
Chippindale, Sam 151
Church Commissioners
 asset distribution **139**
 and Chapter Property Pool 173–4
 direct developments 149, 153
 equity participation arrangements 151–2, 156
 investment policy 113–14, 137–41
 problems following 1990 crash 234–5
 and rent review 134, 137–8
Church Estates Development and Improvement Company 140, 151–2
Citadel Property Co. Ltd 125
City Centre Properties 151, 152, 154
 see also Clore, Charles; Cotton, Jack
City Corporation 223–4
City Gate Estates 231
City Merchant Developers 225
City of London
 damage during Second World War 104
 development to 1914 21–6
 office development on fringes during 1980s 228
 office floorspace 141, 159
 office over-supply during early 1960s 174
City of London Real Property Co. Ltd 22–6, 81, 146
City Offices Co. Ltd 22–6, 81
Civic Trust 144
Civil Defence Act (1939) 99
Clegg, Tony 221
Clerical, Medical and General Life Assurance Society 90, 171

INDEX

and Arndale Property Trust Ltd 126
and Brixton Estate Ltd 84–5, 123–4
developments and development finance 84, 86–7, 88–9, 180
equity participation arrangements 123–6, 149, 155, 156
and First World War 33
inter-war property investments 68–9, 73
investment policy 1945–1951 110
and leaseback finance 53, 88
market commentary 169–70, 172, 194, 201, 206, 207–8
non-property subsidiary companies 127
pre-1914 investment policy 31
property company share-holdings 180
property portfolio rationalization 206
residential property investments 82
and Second World War 99–100, 101
Clore, Charles 51, 109, 117, 122–3
Coal Exchange 144
Cohen, Jack 51
Cole, J.J. 22, 25
Collenette, D.A. 140
Collins, Michael 300
Commercial Union 90, 182
Community Land Act (1976) 203, 205
Companies Act, 1967 186
Competition and Credit Control 186–7
Construction
  orders, value of 184, 202
  output, value of 141
Co-operative Insurance Society 171
Cornhill Consolidated Group 197
Cornwall Properties 182
Corporate growth
  and property finance 54–8, **56**, **57**, 63–4
Corporation Tax 167, 173
  and concentration of property sector 181
  and property company gearing 185
Corruption 120, 144–5
Cotton, Jack 117
  Monico Piccadilly Circus project 144, 149

Counter Inflation (Temporary Provisions) Bill (1972) 171
Covent Garden Properties Co. 74
Coventry 102, 132
Credit squeeze
  1955–1957 147–8
  1960 154
Credit Swiss First Boston 228
Crickmay, John 109
Cross, George 19, 44–5, 77–9
Cubitt, Thomas 12

Dangerous structure notices 125
Davidson, J. 115
Davies, R.L. 179
Debenham, Tewson & Chinnocks 225
Deep discount bond 226
  defined xiv
Derby Estate 13, 74
Development Gains Tax 202–3
Development Land Tax Act (1976) 202–3
Devonshire, Duke of 12
Docklands 218–19
Dodge City 220
Douglas Kershaw & Co. 44
DTZ DebenhamThorpe Research 231, 259
Duffy, Frank 217
Dyos, H.J. 29

Eagle Star 74, 90, 133, 141, 226
  direct developments 181
  and leaseback finance 53
  and Odeon Theatres Ltd 89
  property company takeovers 181
  property investments, 1945–51 112
East, Barry 15
Ecclesiastical Commissioners 14, 113
Economist, The 75–6, 209, 217, 218, 226, 229, 300
Edgson, Stanley 134
Edward Erdman 44–6, 173
Einstein, R. 196
Elephant and Castle Centre 178
Embankment Place 217
English and Continental Property Co. 182
Enterprise Zones 218–19, 221
Entrepreneurship 90–5, 117–21, 127, 142–3, 148, 172, 226

INDEX

Equity & Law 89
Erdman, Edward, *see* Edward Erdman
Erostin 231
Estate agents, *see* Chartered surveyors
Estates Exchange 20
Estates Gazette 20, 70, 207, 266
Estates Property Investment Co. 226
Estates Times 216, 225
Eton College 15
Euston Arch 144

Farrell, Terry 217
Federation of Small Businesses 239
Fenston, Felix 144, 153
Feu duty, defined xiv
Finance Act (1940) 99
Finance Act (1965) 167
Finance Act (1974) 202
Financial institutions 2
    acquisition of City brokerage firms 225
    acquisition of residential estate agents 225
    see also Institutional property investment; Insurance companies; Pension funds
Finsbury Avenue, No. 1 217
First World War 33–4
Fisher, Archbishop 113–14
Flats, 1930s boom 81–2
    see also Property, residential
Fortress Property Co., *see* Bernstein, W.
Foster, Norman 217
Foundling Hospital Estate 12
Fountain House, Fenchurch Street 146
Freeson, Reg 195
Friends Provident 133
Full repairing and insuring lease 12
    defined xiv

Gaul, John 144
George Wimpey & Co. 152
Giltvote 226
Goddard, Claude 74
Goddard & Smith 74
Goldman Sachs 225
Goldsmith, James 182
Goobey, Alistair Ross 209, 216, 221, 299
Gordon, Charles 5, 160

Government policy and the development cycle 166–7, 174–7, 183–4, 201–3, 217–19, 223, 229, 301–4
Graham, R., *see* Citadel Property Co. Ltd
Great Universal Stores 122
Green, Arthur 109, 135
Gresham Chambers Co. Ltd 23
Greycoat group 222, 223, 226
Gropius, Walter 215
Ground rents 29–30, 260–1, **260**
    defined xiv
Grovewood Securities Ltd 181
Grunwald, F. 145
Guardian Properties (Holdings) 197, 200
Guingard, Richard 222

Hagenbach, Arnold 151
Hammersons 105, 177, 196, 227, 236
Hamnett, Chris 81
Hanover Property Unit Trust 173
Haslemere Estates 225
Healey & Baker 42–4, 46, 85, 122, 171, 251, 256–8
    and Montague Burton 93–4
Heim, Carol E. 80
Hemens, Arthur 42
Hepper Watson 222
Heseltine, Michael 218
Hill, Philip 74, 75
Hill Samuel 173
Hillier, Parker, May & Rowden 40–2, 204, 205, 234
    and Marks & Spencer 59
    business conducted, 1922–1937 **42, 43**, 43
    market commentary 70
    and rent review 134
Hillier, Parker Superannuation Fund 92–3
Hillier, William 40
Holford, Lord 143
Horne, W. Edgar 33, 72, 82
Horton, A. E. 115
Houlder, Vanessa 237
Housing and Town Planning Act, 1909 32
Howard Estate 12
Huddersfield Building Society 83
Humphries, Jane 91
Humphrys, Frances 112

INDEX

Hyams, Harry 153, 195

Imry Merchant Developers 226
Income conversion factor 160–1, **161**
Industrial estates 85–6
Industrial Finance and Investment Corporation Ltd 59
Inflationary expectations 110, 113–14, 116, 133, 149, 169–70, 214
Inner-city decline 217–19
Innes, James and John 24
Innovation, *see* Entrepreneurship
Institutional lease, undermining of following 1990 crash 235
Institutional property investment
   decline during 1980s 213–15
   and the development cycle 304–5
   distribution by geographical region 262–5, **264, 265**
   distribution by property type 260–2, **260, 261**
   pattern, 1974–1980, 193, 207–9
   volume of
     determinants 294–6, 295
     statistical series 247–59, **248–50, 252–3**, 254, 255, 259
   see also Insurance companies; Pension funds
Insurance companies
   asset distribution **27**
   boards, conservatism of 154
   as developers 30, 82, 90, 149, 180–1
   during nineteenth century 14–18, 26–7
   investment distribution targets 168–9
   mortgage distribution **155**
   property company takeover bids 181
   and 1974 property crash 199–200
   and property developers 15–18, 27–8, 83–90, 147–62, 205
   property investment activities
     post-1945 108, 133–41, 167, 234, **256**
     pre-1945 27–34, 68–73, **69**, 105
   and Second World War 99–101
   solvency margins 198–9
   with-profits business 127
   see also entries for individual offices; Institutional property investment

Interest rate cap 225
   defined xiv
Interest rate swap 225
   defined xiv
Internationalization of investment markets 214, 224
Investment Property Databank 271, 275–84
Investors Chronicle 158, 162
Investors Chronicle/Hillier, Parker rent index 207–8
Isle of Dogs, *see* Canary Wharf

J. Lyons & Co. 31, 51, 59
J.M. Jones 230–1
J. Sears & Co. 51, 122–3
Jackson, Alan 77
Jasper affair 145
Jefferys, J.B. 51
Jeger, Lena 195
Jencks, Charles 217
John D. Wood & Co. 50
Jones Lang Wootton 209, 222
Joseph, Maxwell 145

Keynes, J.M. 305–6
Keyworth, J.M. 74, 116
Kings Cross, development of 226–7
Kirzner, I.M. 120
Kleinwort Benson 222, 226
Knight, F.H. 117
Knight, Frank and Rutley 173
Kuwait Investment Office 201

L'Anson, Edward 22
Laing's 177, 178
Land & Property Trust 231
Land Commission Act (1967) 167, 184
Land Investors Ltd 158, 225
Land Securities 105, 121, 149, 154, 225, 227
Landmark House, Berkeley Square 227
Law Union & Rock Insurance Co. 133
Lawrence, Jon 21, 22
Lawson, Nigel 227
Lazards 173
Le Corbusier, Henri 215
Lee, S. 275

Leeds 184
Legal & General Life Assurance Society 209, 227, 234
   and Aynsley Bridgeland 123
   and Birkenhead estate 18
   and Charles Clore 122
   development finance **150**, 206
   direct developments 149, 171, 180
   equity participation arrangements 149, 158
   Estate department administrative structure 135, 136
   investment policy 71–2, 137
   investment targets 170–1
   investment yields 135, **137**
   market commentary 194, 199, 205
   overseas property investments 135
   property company share-holdings 180
   property company takeovers 182
   property investment 1944–51, 109–10
   property portfolio decentralization 175
   property portfolio rationalization 206
   and rent review 134
   residential property investments 82
   and Second World War 100
Legenland 123
Leigh, Vivian 144
Leigh-Pemberton, Robin 229
Lessor schemes 125–6
Lester, R.M. 50, 71
Levereged buy-out 226
   defined xiv
Liberal Land Campaign 32–3
Lift, hydraulic 25–6
Limited-recourse loan 222
   defined xiv
Lintang Investments 145
Liverpool, as an office centre 22
Lloyd George, D. 32
Lloyds Bank 227–8
Lloyds Building 217
Local authorities 177, 178–9, 204–5
Local Government, Planning and Land Act (1980) 218
Location of Offices Bureau 174–5
London, Bishop of 13
London and County Securities 197

London and Edinburgh Trust 221, 222, 227
London & Metropolitan 231
London County Council 79, 143, 174
London Docklands Development Corporation 218–19
London Hydraulic Power Co. 25
London Life Association 180
London Merchant Securities, *see* Rayne, Max
Lotery, Edward 51–2
Lundin, Fife, development of 15
Lyon group 197, 200

Macgregor, B.D. 275
Macmillan, Harold 132
McWilliams, Douglas 238
Market value, defined xiv
Marketchief 226
Marks & Spencer
   and Hillier, Parker 59
   inter-war expansion 58–63, **61**
   and Prudential 59, 60, 62
Marks, Michael 58
Marks, Simon 51, 58, 59
Marlborough Property 225
Marriage value, defined xiv
Marriott, Oliver 5, 110, 117–18, 120, 121–2, 144, 145–6, 152, 158–9, 162, 178, 195
Marris, Robbin 54
Marx, Karl 306
Marxist theory of crisis 166
Mason, Sidney 196, 227
Massey, D. 138
May, George Ernest 28–9
Meighar-Lovett, Percy 124
MEPC 105, 197, 225
Mercantile House 226
Mercury Asset Management 226
Merry Hill Centre, Dudley 221
MetroCentre, Gateshead 221
Middle class investment patterns 18–19
Milford Dock & Railway Co. 18
Milford Haven, development of 17–18
Millward, Richard 222
Monetary policy and the development cycle 195, 229–30, 236–7

INDEX

see also Credit squeeze
Montague Burton Ltd 51, 52–8, 63, 85
   and Healey & Baker 93–4
   and leaseback finance 53
   property developments 79–80
   subsidiary property companies 58
Morgan, E.V. 251
Morgan Stanley 228
Morley, G.C. 170
Mortgage lending 14–18
   limits of availability 87–8
   see also Insurance companies and property developers
Mountain, Edward 74, 89, 90
Mountleigh 221, 231, 232
Multiple Traders Federation 103
Murray, H.H. 145
Murrayfield 177
Mutual Property Unit Trust 173

Naish, H.W. 115–16
Nanthakumaran, N. 275
National Building Society 83
National Provident Institution 169
   and Allnatt Ltd 84–5
   and First World War 33
   and Second World War 100
   development finance 86, **87**
   direct developments 180
   investment policy 15, 17–18, 30–31, 91, 112–13
National Westminster Bank 226
NCB Superannuation Funds 114–16, 134, 201–2
Non-recourse loan 222
   defined xv
Norfolk, Duke of 12
Norwich Union 90, 209

Odeon Theatres Ltd 89
Off-balance sheet finance 222–3, 227–8, 298
   defined xv
Offer, Avner 31
Office Development Permits 175–7, 183–4
Offices
   building quality 145–6

construction technology 80
design innovation 146
design, requirements of computer technology 217, 223
employment 22, 105, 168, 223
floorspace distribution 142
inter-war development 80–1
obsolescence 217
nineteenth century development 21–6
rental growth 141
vacancy rates 224, 229, 230, 231, 237, 238
see also City of London
Olsen, D.J. 12, 15
Olympia & York 228–9, 232
Orchard-Lisle, Aubrey 42, 44, 116
Oriel Property Trust Ltd 125
Oversesas investors 224, 237, 258–9, 259
Oxford college investments 14

P&O Property Co. 225
Paddington Estate 13, 151
Parker, Thomas 40–1
Parry, John 235–6
Pearl Assurance 153
Peel Holdings 221
Pegler, Jim 124, 151
Pelli, Cesar 229
Pension funds 92–3, 105, 169, 234
   equity participation arrangements 157
   pattern of investment 172, **256**
   property company takeover bids 182
   and property development funding 182–3
Pension Fund Property Unit Trust 172–3
Peppercorn rent, defined xv
Percy Bilton Ltd 125
   see also Bilton, Percy
Perpetual Investment and Building Society 17
Peter Angliss and Yarwood 173
Phillips, Jackie 101
Planning and Compensation Act (1991) 236
Planning policy, liberalization during 1980s 214, 217–20
Plender, John 182–3, 187
Plot ratio restrictions 143

Portman Estate 12
Post Office pension fund 182–3
Prevention of Fraud (Investments) Act (1958) 172
Privity of contract 235
Property
  commercial
    1902–1914 market depression 31–2
    and bank lending 1
    and corporate finance 1, 238
    decentralization from urban centres 214
    and employment 2
    and First World War 33
    factors influencing investment in 21, 38–9
    market rents 207–8, 224, 230, 271–4, 272, **272–3**, 273, **274**, 289
    obsolescence 214
    rent control 171, 201
    returns to investment in 70, 105, 168, 193–4, 288
    returns to investment, statistical series 274–84, **277–9**, 279, **280–3**
    returns statistics, data smoothing problems 274–5
    and technical innovation 1–2
    transactions costs 214
    values, determinants 291–3, 293
    values, economic impact of changes in 238, 303–4
    yields on, *see* Yields
  companies
    accounting standards 302
    amalgamations 225
    assets 298–9
    bankruptcies after 1990 crash 231
    contraction of sector following 1974 crash 199–200
    ease of entry into sector 298
    equity participation arrangements **156**, **157**, 167
    gearing 23, 233, 289
    growth of sector 19, **20**, 73–5, **73**, 157–8
    investment profile 108
    and property cycle 298–9
    and Second World War 101
    share prices 158–9, **159**, 227, 237
    use of research 298–9
    valuation by stockbrokers 185
    volume of investment 257–9, **257**, 259
  crashes 291
    1974 194–200
    1990 219–36, 297
  developers 117–22, **118**
    and local authorities 177, 178
    low public esteem 143–6
    merchant 221–3, 230
    switch from offices to shops during early 1960s 176–7
  development
    1955–1964 boom 132–62
    1971–1973 boom 183–7
    1985–1989 boom 223–9
    cycle 2, 289–91, 290, 297–306
    finance 83, 146–62
    speculative 15, 289
  industrial 4–5
    as collateral 85–6
  industry, as proportion of GDP 1
  investment
    by individuals 108, 257–8, **257**
    factors influencing growth 21
    profile of investors 106–9, **107**
    see also Institutional property investment
  investment banking 225–6
  market press 20
  overseas 4
  prime, defined xv
  residential 4, 29, 112–13, 230, 237–8
  retail, boom in development 1965–1973 178
    see also Retailers; Shopping centres
  secondary, defined xv
  speculation 29, 47, 101
Property bonds 173, 200
Property income certificates 226
Property unit trusts 169, 172–3, 200
Provincial Insurance Co. 89
Prudential 46, 90, 135, 171, 209, 225
  and Marks & Spencer 59, 60, 62

development finance 89–90
flat developments 82
investment policy 28–9, 33, 72, 91–2
property company takeovers 182
Purdy, D.E. 302

Quantum Fund 237
Queen Anne's Bounty 113

R. Hitchins 222
Rack rented property 260–1, **260**
Randolph, Bill 81
Randsworth 232
Ravenseft 132, 177
Rayne, Max 151–2
Raynes, H.E. 134
Real Estate & Commercial Trust Ltd 151
Redden, R. 5, 251
Rees, G. 60
Regentcrest 221
Reichman, Albert, Paul and Ralph 228
Reid, Margaret 186
Reith, Lord 102
Rent review
  decline in intervals between 141
  introduction 134
  statistics 286–8, **287–8**
Rent-seeking behaviour 128
Retail warehouses 220–1
Retailers
  Co-operative 48, 177
  Department store 48
  multiple
    and property investment market 50–63
    inter-war growth 40, 48–63, **49**
  self-service 177
  variety chain stores 48, 58
Return, rate of, defined xv, 275
  see also Property, commercial, returns to investment in
Ribbon Development Act (1935) 102
Richard Ellis 209, 226
Ridley, Nicholas 219, 236
Rippon, Geoffrey 196
Robins, Julian 239
Rodamco 225
Rogers, Richard 217

Rolling-up interest payments 184–5
Rose, Jack 5, 141, 146, 158, 185
Rose, Philip 158
Rosehaugh 222–3, 227, 231–2
  Rosehaugh-Stanhope 226–7, 232
Rowe and Pitman 199
Rowell, Andrew 90, 101, 110–12, 123
Royal Exchange Assurance 17, 30
Royal Institution of Chartered Surveyors 225
Royal Insurance 182
Russell Estate 12

Saachi & Saachi 227
Sackville Estates 89
Sainsbury, J.B. 78
Sainsbury's 78, 79
  Homebase 220
St Martins property group 201
Sale and leaseback 109, 122, 167
  defined xv
  early development 31, 53–8, 93–4
  see also Insurance company development finance
Salisbury, Marquess of 12
Salmon, B. 59
Salmon, Harry 51, 59
Salomon Brothers 225
Samuel, Harold 105, 117, 119–20, 121, 149
Samuel Montagu 173
Samuel Properties 225
Savills 225
Scarcity premium 26
Schiller, R. 179, 204
School of King Edward VI 13
Schumpeter, J.A. 117
Scott Report on land utilization in rural areas 104
Second Covent Garden Property Co. Ltd 75, 101, 116
Second World War 99–104
  and town planning legislation 101–4
Secondary banking crisis 197–200
Securitization 226, 300–1
  defined xv
Seifert, Richard 143

Share prices and institutional property investment 294–6
Sheraton Securities 231
Shop Investments Ltd 125, 206
Shopping centres
  arcade 103
  covered 178
  development, 1974–1980 203–5
  effects of 1990 crash on development of 235–6
  high street 49–50
  hypermarkets 205
  out of town 204
  precinct 102–3, 176–7
  regional 205, 221
  suburban 51–2, 77–9
Silsoe, Lord, see Trustram Eve, Malcolm
Smyth, Hedley 5
Society for the Protection of Ancient Buildings 144
Solicitors 15, 19, 108–9
Soros, George 226, 237
Southampton 102–3
Southend Estates 225
Speyhawk 221, 227
Standard Life
  during nineteenth century 15–16, 27–8, 30
  market commentary 208
  property development and development finance 180–1, 206–7
  property investment during 1974 201–2
Stanhope 222, 227
  see also Rosehaugh-Stanhope
State Building Society 145
Sterling commercial paper 234
  defined xv
Stern group 197, 200
Stock Conversion 225
Stock market, short-termism of 234, 298
Suburbanization 29
Sun Real Estates 144
Superannuation Fund Property Unit Trust 173

Takeover boom, 1952–1956 122
Tesco's 51
Thompson, F.M.L. 20

Times, The 266
Tovey, Douglas 122
Town and City Properties 151, 177, 197
Town and Commercial Properties 197
Town and Country Planning Act (1932) 102
Town and Country Planning Act (1944) 103
Town and Country Planning Act (1947) 104, 119–20, 143–4
Town and Country Planning Act (1953) 132
Town and Country Planning Act (1962) 174
Traditional landowners 11–14, 257–8, **257**
Travelstead, Gooch Ware 228
Trustram Eve, Malcolm 137
Turvey, Ralph 25–6
TV-AM building 217
Twentieth Century Banking 197

Uniform business rate 230
United Real Property 225
Unitization 226
  defined xv
  early attempts 19, 172–4
Urban Development Corporations 218
Uthwatt Report on compensation and betterment 104

Value added tax on new buildings 230
Van der Rohe, Mies 215
Veblen, Thorstein, theory of commercial crises 166–7, 183, 194, 197–8
Verry, William 24
Victorian Society 144

W.H. Smith 78–9
Walker, Peter 184, 196
War Damages Act (1943) 99, 103
Warren, Mortimer 113–14, 138–9
Wates Ltd 152
Weatherall, Green and Smith 173
Western Ground Rents 89, 182
Whitehouse, Brian 155, 161
Whitley, John 291
Willet's 178
Wingate, Stephen 226
Wolfson, Isaac 122
Woolworth's 50, 53, 60, 204
  and the Prudential 89–90
Wyatt, George 17

INDEX

Years purchase, defined xv
Yields
  during Second World War 99–100, **100**
  equivalent 169–70, **170**
  defined xiv
  initial
    decline during 1930s 70
    decline during 1964–1973 167–8, 184
  defined xiv
  during 1970s 193
  during 1980s boom 224
  reverse yield gap 168, 184–5
  statistics 266–71, **267–8**, 269, 270
Young, Eric 134

Zero-coupon bond 225
  defined xv

Printed in Great Britain
by Amazon